A HISTORY
OF ISRAEL

HOWARD M. SACHAR

A HISTORY

OF ISRAEL

Volume II

FROM THE AFTERMATH OF THE YOM KIPPUR WAR

New York   Oxford
OXFORD UNIVERSITY PRESS
1987

Oxford University Press

Oxford   New York   Toronto
Delhi   Bombay   Calcutta   Madras   Karachi
Petaling Jaya   Singapore   Hong Kong   Tokyo
Nairobi   Dar es Salaam   Cape Town
Melbourne   Auckland

and associated companies in
Beirut   Berlin   Ibadan   Nicosia

Published by Oxford Univesity Press, Inc.,
200 Madison Avenue, New York, New York 10016

Oxford is a registered trademark of Oxford University Press

Library of Congress Cataloging-in-Publication Data

Sachar, Howard Morley, 1928–
A history of Israel. Volume II.

Bibliography: p.
Includes index.
1. Israel—History.   I. Title.
DS126.5.S152   1987   956.94'05   86-16182
ISBN 0-19-504386-3

The maps appearing in this volume are by Jean Paul Tremblay

9 8 7 6 5 4 3 2 1

Printed in the United States of America
on acid-free paper

In Memory of Edward Joel Sachar

ACKNOWLEDGMENTS

This volume is intended as a sequel to *A History of Israel: From the Rise of Zionism to Our Time*, which concluded with the Israeli-Egyptian disengagement agreement of September 1975. For readers who prefer to concentrate on a more recent period of Israeli history, the opening chapter offers a brief reevaluation of events since the 1973 Yom Kippur War.

Even as the format of Volumes I and II is identical, so not a few of the individuals who enriched the earlier book proved equally generous with their time and information for this one. Appreciation is offered here only to those friends whose insights relate specifically to the latest years of Israel's development: Dan Pattir, former press spokesman for Prime Ministers Rabin and Begin; Erella Hadar, cultural affairs department, foreign ministry of Israel; Seth Carrus, senior military analyst of the *Near East Report*; Dan Halperin, economic counselor of the Embassy of Israel in Washington; Eli Eyal, former Washington correspondent of *HaAretz*, and Ze'ev Schiff, military editor of the same newspaper; Ya'akov Erez, military affairs correspondent of *Ma'ariv*; Wolf Blitzer, Washington correspondent of the *Jerusalem Post*.

Gratitude must be expressed, as well, to the late Professor Yigael Yadin, department of archaeology, Hebrew University; Dr. Amnon Sela and Dr. Eytan Gilboa, department of political science, Hebrew University; Professor Shlomo Shoham, department of criminology, Tel Aviv University; Dr. Itamar Rabinovich, Daniel Dishon, and Aluph Hareven, Shiloach Institute, Tel Aviv University; Dr. Jack Cohen, director emeritus, and David Breslau, former associate director, Beit Hillel, Jerusalem; Rabbi Philip Goodman, representative in Jerusalem of the Israel Endowment Fund; and Dr. Avraham Avihai, director-general, Keren HaY'sod.

For their bibliographical assistance in this and earlier projects, my warmest thanks as always to Bruce Martin, director of the research facilities office of the Library of Congress, and to Quadir Amiryar, interlibrary loan director of George Washington University's Gelman Library. Notwithstanding the pressures of children and law studies, finally, my wife Eliana once again gave my manuscript the fullest collaboration of mind and heart.

Kensington, Maryland   H.M.S.
May 1986

CONTENTS

LIST OF ILLUSTRATIONS

A HISTORY
OF ISRAEL

THE RAMIFICATIONS

OF OCTOBER

THE FALL OF GOLDA MEIR

On April 11, 1974, only three months after the Labor Alignment had won a renewed mandate in a close national election, Prime Minister Golda Meir submitted her government's resignation. As much as any soldier fallen in battle, Mrs. Meir was a casualty of the Yom Kippur War. The cost of her nation's victory in that October 1973 struggle was far higher than in any of Israel's previous wars. Over 6,000 troops had been slain or wounded in eighteen days of fighting, and on the Syrian front the toll would continue to mount. The economic losses were only beginning to register. The expenditure of equipment, the decline in production and exports as a consequence of mobilization, would soon reach $7 billion, the equivalent of Israel's Gross National Product for an entire year. It was in the aftermath of this somber "triumph," on December 31, 1973, that voting took place for the Eighth Knesset. The results produced no major political transformations. The right-wing Likud opposition gained 8 seats, to be sure, even as 6 seats were lost by the "Alignment"—the three united Labor parties, in electoral confederation with the left-wing Mapam faction. Yet, while duly punished for the initial reverses of the war, Labor disposed of 51 out of 120 members. It still remained the dominant bloc in the Knesset.

As matters turned out, Mrs. Meir's new cabinet experienced the shortest life span in Israel's history. Its tenure was undermined by a commission, chaired by Supreme Court Justice Shimon Agranat, that had been established shortly after the war to determine responsibility for the "blunder," the government's palpable unpreparedness for the Arab offensive. Issued in early April 1974, the Agranat Commission's report was a serious indictment of the military intelligence service, of the Sinai field commanders, even of the chief of staff, for their sluggish reaction to advance warnings of a possible enemy offensive. Where the commission refused to speculate was in the murky area of Defense Minister Moshe Dayan's personal responsibility for Israel's early battlefield setbacks. On this point, however, public reaction was much less equivocal. An explosion of outrage greeted the report. It appeared unconscionable that career officers were being punished, while Dayan, the self-proclaimed architect of Israel's military supremacy, was being absolved. The Likud opposition accordingly requested a special Knesset session to debate a motion of nonconfidence in the government. Anticipating that debate, which was fixed for April 11, Mrs. Meir was confident that the Labor Alignment would close ranks behind her. In fact, it did not. Hence the prime minister submitted her resignation, and thereby caused the fall of her cabinet.

Golda Meir had acceded to the premiership in 1969, in the aftermath of Israel's most glorious moment, the 1967 Six-Day War victory. The Arab states lay at her feet, broken and stupefied. Washington was providing massive diplomatic and military support. Jewish millionaires were arriving from overseas to plan important development projects. A powerful renaissance of solidarity with Israel was infusing Soviet Jewry. And within four years of this apotheosis, Israel's circumstances plummeted from euphoria to depression, from international prestige to diplomatic isolation, from economic boom to galloping inflation, from large-scale immigration to substantial emigration. The Yom Kippur War was the watershed. It destroyed the myth of Israel's invincibility and eroded the nation's self-confidence. Some of these failures unquestionably developed independent of Mrs. Meir. Yet the doughty old woman could not escape responsibility for exacerbating them. The main source of her considerable impact was her personality, a synthesis of iron will, courage, and warm sensitivity to human and social problems. But the converse of this strength was inflexibility. The veteran diplomat Gideon Rafael, who knew Mrs. Meir well, provided an astute evaluation of her. "To hold on doggedly to concepts, even if they had long been overtaken by events, was Golda's perception of leadership," he recalled. "It endowed her with a singular strength of resistance—a quality much needed by an embattled people—but it impaired her resilience to adapt her policies to changing circumstances."

Thus, Mrs. Meir may have rejected the very notion of permanently ruling a million Arabs. In her insistence on "direct negotiations" with the enemy, however, her opposition to any form of outside mediation, to any interim settlement and to any military withdrawal from the cease-fire borders before the conclusion of a peace treaty, the prime minister was as intransigent as the most militant of Israel's right-wing annexationists. She refused to take seriously a number of important exploratory proposals that Egypt's President Anwar al-Sadat had been floating since 1971. In February of that year, Sadat assured the United Nations emissary, Gunnar Jarring, that he was prepared to reach an accommodation with Israel, that he would contemplate negotiating an interim agreement based on Israel's withdrawal from the Suez Canal. The Egyptian leader twice repeated this offer in the next eighteen months, but Mrs. Meir remained unimpressed by the overtures, and rejected them. It was a tragic irony that Israel was obliged to accept a much less favorable interim arrangement after the 1973 war.

Worse yet, Mrs. Meir failed to grasp the shift in her nation's public morality. Following the initial 1967 victory, the reservoir of tens of thousands of cheap laborers from the occupied Arab territories encouraged Jewish workers gradually to move from production to services. The influx of vast sums of money from abroad encouraged stock market speculation and banking at the expense of industry and agriculture. The exemption of defense expenditures from budgetary restraints or even from the rules of competitive bidding fostered a corruption that infected the Labor establishment itself. Golda Meir witnessed the process with chagrin and sorrow. Yet her effort to stop it by preaching loyalty to "old fashioned" values betrayed a characterological inability to grasp the authentic causes of the rot, namely, the ongoing occupation of the territories and the ongoing state of war. Only a peace settlement could have blocked the social deterioration—the key issue on which Mrs. Meir failed utterly.

In any case, upon the prime minister's departure in April 1974, Labor's priority was not one of introspection but of succession. Within days, rivalry devolved into a contest between David Ben-Gurion's former protégé Shimon Peres, then minister of information, and Yitzchak Rabin, chief of staff during the Six-Day War, later ambassador to the United States, and most recently minister of labor. Peres was the candidate of the party's Rafi (right) wing, but also of many others within Labor. Rabin, although of Achdut HaAvodah background, had not played an active role in this leftist faction, or in any other faction. Indeed, his career was all but apolitical. It was with the credentials, rather, of a military hero that the fifty-one-year-old Rabin, a native-born *sabra*, eventually was picked by Labor's central committee in a tight election. Rabin's foreign minister, Yigal Allon, also came from Achdut HaAvodah. Peres in turn was awarded the key portfolio of defense minister as a conciliatory gesture to Labor's right wing. For the first time in Israel's history, then, the cabinet was led by a member whose roots in his own party were not deep, even as others of its key ministries were dominated by figures who similarly were not products of Mapai—of Labor's veteran centrist faction. As shall be seen, the structural weaknesses of this hybridization would fatally erode the Labor government's political authority, and contribute to its electoral loss in 1977.

A SEARCH FOR DISENGAGEMENT

Prime Minister Rabin's initial concern now was to press ahead for additional military disengagement with the Arab enemy. The Egyptian-Israeli cease-fire of October 22, 1973 had called for immediate negotiations leading to a "just and durable peace." But, afterward, direct Egyptian-Israeli talks at Kilometer 101 on the Suez-Cairo highway failed to produce results. Each army remained ensconced precariously behind the lines of the other. A subsequent "peace conference" in Geneva on December 22, between Israel, Egypt, and Jordan, was adjourned within a single day amidst a farrago of mutual recriminations. It was accordingly in January 1974 that U.S. Secretary of State Henry Kissinger flew to the Middle East in an effort to broker an agreement through personal mediation. American "shuttle diplomacy" between Cairo and Jerusalem hardly represented the kind of direct negotiations the Israelis had anticipated, yet both governments accepted it to defuse the threat of renewed warfare.

Kissinger's efforts were successful. Under the agreement he mediated on

January 17, 1974, the Israelis agreed to pull back 7 miles into Sinai; the Egyptians, to reduce their military presence on the east bank to a thin line of troops and a few score artillery pieces and tanks. In letters to President Richard Nixon, moreover, Sadat committed his government to dredge the Suez Canal and rebuild its cities—in effect, to make the Suez area a hostage to peace—and to allow the passage of nonmilitary cargoes through the waterway to and from Israel (although not in Israeli vessels). The agreement was honored punctiliously by both sides. A certain atmosphere of cautious trust even began to stir between Egyptians and Israelis. Admittedly, Kissinger's efforts to negotiate a disengagement between Damascus and Jerusalem were far more difficult. Syria's hatred of Israel was more implacable than that of any other Arab nation. But here also, following thirty-two days of exhausting shuttle discussions, the secretary of state managed to broker an agreement. On May 31, 1974, a new cease-fire line was established on the Golan Heights, this one slightly west of the post–Six-Day War position. The town of al-Quneitra was returned to Syria, Israel retained control of the adjacent hills. Provision also was made for a United Nations buffer zone between the two enemies. As along the Suez Canal, the retreat effected no serious erosion of Israel's defensive posture.

Except for this last accord with the Syrians, the disengagement agreements were a legacy of the Meir government to the new Rabin government. Thus, once assuming office, Rabin was intent upon sustaining the momentum to peace. So was Kissinger, and so was Sadat. Like Rabin, the Egyptian president by then was opposed to the Geneva format. As he saw it, a reconvened conference under Soviet-American sponsorship would have raised the explosive issue of Palestine Liberation Organization (PLO) participation, and ensured a common Arab stance of noncompromise. Triangular diplomacy through Kissinger still offered the best likelihood for reaching accommodation. Rather, it was the substance, not the procedure, of disengagement that created the impasse between the two sides. The Rabin cabinet anticipated linking territorial withdrawal to peace; that is, of calibrating the stages of Israeli evacuation in the Sinai to the range of Egyptian political concessions. Sadat rejected the formula. Through Kissinger, the Egyptian leader demanded an unqualified Israeli commitment to withdraw from approximately 40 percent of the Sinai; any Egyptian concessions on demilitarization, on nonbelligerency, on additional rights of passage through Suez, would have to await full Israeli evacuation from all occupied territories on all fronts, Egyptian, Syrian, Jordanian. It was a nonstarter for the new Israeli government. By March 1975, the effort to achieve a wider-ranging disengagement appeared to have foundered.

In fact, it had not. Only three days after Kissinger's chagrined return to Washington, Sadat decided to gamble on economic recovery. He announced that he would open the Suez Canal on June 5, 1975, and proceed forthwith to rebuild the canal cities. As Rabin sensed better than had Kissinger, the Egyptians no longer possessed the strength or will power to renew the war. Triangular diplomacy then quietly resumed in mid-June, this time in Washington between Kissinger and the Egyptian and Israeli ambassadors. Finally, on September 1, agreement was reached. Israeli forces would pull back an additional 18 miles into the Sinai. Although the Mitla and Gidi passways across the peninsula would be abandoned, as well as the Abu Rudeis and Ras Sudr oil fields in the southern Sinai, the nature of this withdrawal was to be tightly qualified, and Israel's defensive position would be affected only minimally. Egyptian forces (much reduced in number) would occupy less than 4 miles of the evacuated territory; the rest of the terrain would be manned by a United Nations contingent in a wide buffer zone, even as a group of 200 American technicians would operate electronic monitoring stations in and near the Mitla and Gidi passes. If Israel failed to win a public Egyptian pledge of nonbelligerency, it extracted other commitments. Each side agreed to avoid "resort to the threat or use of force or military blockade against the other." Sadat reconfirmed opened passage through the canal for nonmilitary cargoes destined to or from Israel in foreign carriers. Unlike its predecessor in the Gaza Strip, the United Nations contingent in Sinai this time would be virtually immune to Security Council cancellation.

Despite these built-in guarantees, no member of the Rabin government was dancing in the streets. Sadat had offered the irreducible minimum for Israel's territorial concessions. Indeed, the Egyptian president emphasized the purely military, nonpolitical, nature of the accord by dispatching officers of modest rank to the signing ceremony in Geneva on September 4, 1975 (the facade of an ongoing peace conference was retained). Afterward, too, he stressed repeatedly that his negotiations at no time had been with the Israelis, but rather with the Americans; that the very notion of a conclusive political understanding with Israel was premature. "I leave to the next generation the trouble of deciding if it is possible not only to exist with the Jewish state but also to cooperate with it," he explained to a *Le Monde* interviewer. Given this apparently unrequited Egyptian animus, it was Washington's, not Cairo's, commitment of security that now assumed a new and decisive importance for the Israelis.

RABIN'S YEARS OF GRACE

Indeed, the developing quasi alliance with the United States may well have been the single most notable achievement of Rabin's prime ministry. It had been largely an arm's-length relationship until the Six-Day War in 1967. A military victory for Egypt then would have established Gamal Abd al-Nasser's hegemony in the Middle East and swept the region clean of any remaining American influence, even in the Arab oil countries. Washington would have been obliged to take drastic action to check Nasser at a time when U.S. armed forces were mired in Vietnam. It was Israel's surgical victory, achieved "cleanly" with its own manpower, that dispelled these threats. From then on, U.S. strategy in the Middle East remained heavily centered on Israel, buttressing an already widespread moral interest in the democratic Jewish state. Earlier American attempts to stabilize the region through a military balance had failed. Now Washington was determined to use Israel's victory to achieve a "final" settlement in the area. The support enabled Israel to resist all diplomatic and military pressures to withdraw from its captured territories short of a comprehensive peace agreement.

Yet, even as the Yom Kippur War fortified this relationship, it exposed certain tactical divergences between Washington and Jerusalem. On the one hand, the United States for the first time intervened openly and dramatically on Israel's side, providing unprecedented quantities of arms and financial assistance. But in order to establish the United States as a credible peace broker between Egypt and Israel, Kissinger also blocked Israel from achieving a total victory. Rather, from then on, more American pressure was exerted on Israel than at any time since the Eisenhower years, and the disengagement agreements with Egypt accordingly offered the Jewish state less than ideal terms. It was thus by way of compensating Israel for accepting those conditions that Washington was prepared to make certain far-reaching new commitments. Through separate letters exchanged between Rabin, Kissinger, and President Gerald Ford, following the second disengagement agreement of 1975, the Americans agreed to provide Israel with substantial quantities of the latest weapons; to meet Israel's oil needs over a five-year period (if those supplies were not available elsewhere); and, in the event of a major threat to Israel's security, to consult promptly with Israel "with respect to what support, diplomatic or otherwise, or assistance [the United States] can lend to Israel in accordance with its constitutional practices." The commitment was additionally strengthened by an American pledge not to recognize the Palestine Liberation Organization, and to veto punitive measures against Israel in the UN Security Council. These were longer-term obligations to Jerusalem than Washington had ever before accepted.

Still another of Rabin's legacies from the Golda Meir era was a dramatic new infusion of immigration. The upsurge came from the Soviet Union. It was the unlikeliest of sources. Following the "black years" of Stalinist terror, this largest of European Jewries normally would have been fearful of bestirring itself on behalf of any communal right, least of all emigration. Yet, as early as the 1950s, Prime Minister Ben-Gurion ordered Israel's embassy in Moscow to develop a subterranean dialogue with Soviet Jews. The effort was productive beyond all expectations or hopes. The ethnic loyalty of substantial numbers of Jews already was being kindled by outrage. Moscow was denigrating them as a "non-people" in the constellation of Soviet nationalities, evicting them from responsible positions in the government and party. Until the 1950s, to be sure, Jews had been willing to accept all the inequities and constraints of Soviet life—so long as they were permitted to maintain their traditional role as an educational elite. But, in later years, a harsh numerus clausus in the universities seriously threatened that role. Moreover, in the aftermath of Israel's 1967 victory, the Kremlin launched a furious denunciation of Soviet Jews as an "untrustworthy," "Zionist-deviationist," "unpatriotic" element. The campaign of abuse did not abate in subsequent years. Convinced that they had little further to lose, then, thousands of Jews decided to run the very considerable risk of formal application for exit visas to Israel.

The emigration movement doubtless would have been suppressed instantly and ruthlessly had it not been for the Soviet leadership's desire for access to American technology. Here the obstacle in Washington was less the White House than the legislative branch. By 1970 a powerful Jewish lobby in the United States had won congressional support for the cause of Soviet Jewish rights, including the right of emigration. The Kremlin understood well now that these congressional misgivings would have to be dissipated, even at the price of a certain limited number of Jewish departures. As it happened, a rationale existed for "family reunions." It was the—intermittently invoked—Leninist tradition of respect for the Soviet peoples' ethnic sensibilities. In the case of the Jews, family reunions in a Jewish homeland, in Israel, might fit this Leninist pattern. Very cautiously, then, the ban on Jewish emigration was relaxed. Between 1969 and 1971, some 15,000 Jews were allowed to depart for Israel. The number more than doubled throughout 1972, reaching 31,652, and edged up to 34,733 in 1973. The process of departure was never easy. The Kremlin was uninterested in encouraging other Soviet nationalities to emulate the Jewish example. In 1974, the number of permitted Jewish emigrants actually dropped back to 20,628; in 1975, to 13,221. Yet the level rose again to 14,261 in 1976, to 16,737 in 1977. Once reaching Israel, moreover, the newcomers were received enthusiastically, even gratefully. Housing provided for them often matched that available to veteran Israeli citizens. Employment literally was manufactured for them, at heavy public expense. The effort, and the cost, were by no means extravagant. Many of the immigrants were well educated. They appeared to be superb human material for Israel's thinly stretched population. Their arrival in these unprecedented numbers was a source of quiet euphoria to the Israeli government and public alike.

Still another achievement could be attributed even more directly to the Rabin government. Early in May 1976, Israeli intelligence learned that a new series of terrorist attacks were being organized by a radical Arab guerrilla faction, the Popular Front for the Liberation of Palestine. Unknown to the Israelis, in fact, one of the PFLP's leaders, Dr. Waddia Haddad, was planning a daring skyjacking operation. In mid-June, Haddad assembled his staff in an Aden apartment to lay out the basic strategy. Among those attending was a German, Wilfried Boese, who had been associated with the German terrorist Baader-Meinhof gang, and who now became Haddad's agent for the recruitment of additional Arabs and Germans. The selected victim was Air France, an airline that was considered immune from terrorist activity due to President Giscard d'Estaing's pro-Arab policy. The takeover was scheduled for Athens airport, where security was notoriously weak. Thus, at 7:00 a.m. on June 27, a German "couple" and two Palestinians disembarked separately in the transit section of the Athens airport, traveling with forged passports. At 11:00 a.m. they joined the passengers of an Air France plane that had just arrived from Tel Aviv, en route to Paris. There was no inspection of hand baggage. At 11:33 a.m. the Air France plane took off again. Eight minutes later the four terrorists seized control of the plane, its crew, and its 254 passengers. The captain was ordered to change course for Benghazi, and complied. Once in Libya, the plane was refueled, then took off again. After nearly six hours of additional flight, it landed at Entebbe, outside the Ugandan capital of Kampala, where another team of terrorists was waiting.

The passengers—83 of them Israelis—were confined in an old, unused terminal. Soon afterward, Idi Amin, the pro-Arab dictator of Uganda, declared the passengers to be his "guests," and appointed himself intermediary between them and the Israeli government. Two days later, on June 29, Amin revealed the hijackers' demand. It was release of 53 Palestinian "freedom fighters" imprisoned in five countries (40 of them in Israel), 6 German Baader-Meinhof members imprisoned in West Germany, 5 in Kenya, 1 in Switzerland, and 1 in France. Within forty-eight hours—that is, by 2:00 p.m. of July 1—the released terrorists were to be brought to Entebbe. If these terms were not met, the passengers would be killed.

The ultimatum presented Rabin and his closest ministerial advisers with a crisis of conscience. It was the traditional view of Israeli governments that compromise with terrorism was unacceptable if there existed the faintest possibility for direct action. On the twenty-ninth, the prime minister queried the military chief of staff, General Mordechai Gur: "Does the Israel Defense Force have any possible way of rescuing the hostages? If not, we shall have to consider negotiating with the hijackers." Negotiation was a route that Shimon Peres was unwilling even to contemplate. On his own, the defense minister pressed General Gur and other senior officers to devise a military alternative.

The "window" for that option suddenly opened when the hijackers made two mistakes. Even before learning of Rabin's impending reply, his willingness to "discuss" the ultimatum (the cabinet's tactical effort to buy time), the terrorists extended their deadline by seventy-two hours, until 11:00 a.m. of July 4. They also decided to free over 100 non-Israeli hostages. The unexpected three-day grace period gave the Israeli general staff planning time. Moreover, 8 of the freed hostages had been returned to Paris and now were available for questioning by the Israeli secret service. It was from them that interrogators learned the precise location of the captives, of their guards, of a unit of Ugandan troops at the old terminal at Entebbe airport. To this information the military in Israel added data gleaned from photographs taken by their air force personnel some years before, when they were training the Ugandan air force. Whereupon General Dan Shomrom, Israel's commander of paratroopers, began to formulate a rescue plan. On the morning of July 3, Chief of Staff Gur informed Peres that a military alternative was both possible and feasible. Rabin's special ministerial "crisis" team then endorsed the plan on the spot.

At 3:00 p.m. of July 3, even as the full cabinet was debating the scheme, four air force Hercules transport planes and a Boeing 707 command plane took off from Sharm es-Sheikh. Peres had given the takeoff order on his own authority; the planes could be recalled if the cabinet vetoed the operation. The cabinet approved. After seven hours of flying, mainly along the East African coast, the five aircraft came within range of the Entebbe radio tower.

The expedition was planned to enable the first Hercules to dovetail behind a scheduled British cargo flight. Thus, as the British plane landed, the lead Hercules glided in immediately behind it without arousing suspicion in the tower. The advance paratroop unit jumped out while the aircraft was still taxiing, and placed mobile landing lights for the other three transport planes. The Hercules then taxied to a dark corner of the field and the rest of the landing party disembarked with a Mercedes automobile and two Land Rovers—vehicles that precisely resembled President Amin's usual entourage. The automobiles drove toward the old terminal building. Challenged there by a Ugandan soldier, the Israelis shot the man, then burst into the terminal, where the hostages were confined. The four terrorists stationed there were promptly cut down. Thus far, the operation had worked to near perfection.

At this point the three other transport planes landed, discharging additional troops and a number of armored cars. The Israeli contingent mopped up at the terminal, killing another two terrorists, then bundled the hostages out of the hall and into the evacuation plane. Meanwhile, General Shomron, directing the ground operation, ordered a unit to "clean out" the new terminal. The assault went smoothly. The Ugandan troops there surrendered. Originally the Israelis had intended to refuel their planes from tanks in Entebbe, but the Boeing 707 command now received word from Tel Aviv that the expeditionary force would proceed to Nairobi. Detesting Amin, Kenya's President Jomo Kenyatta had quietly agreed to cooperate with the Israelis. Only fifty-seven minutes after the initial landing, then, the first Hercules took off, loaded with freed hostages. Forty-two minutes later, the fourth and last aircraft departed. All planes landed safely in Nairobi, where they were refueled, and then proceeded to Israel. Of the 104 hostages evacuated, 3 had been killed in the operation. One Israeli officer, Lieutenant Colonel Yonatan Netanyahu, was the sole military casualty. A final hostage, an aged Israeli woman, who was in the hospital at Kampala at the time of the raid, was subsequently murdered by vindictive Ugandan soldiers.

Once news of the rescue was announced, the entire population of Israel appeared to explode with joy. A huge crowd descended on Ben-Gurion Airport to welcome the returning commando units and passengers. The nation's thrill of pride and relief evinced more than gratification at a remarkable military feat. It was a collective psychological release after the trauma of the Yom Kippur War, after endless economic problems, strikes, and unrelieved Arab hostility. The reprieve was an important one for Rabin, too, whose political tenure was becoming shaky. Not least of all, the example of Entebbe inspired an upsurge of antiterrorist efforts well beyond Israel. Thus, on October 18, 1977, in Mogadishu, Somalia, a West German commando unit flew in to rescue a Lufthansa plane that had been diverted there by four PFLP terrorists. Ninety hostages were freed. In Larnaca, Cyprus, on February 19, 1978, an Egyptian commando unit similarly was flown in to liberate eleven hostages on a Misrair passenger plane, hijacked by Palestinian terrorists. Tragically, the commandos were wiped out by the Cypriot National Guard; but the Cairo government at least had not been intimidated from taking action. At Orly Airport on May 15, 1979, French police shot three armed Palestinians who had been preparing to massacre El Al passengers en route to Tel Aviv (in 1975, the same French police put an airliner at the disposal of terrorists who had just wounded twenty persons in an attack on an El Al plane). Once again, little Israel projected a larger-than-life image throughout the world.

THE ALBATROSS OF THE TERRITORIES

As matters developed, the 1975 disengagement agreement with Egypt and the sensational Entebbe operation climaxed Labor's historic domination of Israeli public life. It was the issue of the occupied territories that ultimately took its toll of the Rabin cabinet, as the Yom Kippur War had eroded the credibility and authority of the Meir government. In the immediate aftermath of the earlier, June 1967 victory, the assumption was widespread that Israel would return the largest part of captured terrain in exchange for peace. Through U.S. intermediaries, the Levi Eshkol cabinet on June 19, 1967 declared its willingness to sign peace treaties with Egypt and Syria based essentially on the former international boundaries, although subject to demilitarization of the evacuated areas and a special agreement for Sharm es-Sheikh; and to be "unbelievably generous" (as Foreign Minister Abba Eban promised) in its territorial accommodation with Jordan. But the "phone call" from the Arab capitals did not arrive. Rather, in late August 1967, a gathering of Arab leaders in Khartoum flatly rejected the very notion of peace with Israel. The Eshkol cabinet thereafter abstained from any further public discussion of returning captured lands.

Instead, over the next few years, Israel carried out a hybrid policy that in practice combined the ideas of Moshe Dayan and Yigal Allon. Dayan, charged with the military government of the territories in his role as defense minister, presided over a kind of informal dual Israeli-Jordanian rule in the West Bank, with the Israeli army responsible for the security of the area, and local administration left to the responsibility of Arab mayors, their councils, and the civil servants originally appointed by Jordan's royal Hashemite government. The arrangement represented functional, as distinct from territorial, partition. By contrast, the government's territorial format was influenced by a document submitted within three weeks of the Six-Day War victory by Yigal Allon, minister of education in the Eshkol cabinet, labor minister under Golda Meir, and subsequently foreign minister in the Rabin government. The "Allon Plan" would have relinquished the main Arab-populated areas of the West Bank to Hashemite political jurisdiction, but retained under direct Israeli military control the strip of uncultivated and thinly populated land along the Jordan River. Arab military forces would have been prohibited west of this security buffer.

For all its liberality, the Allon Plan evoked little interest in the Arab world, even from King Hussein of Jordan, normally the most moderate of

Arab leaders. As Hussein observed in a private meeting with Abba Eban in London in 1969, the formula still would have left Israel in possession of nearly one-quarter of the West Bank—however barren and underpopulated—as well as of Arab East Jerusalem. Such a map, explained the king, would have been politically unacceptable in the eyes of his people and the Arab world. In ensuing years, therefore, it became Israel's de facto policy to carry out the Dayan version of benign military government, on the one hand, and to use the Allon boundaries as guidelines for Israeli settlement, on the other. In East Jerusalem, Jewish residential clusters went up both within the traditional city limits and in a network of outlying suburbs. Beyond the Etzion Bloc, a picket line of Nachal—farmer-soldier—outposts took shape along the Jordan Valley. Later, as shall be seen, certain exceptions were made to this carefully delimited settlement. But it was significant that, more than six years after the June war, the pattern of Jewish habitation in the West Bank was quite skeletal.

This minimalist holding action was not accepted with equanimity by all factions within the Labor Alignment, let alone by other political parties. To be sure, many in the Alignment discerned in the 1967 triumph a historic opportunity to prove Israel's desire for peace by renouncing all territorial claims. Even such Labor centrists as Abba Eban, Pinchas Sapir, and David HaCohen regarded the very notion of ruling a million captive Arabs as a political and moral threat to Israel's democracy. There were others, however, with roots equally deep in the Labor mainstream, who adopted a far tougher stance on the occupied areas. They included such eminent Socialists as Rachel Yanait Ben-Zvi, widow of Israel's second president and a veteran Labor leader in her own right; Eliezer Livneh, a respected Labor theorist; and Israel Galili, elder statesman of Labor's Achdut HaAvodah faction. Among these "territorialists," security was an unavoidable consideration, in the need to gain additional defense in depth.

But eventually psychological, even mystical, currents were tapped. These welled from classical Zionism itself. After the 1967 victory, hundreds of thousands of Israelis came to regard the Six-Day War less as a struggle for survival than as a direct extension of the 1948 war of independence, a culmination of Zionism's "unfinished business." The historic mythos of the "Promised Land" was accepted by every Jewish citizen, after all, religionist and secularist alike. Biblical studies were a central feature of the nation's school curriculum. No one needed reminder that the early Zionist leaders had rejected all thought of a Jewish homeland anywhere but in Zion. Whenever Chaim Weizmann, David Ben-Gurion, Moshe Sharett, and other senior Zionist statesmen had acquiesced in an attenuation of the homeland—by the Cairo Conference in 1921, the Peel Report in 1937, the United Nations Partition Resolution in 1947—they had done so under heavy pressure, and for purely tactical purposes. Now, at last, since the June victory in 1967, that pressure could be resisted. The West Bank had been governed by the Hashemites without de jure international recognition, after all. The Arab inhabitants of the Gaza Strip had been ruled—harshly—by an Egyptian military government. For all the moral resonance of the Arabs' case, the vulnerability of their legal claim to Palestine appeared further to vindicate the conviction that the moment of Zionist fulfillment had arrived.

There were, as well, economic factors that militated against "repartition." By the early 1970s the interaction between Israel, the West Bank, and Gaza had become that of a common market, with the agricultural Palestinian territories all but sucked into the highly developed Israeli industrial economy. Perhaps the most visible evidence of this integration was the influx of Arab workers commuting to Israel each morning from Gaza and the West Bank. By the mid-1970s, Arabs both from Israel and the territories provided nearly a quarter of Israel's factory labor, and half the workers in construction and in service industries such as hotels, garages, and municipal sanitation. It was the supply of cheap labor, in turn, that relieved the pressure for higher wages in Israel, thereby reducing the inflation rate and the need for more extensive capital investment. Moreover, with their geographical proximity, their negligible transportation costs and high tariffs, the territories became virtually a captive outlet for Israel's manufactured goods. By 1975 not less than 16 percent of all Israeli exports were sold in the territories, a larger market for Israeli goods than that of Britain and Germany combined, and more than half the size of the entire American market. With Israel's initial administrative expenses in the West Bank and Gaza essentially covered by revenues, the occupied areas soon became so integral to Israel's own economy that the term applied to them, "territorial administrative entity," eventually became as irrelevant as the term "Beersheba administrative entity."

Little wonder then that the problem of the West Bank and Gaza should have confused and tormented the Labor Alignment, arousing more contention among its members than did any other single public issue. Prime Minister Golda Meir managed the feat of accommodating these adverse viewpoints simply by refusing to define a position on the territories at all. Her personal immobilism, it is recalled, was translated into a kind of national posture, and her approach to the Palestinian question was to deny its existence, or even the existence of a historical entity such as the Palestinians altogether. Yitzchak Rabin, who assumed the prime ministry as a member of a minority component within Labor, enjoyed even narrower leeway on the issue. Temperamentally, he himself was close to the partitionist approach of his friend and mentor, Yigal Allon; but he dared not provoke the further hostility of the Rafi group, whose leader, Shimon Peres, was emerging as a bitter political rival. Immobilism accordingly remained the prime minister's stance, as it had been of his predecessors. That posture may well have hardened during Rabin's disengagement negotiations with Egypt. In the eyes of this former military chief of staff, it was the relationship with the largest and most powerful of enemy states, not with the Palestinians, that offered the key to Israel's security.

It happened that, in earlier years, Jordan's King Hussein had expressed a cautious tactical willingness to regain the West Bank in stages. Thus, immediately following the Yom Kippur War, the Hashemite ruler had proposed his own version of a separation-of-forces agreement under which Israel would pull its troops back from the Jordan River along a strip 7.5 miles wide. The Meir cabinet initially rejected the notion. A year later, however, Yigal Allon, foreign minister in the new Rabin government, floated the idea of returning the much smaller Jericho area and linking it to Jordan. This time it was Hussein who was reserved; the proposal was too grudging for his taste. Afterward, then, as the Rabin government became increasingly preoccupied with the task of defusing the Egyptian front, of mobilizing national support for possible additional Israeli withdrawal in Sinai, it simply dared go no further on the much touchier issue of Palestine. In August 1974, Hussein arrived at a Tel Aviv suburb by helicopter for yet another secret meeting with Israel's leaders. In the ensuing discussions, Rabin and Allon ruled out unilateral withdrawal. Instead, they offered a federative arrangement on the West Bank, with Israel and Jordan sharing in security and economic arrangements. Graciously, but firmly, Hussein rejected the palliative, and Rabin was fearful of offering more. Two subsequent meetings with the king, in late 1974 and early 1975, failed to produce movement. Beyond his fixation with a Sinai disengagement agreement, Rabin had a new worry. The National Religious party, negotiating to enter the government coalition, insisted that any future withdrawal proposal for "Judea and Samaria" be submitted to the referendum of a national election. To Rabin, the prospect of facing the voting public was distinctly unappealing.

THE COMMUNITY OF BELIEVERS

Labor's indecisiveness, its structural vulnerability on the territorial issue, was revealed as early as April 4, 1968, in the arid little West Bank town of Hebron. It was the eve of Passover, and ten Jewish families, all Israelis but masquerading as Swiss tourists, registered at the local Park Hotel. The next day the group's leader, Rabbi Moshe Levinger, son of a Munich-born physician, announced that he and the others were the vanguard of a revived Jewish settlement in Hebron (the previous Jewish community had been slaughtered by Arabs in 1929). Thereupon Levinger and his followers barricaded themselves in the hotel and refused to be moved. The challenge caught the Eshkol government by surprise. Hebron was without strategic importance. A devoutly Moslem enclave, it was the last place even Defense Minister Dayan would have chosen to implant a Jewish outpost. Before the army could move to evict the squatters, however, other Israelis began speaking out on their behalf. In addition to religionists of the Levinger stripe, supporters of the Hebron group included members of Labor, even of kibbutz movements. Reports that Jews were returning to this ancient Palestinian town evidently kindled a dormant spark of half-forgotten biblical memories. Here, after all, the forefather Abraham bought his first piece of land in Canaan, the field of Ephron the Hittite, and later buried his wife Sarah in the Cave of Machpelah. If ever there were terra irredenta for mystics and romantics, Hebron fitted that profile.

Now, therefore, in April 1968, the religionists' challenge swiftly provoked a nationwide debate. The Eshkol cabinet in turn was so taken aback by the public clamor that it decided not to risk a forceful eviction of the zealots. Rather, it devised a "compromise" under which Levinger, his American-born wife, and their retinue would be moved to an Israeli army compound in Hebron. But when the squatters disobeyed military orders by conducting aggressive prayer sessions at the Tomb of the Patriarchs—a Moslem holy place, too—five of them were ordered out. Here again a political storm was touched off. And again the government backed down, countermanding the eviction order. Whereupon Levinger and his group launched into the construction of a settlement, "Kiryat Arba," on the outskirts of Hebron. In ensuing years, the number of settlers and homes continued to grow. Their sewage facilities, electricity, and other town services in fact were provided by the government of Israel itself. Far from merely tolerating a Jewish community that had no imprimatur in Israeli law or military purpose, the government between 1969 and 1973 spent over $10 million making Kiryat Arba viable (fifteen years later the town's population approached 2,000). A pattern of defiance was set. The Hebron squatters had created "facts."

As it happened, Levinger and his band of fanatics had been prefigured, and eventually would be transcended, by a group to be known as the Gush Emunim (roughly translated, the Community of Believers). Most of its initial members were young men and women who had been educated in state religious schools and in the pioneering religious B'nai Akiva movement. Few were zealots of the Levinger ilk. Indeed, before 1967 B'nai Akiva had been a generally restrained and courteous group, associated with the moderate Mizrachi Workers faction of the National Religious party. Their personal religiosity had not obtruded on their neighbors' freedom of action, and still less on the government's foreign and security policies. One of their components, however, included graduates of the famous Rabbi Kook Yeshiva, a religious seminary named in honor of the beloved (and generally moderate) Avraham Kook, chief rabbi of Palestine Jewry's Ashkenazic (European) community during the British mandatory period. The yeshiva's director and guiding light was Rabbi Zvi Kook, son of Avraham. In May 1967, on the eve of Israel's independence day, it was Zvi Kook who addressed his students, lamenting the sins of the nation for "abandoning" Hebron and Bethlehem to Arab rule. In the light of the sensational Six-Day War victory a month later, Kook's observations retrospectively appeared prophetic. Thus, in ensuing months and years, they nurtured an annexationist spirit among his followers and among other young religionists. Under the covenant between God and Abraham, insisted the rabbi, the right of the Jewish people to Samaria and Judea was incontestable and nonnegotiable. It was a curiously unbudging stance for Israel's pietists. Until the early 1970s, the National Religious party, to which most of the youthful B'nai Akiva and Rabbi Kook disciples belonged, still remained less preoccupied with the messianic obligation of settling Judea and Samaria than with the purely domestic role of Orthodoxy in Israel's national life. It was specifically to transcend this earlier, parochial association, therefore, that the young activists gathered in February 1974 to organize themselves as the Gush Emunim, and to sever their formal organizational links with the NRP.

The move was a shrewd one. Henceforth the group was open to anyone, religious and nonreligious. Their single-minded preoccupation with settling the historic sites of the biblical Land of Israel touched wellsprings of a Zionist idealism that attracted kibbutz and moshav members, native-born *sabras* and recent immigrants from the United States and the Soviet Union, marginally educated North Africans and learned academicians. Soon the Community of Believers picked up momentum almost in direct ratio to the Labor government's confusion and equivocation. In 1974 and 1975, it launched several new settlements in the heart of Arab-inhabited territory in Judea. And once again, as in the case of the Hebron squatters, the Labor cabinet was ambivalent in its reaction. Rabin and Peres were vying then for leadership of the party, and each kept a prominent Gush Emunim spokesman in his ministry. Rabin had Ariel Sharon, the hero of the 1973 Suez Canal crossing. Peres had Yuval Ne'eman, a distinguished scientist with plainspoken annexationist views. At Peres's orders, army bases provided electrical and water utilities for the Gush Emunim settlers. Rabin, in turn, although repelled by the group's aggressiveness, similarly found it useful to express admiration for their "pioneering zeal."

A particularly flagrant Gush Emunim challenge to the hapless Labor confederacy occurred in June 1974. The site this time was a tract of land near the West Bank city of Nablus, "capital" of Samaria and a traditional seedbed of Arab nationalism. Here some twenty young Israeli families ensconced themselves on a plot near the ruins of ancient Sebastia and proclaimed their own settlement to be the revived biblical community of Elon Moreh. The army promptly ordered them out. The settlers ignored the order. Soon the roads to Sebastia were crowded with thousands of Gush Emunim members and supporters. Here the government again split into two camps. Foreign Minister Allon insisted upon the immediate and unflinching eviction of the squatters. Rabin counseled restraint. So did Peres. Finally, in January 1975, after months of indecision, the government came up with an apparent compromise. If the settlers agreed to depart Sebastia, they would be transferred to a nearby army camp outside the Arab village of Qadum until the fate of their project was decided.

Yet, when the group accepted the "compromise" as the victory it was, Foreign Minister Allon angrily demanded that the agreement be repudiated; and Rabin, taken aback by this reaction from his closest political ally, compounded his earlier indecision by declaring that the settlers would be evacuated from Qadum within a few weeks. But they were not. The weeks passed, then the months and years, and there was no evacuation. Again, a Gush Emunim settlement had become a fait accompli, the government had been exposed as spineless. By early 1977 there would be five Gush Emunim settlements on the West Bank. By then, too, it was uncertain who entertained the greater contempt for Rabin and his cabinet: the "doves," who favored eventual withdrawal from the Arab-inhabited areas; or the "hawks," who

divined unerringly the indecision and confusion of a Labor government possessed of no collective vision, not even a clue, to the disposition of "reclaimed Judea and Samaria."

A DETERIORATION OF PUBLIC TRUST

Labor's record on the economy was not significantly more brilliant than on
the occupied territories. Many of the nation's financial burdens unquestionably represented a force majeure. To compensate for a vast infusion of Soviet
and French weaponry to Egypt and Syria after the Yom Kippur War, the
government was obliged to double its security budget. Tens of thousands of
Israelis were pulled out of the economy for extended terms of military service. The share of defense in Israel's GNP soon more than doubled, from 21
percent in 1972 to 47 percent in 1976. The escalating (worldwide) cost of oil
and of other imported goods added to a national budget deficit of over $4
billion by 1976. With only $1 billion of that deficit made good by U.S. grants
in aid, by German restitutions and Jewish philanthropies, the rest had to be
covered by loans from Washington and other sources. The expense of servicing the debt, in turn, of covering a wider balance of payments gap, was
reflected in the cost of living. Inflation rose by 30 percent in 1973, 40 percent in 1974, and again 40 percent in 1975. If Rabin did not create this
situation, neither did his administration manage to resolve it. Although his
able finance minister, Yehoshua Rabinowitz, ruthlessly cut imports and food
subsidies, and eventually achieved a modest replenishment of Israel's hard
currency reserves, the premier himself issued no forceful call for the nation
to accept additional necessary sacrifices. Rather, in this period, Israel was
plagued by waves of strikes, most of them organized not by blue-collar workers but by salaried professional groups—engineers, physicians, teachers—
who proved remarkably adept at immobilizing vital public services. The government for its part seemed to look on helplessly, and in the end to surren-
der to the strikers' demands.

Israel's social economy was further distended by an inequitable fiscal
structure. Affluent citizens often succeeded in avoiding the heaviest taxes.
Indeed, many of the self-employed and well connected managed to avoid
taxes altogether. The rot infected not only the business and professional
communities (Chapter X), but the highest echelons of public life. A particularly notorious scandal involved the Israel Corporation, established in 1968
by the Eshkol government and funded by Jewish investors throughout the
world. In February 1975, the corporation's former managing director was
indicted (and eventually convicted) of fraud and theft for having transferred
millions from the company to a number of shaky enterprises in Europe. A
year later, the director of customs was indicted for having amassed extensive
cash payoffs from a businessman whose imports had been allowed in duty-
free. In March 1976 the deputy director of the Bank of Israel's examiner of
banks department was arrested for having used his position to speculate and
earn vast profits in bonds. Meanwhile, the former manager of Netivei Neft,

the Israel-operated oil wells in the occupied Sinai, was sentenced to prison for having privately operated oil-drilling equipment he had bought with public funds. Then the director-general of the Histadrut—labor federation—sick fund was indicted for accepting bribes from potential suppliers. Some months later the minister of housing was placed under investigation on suspicion of having illegally dispensed funds to the Labor party. However widespread, the atmosphere of permissiveness and moral decay was increasingly identified with the Labor leadership.

In the end, it was the sapping guerrilla warfare within that leadership that delivered the coup de grâce to the government. Since early 1974 the struggle for power among Rabin, Peres, and (to a lesser degree) Allon led to a near paralysis of executive responsibility. A political newcomer, Rabin often functioned as little more than chairman of the board at cabinet meetings. The animus between himself and Peres became all but uncontrollable, intruding in ministerial discussions, even undermining the line of command in the defense establishment. Thus, to outflank Peres, Rabin in 1975 appointed General Ariel Sharon, closely associated with the right-wing Likud bloc, as his special military "adviser." Dropping all restraints in fighting back, Peres in turn began consulting his old comrade-in-arms Dayan on defense matters, even letting him in on highly secret information. Whereupon Rabin issued orders that Peres, the minister of defense, should be denied access to military intelligence reports. "It was no secret that there was a considerable degree of friction between the defense minister and myself," Rabin wrote later, with much understatement. By 1976 Peres had decided to challenge Rabin within the party central committee before the scheduled November 1977 elections.

Rabin, in turn, aware that opinion polls indicated a national preference for Peres, resolved to fight back. In December 1976 he thought he saw his opportunity. He had attended an airport welcoming ceremony that month for a newly delivered contingent of American jet fighter planes. The event intruded into the Sabbath. Learning of this "desecration," the ultra-Orthodox Agudat Israel faction introduced a motion of nonconfidence in the government. The motion was easily defeated in the Knesset, and there the matter might have ended. Yet, it developed that several members of the National Religious party, partners in Rabin's coalition government, had abstained on the vote, thereby violating cabinet discipline. Hereupon Rabin decided to tender his resignation, together with his cabinet's; and a special Knesset resolution then advanced the 1977 elections from November to May of that year. In the interval, Rabin would stay on as premier of a caretaker government, and presumably the crisis of an accelerated election would induce Labor to rally around its current leader. The tactic proved less than inspired. At the central committee meeting in February 1977, Rabin was indeed re-elected as party leader, but by the hair-thin margin of 1,445 to 1,404. His standing this time was publicly shaken, and Peres's was augmented even further.

In the ensuing months of the campaign, too, more of the national financial scandals broke, including the conviction of the former director-general of the

Histadrut sick fund and the suicide of the minister of housing. Then, on March 15, only three weeks after Rabin's reelection as party leader, an apparently innocuous item appeared in *HaAretz*, the nation's leading newspaper. It disclosed that Rabin's wife Leah had maintained a secret dollar account in a Washington bank, well after the couple's return to Israel following Rabin's stint as ambassador. As a joint signatory of the account, the prime minister quickly admitted the violation of Israel's currency laws, but insisted that the sum involved was a mere $2,000, and was simply an oversight. It was rather more than that. The attorney general subsequently learned that the Rabins possessed a second Washington account, this one for $23,000. Into that account Rabin had deposited a portion of the tidy nest egg he had built for himself during his tenure as ambassador, when he had accepted fees for lectures around the United States. Once this information was revealed, the prime minister could not escape public obloquy. On April 7 he announced his intended resignation as head of government and chairman of the Labor party. Yet even this gesture was accomplished clumsily. It turned out that, as head of a caretaker government, Rabin was unable to resign. With his only alternative a departure "on leave," he then asked Peres to take over from him as "chairman of cabinet meetings." It was not until April 21 that a special meeting of the party central committee confirmed Peres's candidacy for the premiership.

By then confusion had exacerbated paralysis. The Israeli people, even many tens of thousands of Labor's traditional supporters, witnessed the public display of scandal and backbiting with growing revulsion. The nation was in disarray, its morale shaken by the belated implications of the Yom Kippur War, its Arab enemies seemingly rejuvenated in their ability to confront Israel on future battlefields, its citizens bitterly divided on the fate of the occupied territories, its economy battered and dependent almost entirely on the goodwill and largess of the United States. It appeared that the erosion of public morality had left both the Labor party and the Labor government not only corruption-ridden but evidently rudderless. If the electoral debacle that followed was a "revolution," then, it was hardly an upheaval without auguries.

CHAPTER 11

ISRAEL TURNS TO THE RIGHT

THE UPHEAVAL OF MAY 1977

Amidst a turmoil of contending parties, the electoral campaign gave every appearance of following earlier precedents, of devolving essentially into a contest between three major blocs. The largest bloc, the Labor-dominated "Alignment," was established in 1968, and comprised a reunited Labor party (itself a consolidation of the old Mapai, Achdut HaAvodah, and Rafi parties) functioning in alliance with the leftist Mapam. At the other end of the political spectrum, the Likud bloc dominated the right-wing opposition. Under the leadership of Menachem Begin, this confederation represented the original Cherut party, the much smaller Liberal party, and even tinier residual factions such as the State List and La'am. Finally, there were the religious parties. Preeminent among these was the National Religious party, a joinder of the former Mizrachi and Poel Mizrachi (Mizrachi Workers); together with the loosely associated, if far more intractably fundamentalist, Agudat Israel faction. The NRP generally had been able to work out an accommodation with Labor over the years, to share in Labor-dominated cabinets. But it was the religionists' defection in December 1976 that made their allegiance problematic in any future Labor coalition. During its opening months, in any event, the campaign seemingly reflected the classical triangular rivalry among Israel's three historic competitors.

By the spring of 1977, the political rhetoric became increasingly shrill. Labor warned that a Likud victory would threaten democracy and workers' rights. Likud charged that Labor would offer unacceptable territorial concessions to the Arabs and fuel inflation. The campaign's bitterness was evident then in a harsh television debate between Peres and Begin. Ironically, on foreign policy issues, the two candidates appeared to be separated more by degree than by kind. Although such Labor stalwarts as Yigal Allon and Abba Eban favored a substantial Israeli withdrawal from the West Bank, Peres adopted an indeterminate "middle" position. No doubt he had been made unduly cautious by the polls, which indicated a slight edge to Labor. Likud's campaign was more shrewdly managed. Ezer Weizman, a former air force chief of staff, returned from his private life as a successful businessman to coordinate the right wing's effort. With the advice of an American public relations firm, Weizman concentrated less on foreign policy issues than on Labor's corruption scandals, its flaccidity on economic matters. When Labor in turn countered with allusions to Begin's early years as a terrorist and inciter to political violence, its leadership forgot that a majority of Israel's voters was young, with memories of Begin not as a fire-eater but of a states-man who had shared in the coalition government of 1967–70. And people knew exactly where Begin and Likud stood on the issue of the territories— as they did not with Peres and Labor.

The pollsters who had forecast a close Labor victory were proved wrong. When the votes were counted, Likud was found to have increased its Knes-set representation from 39 to 43 seats. The National Religious party similarly enlarged its delegation, from 10 to 12 seats (the Agudists essentially held their own, with 5 seats). Yet the true surprise was not the impressive show-ing of the Right but the collapse—it was nothing less—of the Alignment, from 51 seats in 1973 to 32 now, a disappearance of a third its strength. Indeed, the humiliation of the leftist parties was all but complete. The In-dependent Liberals, formerly the "progressive" wing of the Liberal party, lost 3 of their 4 seats in 1977. Shulamit Aloni's egalitarian Citizens Rights Movement lost 2 of its 3 places. Sheli, an intensely dovish faction, lost 3 of its 5 seats. Altogether, between the elections of 1973 and 1977, about half the nation's voters changed their political allegiance. Never before in Israel's history had there been such a shift away from the Left.

To the extent that the election signified a repudiation of Labor, it was a new centrist group, the Democratic Movement for Change (DMC), that played a decisive role in the political upheaval. The DMC had its origins in the protest demonstrations that welled up in the aftermath of the 1973 Yom Kippur War. Among the civilians and demobilized soldiers who indicted Dayan and the entire Meir government for the "blunder," there were some who perceived a more fundamental dysfunction in Israel's basic political sys-tem. These were the critics who subsequently focused on the domestic pic-ture, and urged a transformation of the party system, a depoliticization of the military establishment, of the economy and government bureaucracy. Their effort finally took organized form in 1976, adopting the forthright title, Democratic Movement for Change.

Although new political configurations outside the main party lines rarely had succeeded in Israel, the DMC seemed to have a chance. It offered its moderate, centrist followers—most of them educated, upper-middle-class Europeans—the opportunity to cast a protest vote equally against the inertia of Labor and the chauvinism of Likud. The caliber of the DMC's leadership was another inducement. Its chairman was Professor Yigael Yadin, the re-nowned archaeologist and military commander of Israel's war of indepen-dence. Its co-chairman was Amnon Rubinstein, dean of Tel Aviv University Law School, and a founder of the Shinui ("Change") faction that now was confederated into the DMC. Equally attractive were the figures Yadin and Rubinstein recruited for their ticket. These were a cross-section of doves and hawks, respected former Laborites and military leaders, intellectuals and industrialists—all determined to wield a fresh broom for reform.

The DMC broke little new ground in foreign policy. Initially, its spokes-men adopted a rather hard line on the territories, rejecting withdrawal in the absence of a formal and enforceable peace treaty. But Yadin and Rubinstein in any case preferred to concentrate on such domestic issues as inflation, labor unrest, social inequities, and government corruption and inefficiency. Above all, the DMC stressed the urgent need for electoral reform, with constituency districts to replace the anachronistic party lists and oligarchical party central committees. The program, and its advocates, made their impact. A postelection survey revealed that two-thirds of the DMC's votes came from Israelis who had supported the Labor Alignment in 1973. Without this defection, the DMC would never have become almost overnight Israel's third largest parliamentary delegation, winning an astonishing 15 Knesset seats.

Yet if the political upheaval represented less an accession of Likud strength than an outward flight of Labor voters, the victory of the right wing was impressive enough to be taken seriously on its own account. Menachem Begin himself plainly no longer frightened as many citizens as in earlier elections. During his three-year stint as a member of the Government of National Unity, from 1967 to 1970, he had kept a low profile, exuding respectability, shedding the label of demagoguery Ben-Gurion had pinned on him. Electoral issues, too, had changed after the Six-Day War. Israel's political debate until then had concentrated largely on domestic matters at a time when the nation's population was quadrupling and grave economic crises had to be faced. Questions of partition, of borders and Jerusalem—those that had brought Revisionism into existence in the first place—seemed to be irrelevant during Israel's early years. All this changed in 1967, of course. Israel henceforth remained in constant turmoil over the future of the occupied territories. If Begin's views on the subject did not differ significantly (except in rhetoric and historical mysticism) from the painstakingly balanced territorial stance adopted by Peres, that fact too accorded his annexationist posture on Judea and Samaria a certain retroactive legitimacy. And at least his position was simpler than the rather tortured qualification of Labor; for it projected clear-cut moral choices and perceived an "undivided Land of Israel" in terms of an ancient vision.

Of transcendant importance, finally, in the massive shift away from Labor was the revised political allegiance of Jews from Moslem countries, and particularly of their children, who were reaching voting age. Postelection studies left no doubt on this point. The defection of educated Ashkenazic (European) Jews may have been to the Democratic Movement for Change. But in workers' and lower-middle-class districts, most of these heavily oriental, the principal beneficiary of the switch was Likud. In the 1950s, Jews of oriental or Sephardic (Ibero-Near Eastern) background had identified overwhelmingly with Labor, and as late as 1969 the Alignment received 55 percent of their vote. Then, in 1973, the figure dropped to 38 percent. And now, in 1977, it sank to an astonishing 32 percent. The rightist parties, on the other hand, claimed 26 percent of the oriental-Sephardic vote in 1969, 39 percent in 1973, and 46 percent in 1977. Neither did these statistics tell the full story. With their larger and younger families, the orientals were becoming Israel's demographic majority.

The non-Europeans' repudiation of Labor in turn was fueled by plain and simple resentment of the Ashkenazic elite. As late as 1977, even after years of government-directed affirmative action programs in education and public welfare, in the Knesset and political parties, in the mayoralties and munici-pal councils of Israel's principal development towns, the orientals—particu-larly Moroccan and other North African Jews—still had failed to achieve a quality of life approaching that of the Europeans. Despite an absolute rise in their real income, their share of white-collar jobs did not exceed half that of Ashkenazic Jews, whose per capita consumption remained 65 percent higher than that among the orientals. Indeed, the galloping inflation of the mid-1970s served only to widen the social and economic gulf between the "two Israels." In the so-called black belt, the poorer, oriental Hatikva quarter of Tel Aviv, in the Musrara and Katamon quarters of Jerusalem, local discon-tent was aimed specifically at Ashkenazic control of the upper echelons of the economy, of the national government, the Histadrut, the political par-ties, the armed forces, the universities.

The Labor Alignment was uniquely vulnerable to this frustration. There had been a time when socialism in Israel was identified with authentic egal-itarianism. By the 1960s and 1970s, this was hardly the case. Not when "Socialists" like Eban, Rabin, and Dayan enjoyed the lifestyles of million-aires, and when other prominent government figures were being convicted of fraud and embezzlement. During the Ben-Gurion era, Labor had offered the oriental communities enough welfare and patronage to ensure a majority of their votes. By the 1970s, Ashkenazic "paternalism" and "tokenism" no longer evoked that loyalty. One manifestation of the new restiveness was the Black Panther movement. Borrowing from the example and terminology of American ghetto dwellers, groups of young Moroccans from Israel's urban slums launched clamorous public demonstrations for improved housing and employment opportunities. Golda Meir failed to grasp the sociology behind these protests, and could only describe the Black Panthers as "not nice boys"— a widely publicized remark whose implications eventually would help bring Menachem Begin to power. In fact, the orientals were by no means an un-differentiated proletariat any longer. Many, particularly the Iraqis, had be-come shopkeepers, building contractors, truck and restaurant owners. Yet it was this ascendant Sephardic-oriental petite bourgeoisie, too, that joined with their blue-collar kinsmen in rejecting the customary tutelage of Labor block captains.

To these social factors was added a deeply rooted suspicion of the Arabs. It was the orientals' families and forebears, after all, who had suffered im-memorially under Moslem rule. Renewed contact with their former perse-cutors after the 1967 war, when thousands of West Bank and Gaza Arabs became commuting laborers in Israel, revived old grievances. Ensconced vocationally a rung above these Arabs, moreover, oriental Jews tended to develop a "poor-white" mentality toward them, to generate a vested interest in the status quo. As late as 1981, the writer Amos Oz quoted the recrimi-nations of a group of North Africans in a development community:

What did you bring my parents to Israel for? Wasn't it to do your dirty work? You didn't have Arabs then, so you needed our parents to do your cleaning and be your servants and your laborers. . . . You brought our parents to be your Arabs. But now I'm a supervisor. And he's a contractor, self-employed. And that guy there has a transport business. . . . If [you] give back the territories, the Arabs will stop coming to work, and then and there you'll put us back into the dead-end jobs. . . . Look at my daughter: she works in a bank now, and every evening an Arab comes to clean the building. All you want is to dump her from the bank into some textile factory, or have her wash the floors instead of the Arab. The way my mother used to clean for you.

The observation may not have been off the mark. If Socialists of European background longed for a return to the traditional value of Jewish self-labor, it was labor performed by oriental Jews. Menachem Begin did not hesitate to exploit that suspicion. In his campaign, he excoriated the "kibbutz millionaires" and their "snobbish indifference" to the non-Europeans. Begin's hawkishness similarly evoked an instant response from the orientals. With little experience of democracy in their family or ethnic background, with their traditional exposure in their lands of origin to strong paternalistic leaders (Labor had produced none since Ben-Gurion), the non-Europeans were uniquely susceptible to Begin's charismatic authoritarianism and florid invective. The 1977 election was their long-awaited moment of vindication against the Ashkenazic power structure.

BEGIN: THE MAN AND THE IDEOLOGY

The moment was even sweeter for Menachem Begin. Now sixty-four years old, a bespectacled, unprepossessing man of frail stature and uncertain health, he was a leader driven less by ambitions of glory than of nationalist fulfillment. These were first and foremost territorialist. Since the 1967 war, Begin had contemptuously rejected a description of the occupied West Bank areas as either "occupied" or "administered," and had insisted that they be identified by their biblical titles of Judea and Samaria. This was no mere gesture of ingratiation to the religious parties, those whose support ultimately would provide the margin of safety in his coalition cabinet. Rather, it bespoke his lifelong commitment to a wholly redeemed Land of Israel.

Polish-born, the son of a prosperous timber merchant, Begin was raised in the heavily Jewish city of Brest Litovsk. There he encountered the full force of government-inspired antisemitism—first tsarist, then Polish—and shared in his family's intense redemptionist devotion to Zionism. For young Begin, the commitment was given renewed fervor in 1935. Completing his law studies at Warsaw University that year, he first met and was overwhelmed by Vladimir Jabotinsky, founder and leader of Revisionist Zionism. Jabotinsky's teachings did not lack in forthrightness: every Jew had a right to enter Palestine; only militant self-defense and retaliation would deter Arab violence; Jewish armed strength alone, not collaboration with the British or diverted effort into Labor Zionist utopianism, would ensure a Jewish state for the historic Land of Israel—that is, Palestine (including Transjordan) in its entirety. Begin was enthralled by Jabotinsky's maximalism. The great man's style also left its impact. As a journalist in Rome, Jabotinsky had been impressed by Mussolini's bravura oratory, which he adopted, even as Begin later adopted Jabotinsky's. Both men similarly shared the rightist propensity for authoritarianism, discipline, uniformed paramilitary youth organizations—all characteristics of European irredentism between the wars. In the case of the Revisionists, the youth group was known as Betar (Beit Trumpeldor), and its membership in Poland reached 70,000 by 1939, when Begin was elected its chairman.

Visibility in a right-wing nationalist cause almost proved the young man's undoing. Once the Soviet army occupied eastern Poland in September 1939, Begin was arrested and dispatched to a Siberian prison camp. His subsequent eight-month ordeal in the "white nights" of Soviet imprisonment may well have toughened his soul irretrievably. So did the discovery later of his parents' death in a Nazi concentration camp. In 1942, as part of a three-way deal between Moscow, the Western Allies, and the expatriate Polish government in London, several thousand Polish prisoners in the USSR were allowed to enlist in the "Polish Army in Exile." Begin was one of these. Less than a year later, the émigré force was shipped to Palestine for advanced training under the British. And there at last, with the tacit approval of his Polish officers, Begin departed his unit and vanished into the companionship of the Irgun Z'vai Le'umi—"Etzel"—the Revisionist underground organization. Almost immediately he assumed a leadership role in the group.

Indeed, during the ensuing years, Begin infused the Etzel with his own grim militancy of purpose, his determination to hound the British out of Palestine and eventually to achieve a sovereign Jewish state. Intensifying its assaults on British military installations, the Etzel subsequently killed and wounded British personnel, extorted funds from Jewish businessmen, "executed" suspected Jewish collaborators and informers. In July 1946, an Etzel unit dynamited Jerusalem's King David Hotel, destroying the wing used by the mandatory government's criminal investigation department, indiscriminately killing ninety-one Britons, Arabs, and Jews. Thereafter, execrated by the Zionist leadership, under threat of hanging or long prison terms at the hands of the British, the Etzel partisans adopted false names and moved from one hiding place to another. Begin himself masqueraded as a bearded rabbi. Nothing deterred him or his followers. Rather, they widened the scope of their attacks, engineering prison escapes, murdering British prisoners, ultimately paralyzing British military communications in large areas of Palestine. Many factors, diplomatic, political, economic, combined to force Britain's abdication of its Palestine mandate, but the dissident guerrilla campaign unquestionably was significant among them.

Begin's reaction to the United Nations Partition Resolution of November 29, 1947, was characteristic. "The amputation of the homeland is illegal," he warned. "It will never be recognized. . . . It will not bind the Jewish people." When the independence of Israel was declared on May 14, 1948, Begin was on the Etzel radio the same night, broadcasting his rejection of the partition boundaries in terms the late Jabotinsky would have approved. "The homeland is historically and geographically an entity," Begin insisted. "Whoever fails to recognize our right to the entire homeland does not recognize our right to any of it. We shall never yield our natural and eternal right." The ferocity of this commitment was also prefigured in the Etzel's independent military operations against the Arabs. On April 9, 1948, their massacre of some 200 men, women, and children in the Arab village of Deir Yassin during the struggle for the Jerusalem highway was a typically brutal response to earlier Arab atrocities against Jews.

Ultimately, the Etzel's most traumatic confrontation was with the emergent government of Israel itself. On June 20, 1948, a 4,000-ton vessel, *Altalena*, approached the Tel Aviv shore. It had been purchased by the Etzel in Europe and carried some 900 Jewish volunteers recruited abroad, with a sizeable cargo of French weapons and ammunition. Prime Minister Ben-Gurion had known of the ship's imminent arrival and had tacitly acquiesced in the violation of the current United Nations truce and arms moratorium. Yet he rejected Begin's demand that the *Altalena*'s weapons cargo be distributed exclusively to Etzel units. When the boat landed, then, and began discharging its passengers and freight, it was attacked and sunk by the Israeli army, with the loss of most of its equipment. Casualties were sustained by both groups. For the shocked public, Etzel defiance of government authority seemed particularly unforgivable at a moment when Israel was struggling for its very life. Nor did Begin's tearful, semicoherent defense of his position on the Etzel radio salvage his reputation for stability. The *Altalena* episode would hang like a cloud over him for years.

When the Palestine war ended, the Revisionists grudgingly agreed to be "civilianized" as the Cherut party. Their showing in the first Knesset election in 1949 was a paltry 14 seats, however, with most of their votes drawn from former Etzel supporters and poorer oriental elements. In fact, Begin had not yet made his peace with the Labor establishment. In 1952, when the Ben-Gurion cabinet sought Knesset approval for a reparations settlement with Germany, Begin threatened the government with physical violence, even mobilized crowds of Cherut partisans in a march on the Knesset building that was driven back only by police reserves. The shocked legislators afterward deprived Begin of his parliamentary seat for three months. Altogether, the man's incendiary conduct helped keep Cherut beyond the political pale for the next fifteen years. Neither did the personal animus between Ben-Gurion and Begin mitigate public suspicion of the Cherut leader. To the prime minister, Begin was a "romantic fool . . . a windbag full of rhetoric and metaphors." By the same token, Begin's endless hectoring of Labor for its "appeasement" of the Arabs, its willingness to "tolerate" Arab guerrilla attacks, its "cowardly" withdrawal from Gaza in 1957 following its recent Sinai victory, more than occasionally transcended the vigorous role expected of a political opposition.

Yet, by the 1960s, Begin appeared to have mellowed somewhat. Unable to invoke the old slogans about "liberating the homeland" in those years of comparative tranquility, he shed his abrasive image and began polishing his skills as a parliamentarian. Thus, he concentrated on the task of negotiating a partnership with the Liberal party. A centrist element, the Liberals traditionally had not supported Cherut's militant foreign policy, but their economic views at least were closer to Begin's. Eventually the two groups joined forces in a common "Gachal" bloc, each maintaining its independent identity, with foreign affairs left to Cherut, domestic issues to the Liberals. In view of his evident pragmatism and flexibility, moreover, it was not entirely a shock that Begin acted with courage, even greatness, during the May–June 1967 crisis by suggesting that his old enemy Ben-Gurion be invited back to replace the indecisive Levi Eshkol as prime minister. Once hostilities began and then ended in victory, Begin's vision characteristically shifted again from national survival to territorial annexation. Nevertheless, he remained on after the war as minister without portfolio in the National Unity cabinet. Although he regarded it as his mission to ensure the "inalienable right of the Jewish people to liberated Judea and Samaria," the stance produced no government crisis. The Labor-dominated coalition was annexing Jerusalem, after all, and building settlements in the West Bank. Begin similarly developed a good relationship with Dayan. Within the government, the two were the nucleus of a hawkish alliance that crossed party lines.

Then came the prolonged agony of the War of Attrition, from spring 1968 to summer 1970. In August 1970 the cabinet decided to accept the "Rogers Plan," which envisaged not merely a cease-fire with Egypt but compliance with UN Resolution 242—that is, the obligation to withdraw "from territories occupied in the recent conflict." At this point, Begin pulled his bloc out of the government coalition. There would be no compromise on the issue of the "Land of Israel." Neither would there be in ensuing years. Preparing for elections in the spring of 1977, Begin and his constituency—a confederation that was further expanded now to include additional, smaller rightist factions, and known henceforth as Likud—submitted a terse and uncompromising foreign policy plank:

> The right of the Jewish people to the Land of Israel is eternal and inalienable [it stated], and is an integral part of its right to security and peace. Judea and Samaria shall therefore not be relinquished for foreign rule. Between the sea and the Jordan, there will be Jewish sovereignty alone.

The program was no less forthright in its claim to eastern Sinai, for that matter. Indeed, Begin revealed that he intended to purchase his own retirement home in Yamit, the Rafa cluster of settlements in the peninsula's northeastern corner. It was with this less than conciliatory approach, therefore, after the May 1977 election, that Likud suddenly emerged as the dominant bloc in a rightist government. Anathematized for decades as an incorrigible demagogue, Menachem Begin now stood at last on the threshold of his nation's prime ministry.

A PUTATIVE ECONOMIC AND
SOCIAL TRANSFORMATION

On June 20, 1977, the new cabinet was sworn in. Its members represented a party coalition encompassing 62 Knesset seats. In addition to its own Likud delegation, the government was supported by 2 members of Ariel Sharon's Shlomzion party; by Moshe Dayan, who now defected from Labor to proclaim himself an "independent"; by the 12-member faction of the National Religious party and the 5 members of the ultra-Orthodox Aguda group. Yet this bare majority was less than sufficient to ensure decisive political leadership. As a result, Begin opened talks with Yigael Yadin, leader of the 15-member Democratic Movement for Change. Discussions would continue for five months, with the prime minister unwilling to commit himself on Yadin's central desideratum of constituency elections as a replacement for the party-list system. Ultimately, in late October, it was Yadin who capitulated. Without a firm Likud promise on any substantive issues, Yadin was given assurance at least of significant cabinet participation at a time of rumored future peace talks with Egypt (Chapter III). He himself was named deputy prime minister, and his DMC colleagues were awarded the ministries of justice, transport, communication, and labor.

Begin was still not breathing easily in his tenure. His own bloc, Likud, remained a loose heterogeneity of contentious factions. Among these, the Liberals spoke for such powerful interests as the Manufacturers Association and the private agricultural sector, and Begin was obliged to defer to this group in economic matters. Thus, the key portfolio of finance went to Simcha Ehrlich, chairman of the Liberal party; while the ministry of commerce and industry, and numerous directorships of state corporations, went to other Liberals. The task they confronted was not an enviable one. During Begin's incumbency, as in Rabin's, oil prices maintained their terrifying upward spiral. So did Israel's defense expenditures. So did the national debt, and the cost of servicing it. Indeed, defense, energy costs, the national debt, and welfare expenditures devoured most of the government budget. None apparently could be reduced. Neither could social welfare costs be tampered with, for fear of alienating the oriental constituency that had brought Likud to power. These chronic outflows were all but structural. It was the cabinet's Liberal faction, nevertheless, that had the responsibility of coping with them, with an economy gone dangerously awry.

The new regime had entered office in 1977 with a commitment to fight inflation, corruption, the tax burden, to halve the cost of living increase, to raise the nation's GNP by at least 40 percent through encouragement of investment, reduction of the adverse trade balance, and reversal of bureaucratic interference in the marketplace. For help in activating this imposing program, the government had even invited the renowned American monetarist and Nobel laureate Milton Friedman to offer counsel and moral support. Accordingly, within months of taking office, Ehrlich and his Liberal colleagues announced a "New Economic Policy," a campaign to liquidate the nation's endemic currency speculation by allowing the free play of supply and demand. As matters developed, the changed approach was substantially effective in drying up Israel's notorious black market in dollars. But in the process, it spawned other, unanticipated, consequences. One was the creation of even larger windfall profits for holders of foreign exchange—essentially the recipients of restitution payments from Germany. Some 100,000 European Jewish families in Israel now found themselves the possessors of buying power augmented well beyond the level of the Rabin years; and their economic strength continued to grow in ensuing years as the value of Israel's currency dropped. There was no parallel flow of overseas funds to oriental Jews, of course. Worse yet, by allowing free convertibility in Israeli and foreign currencies, Ehrlich unwittingly encouraged the lavish exchange of American dollar loans into shekels (the new Israeli monetary unit). The unanticipated flood of shekels in turn created a printing press image that fueled the inflation even more alarmingly.

With fresh enterprise the theme, yet another component of Ehrlich's master plan was the envisaged sale of government corporations. Since Israel's earliest years, these companies were prominent in energy production, banking, transportation, mineral extraction, water development, and tourism, and a brisk public trade in their shares often had allowed the government to make quick profits. Ehrlich was determined to reverse this practice, to dispose of the public companies altogether. He failed. Few potential buyers were impressed by these firms, with their less than efficient work forces and suspect bookkeeping. Only one corporation, a mortgage bank, ever made it to Ehrlich's sell list, and by 1980 the effort to auction off the rest had to be relegated to the back burner. "Postponement" also was the fate of Likud's promise to nationalize health care, to remove the country's single largest medical program from Histadrut domination. The plan was by no means a radical one. Ben-Gurion and his Rafi followers had advocated it in the 1960s. Even so, with little public support behind it now, the measure died in committee. So did the government's plans to cut back on food, transportation, and housing subsidies. Here Ehrlich also was given pause by strong cabinet opposition, particularly by Housing Minister David Levy, a populist of Moroccan extraction and Begin's key link to the oriental communities. When a few, limited cuts in subsidies were introduced, the price of milk, bread, cooking oil, sugar, rice, and public transportation rose sharply. And so, even more dramatically, did public protest. After twenty months, a halt was called to further reductions. By then the damage was done. The lower classes had lost further ground to the escalating cost of living.

For years, too, Labor had been criticized for the high tax rates that stifled initiative and encouraged evasion. As it happened, major tax reforms had been in preparation as far back as the last months of the Rabin government, including a reduced rate structure, self-assessment of income tax, and a value-added (sales) tax. The margin for innovation accordingly was small. Nevertheless, Ehrlich did attempt three new measures. One succeeded. This was a modest reduction in the income tax, and a concomitant assessment on previously untaxed employee fringe benefits. The two other planned innovations were heavier taxation of the thriving kibbutzim, and a general tax amnesty for earlier evaders who now agreed to pay up. Neither of these proposals got off the ground. A government commission soon discovered that the kibbutzim actually were paying more than the amount required by law. As for the tax amnesty, public debate on this issue was so acrimonious, Histadrut opposition so fierce, that the proposal eventually was dropped. By and large, taxes remained as high under Likud as under Labor, climbing to 66 percent of the GNP. This was twice the rate even of the Scandinavian welfare states. Meanwhile, Likud's policy of devaluation and massive currency convertibility, together with educational and housing innovations (discussed later), continued to fuel the nation's inflation.

Altogether, the results of the New Economic Policy were calamitous enough, in soaring costs and palpable social inequities, to force Ehrlich's resignation in 1980. His successor, Yigal Hurevitz, a member of Likud's smaller La'am faction and a self-made millionaire, appeared to be a much tougher character. As he turned his full attention now to the inflationary crisis, Hurevitz demanded across-the-board cuts in subsidies and capital gains alike. He did not get them. Perennially threatening to resign if his austerity program fell through, Hurevitz held on for a year, the government's Cassandra, until January 1981. At that point, the cabinet faced a nationwide strike of teachers, who demanded a pay hike. The ultimatum conflicted with a major Likud election pledge, to introduce compulsory arbitration of labor disputes in essential services. Indeed, by then the public had been sufficiently exasperated by the continuous train of strikes and government capitulation under Labor that it was quite prepared for stringent measures of self-sacrifice. At first, too, the Begin cabinet had seemed to respond to that mood by firmly holding the line against a wildcat strike of El Al employees. But this single act of resolve did not become the norm. Preoccupied increasingly with foreign policy issues, the government followed the vacillating example of its Labor predecessor. When the teachers struck in 1981, and Hurevitz refused to compromise, he was overridden. The cabinet voted to give the teachers their wage hike, and Hurevitz himself immediately resigned. Moreover, pulling his two La'am colleagues out of the government coalition, the finance minister also deprived Begin of his majority and precipitated a new election (Chapter VII).

Were there specific populist measures the cabinet was prepared to take to accommodate its lower-class oriental supporters? In fact, there were three. Likud had promised a "revolution" in housing for young couples, by providing financial guarantees to private builders for the large-scale construction of rental apartments. Through the initiative and persistence of Minister of Housing David Levy, the bill was passed, and turned out to be modestly successful. The Likud platform also had made a commitment to three additional years of tuition-free education through high school. That promise too was honored. Although both these measures were put into operation at near surrealistic expense to the public treasury, they proved to be among the few genuinely popular domestic innovations of the Begin years. So also did a promising effort to upgrade the nation's deprived neighborhoods. Earlier attempts had been made to deal with family breakdown, juvenile delinquency, poor school performance or truancy; yet no integrated approach to the problem had been devised on a neighborhoodwide basis. Begin cared about these slum dwellers. They had put him in office. His imaginative approach to their needs now was to call on the Diaspora to join Israel in "Project Renewal," a common undertaking of redevelopment.

Gathering in Jerusalem in the summer of 1978, the assembly of the Jewish Agency responded to Begin's appeal. A comprehensive, even ingenious, blueprint "twinned" eighty-two renewal neighborhoods with thirty-two Jewish communities abroad. Within the neighborhoods themselves, joint committees of local inhabitants, of government and Jewish Agency personnel, met to weigh project recommendations and approve budgets. And, over the years, a systematic effort was carried out to rebuild plant and housing, to provide loans and mortgage funds, to renovate community centers, schools, day-care centers, clinics. Special programs were launched or enhanced: academic, geriatric, vocational training, antidelinquency. By November 1983 some $130 million had been spent on Project Renewal. Progress was uneven. Much of the program was eroded by inflation and chronic budgetary restrictions. Nevertheless, the venture could be rated a qualified success. In numerous urban slums and outlying development communities the quality of life unquestionably was improved. And like the Begin government's other domestic successes—rental housing and free extended secondary education—Project Renewal was the very antithesis of the New Economic Policy. In truth, an authentically right-wing economic policy did not have a ghost of a chance in Israel—under Begin any more than under the previous Labor governments. Not in a state devoted to the "ingathering of the exiles" and the rejuvenation of the Jewish people.

"THE ENEMY WITHIN"

Over the years, the Labor Alignment normally had been able to count on the bulk of Israel's Arab votes. As recently as the 1973 Knesset, three Arabs were elected on lists "affiliated" with Labor. But in 1977 the Alignment's "Arab" list won only a single seat. The sudden shift of Arab Israelis from the Alignment even further to the left was as significant as the wider defection of Jewish Israelis to the right. Here too the change was partly generational. In earlier years, those Arabs who had been "coopted" by Labor for political leadership as a rule were their community's respected older notables, heads of clans, often mukhtars—village leaders. It was specifically this element that was repudiated in 1977. A younger, disenchanted Arab generation was turning now to the Rakach party. At no time did Rakach enjoy official status as an Arab list. In law, it was the Israeli branch of the Communist party, whose leadership still were Jews. By the mid-1970s, however, Rakach's appeal was almost exclusively to Arab voters; for the party represented the one available vehicle of protest against Zionist rule. Thus, in 1965, Rakach won 24 percent of the Arab vote; in 1969, 30 percent; in 1973, 37 percent. In 1977, it won

50 percent. Indeed, Rakach polled 62 percent of the vote that year in the Christian Arab city of Nazareth; and in Shfar'am, another Arab town, the Rakach vote was 90 percent!

This spectacular transformation in political allegiance among non-Jews was not entirely monolithic, no more than the treatment that provoked it. Israel's Druze and Circassian subcultures had long since reached a modus vivendi with the Jews. Arab by blood and language, the Druze sect had broken from Shia Islam in the Middle Ages to embrace certain elements of Ismaili theology. For their heresy, they were persecuted by their Moslem overlords and eventually driven for refuge to the mountainous terrain of southern Syria, Lebanon, and northern Palestine. By the 1970s their numbers in Israel reached approximately 40,000, most of them still concentrated in the north in some eighteen villages of their own. The Circassians were a far smaller group, hardly exceeding 2,000. Sunni Moslems of partly Turkic origin, the Circassians were natives of the Caucasus Mountains. Because they were renowned as tough fighters, the Ottoman government in the nineteenth century transplanted several thousand of them to the wild border regions of southern Syria and northern Palestine, where they served as a frontier gendarmerie. Under Israeli rule, the Circassians were farmers. Living in the two northern villages of Kafr Kama and Rihaniyya, they continued to speak their own Caucasus dialect and to shun social contact with the neighboring Arab population. Like the Druze, they maintained their family-based traditions, and married exclusively among their own.

Israel learned to trust both groups. From the earliest years of statehood, Druze and Circassians were conscripted (at their own request) into the Israeli armed forces, and proved to be formidable soldiers. Indeed, the Druze all but monopolized Israel's crack border police units, and participated in several important military operations (among them, the Entebbe rescue). The government made a point of reciprocating their loyalty. Druze and Circassian villages were provided with paved approach roads to main highways, assured bus routes nearby, full access to electricity, water, telephone, and health facilities. A Druze notable, Sheikh Jabr Mj'adi, was a perennial Knesset member and served briefly as a deputy minister. Other Druze were appointed consuls and members of Israel's United Nations delegation. Eventually, their sect was granted official recognition as a separate religious community—a status no Moslem state was prepared to concede—with the right to maintain their own religious courts. The government was fulsome in its praise of Druze loyalty, and important public officials made a point of attending Druze festivals. For their part, the Druze tended to vote for Knesset lists sponsored by Labor.

But Druze and Circassians represented less than a twentieth of Israel's non-Jewish population, and the emergence of the West Bank issue after 1967 belatedly exposed a host of resentments nurtured by Israel's local Arab minority. Those grievances were not trivial. Since the early days of statehood, they included the systematic, often inadequately compensated, appropriation of Arab lands for security purposes. Thus, by 1973, there were 149,000 less dunams (approximately 38,000 acres) of Arab-owned agricultural land in

Israel than in 1953. The value of Arab agricultural production in 1975 represented a bare 4 percent of the national farm output, although Arabs comprised 23 percent of Israel's agricultural population. The shrinkage of Arab-owned land similarly narrowed the tax base of Arab villages, in turn reducing the local funds available for roads, schools, and other public services. As late as 1977, less than half the Arab villages in Israel enjoyed access to electricity. Arab agriculture received little of the government or Jewish Agency support available to Jews. Neither did Arab commerce or industry. Indeed, industrial development hardly existed among the Arab community. Jewish capital, public and private, went almost exclusively into Jewish projects. The last Arab-owned factory in Israel, the Arab Cigarette and Tobacco Company Ltd., of Nazareth, failed to secure loans on the same terms as did its Jewish competitors, and eventually had to be liquidated. For years, Arabs males who commuted for employment to the Jewish cities endured discriminatory working conditions and wages. Although the gap narrowed in the 1970s, Arab per capita income remained almost precisely half that of the Jews.

The imbalance between the two peoples was evident at many levels. As late as 1975, infant mortality among Israeli Arabs remained twice as high as among Jews. Although Israel provided its Arabs with free, compulsory education, the quality of that schooling often was eroded by the depleted tax base of Arab villages and by the altered school curriculum. For years secondary schools required more Hebrew and Zionist studies than Arabic studies. So did the matriculation examination for university qualification. While approximately 2,000 Arabs were attending Israeli universities by 1978, their student lives were not easy. Beyond the pressure of study in a "foreign" language, it was difficult for them to secure lodgings from Jewish landlords, or access to appropriate counseling services. Upon graduation, too, they found meaningful employment all but closed to them in the principal—Jewish—economic sector. If they returned to their parents' villages, they faced painful generational conflicts. If they settled in Arab urban areas, even the best trained professionals among them encountered a narrow market for their skills.

Their discontent finally boiled over in the aftermath of the Yom Kippur War. With unprecedented vehemence now, Arab newspapers criticized the expropriation of Arab lands, the foreclosure of Arab national affiliations, the limitation of Arab economic opportunities. An upsurge of poems and stories elaborated upon the loss of the ancestral homeland, the plight of the refugees, the travail of social and economic discrimination. In "The Imprisoned Poet," Samiah al-Qasim dedicated his verses to his aggrieved compatriots:

> You were imprisoned,
> But is it possible to imprison the spirit of defiance?
> Cry unto the arrogant jailer!
> Torture my body with whips,
> Paint my ribs and my uplifted brow with my own blood,
> Smash my arm and my breast, you son of dogs!
> For my spirit continues to search for freedom,

The rivers of rebellion wash out the walls of the conqueror.
(Translation by Avraham Yinon, in *HaMizrach HeChadash,* 1975)

In 1974 a novel, *The Strange Case of Said Pessoptimist*, was published by
Emile Habibi, a three-term Knesset member and editor of the Rakach journal *al-Ittihad*. Almost immediately it became a runaway best-seller among
Israel's Arab community. The book took the form of a long anecdotal letter
addressed to an unnamed "friend," a radical leftist journalist (the author
himself, of course). Mysteriously departed from the earth and now floating
above it, Said was free at last to reveal his long-suppressed secret, the painful circumstances of his life in Israel. Much of Said's contempt was directed
at his fellow Arabs for their passivity. Yet he aimed his bitterest diatribes at
the Jews: for discriminating against the Arab minority, for denying fugitives
the right to return, for ignoring—and thereby obliterating—Arab national
dignity. Said ended his detailed litany of frustration and humiliation by con-
fessing that the only protest emerging from beneath his moustache was a
"miauowing sound, like that of a cat." A brilliant tour de force, *Said Pessoptimist* often was cited as the single most important work of fiction ever to
emerge from Israel's Arab community. It was unquestionably the most de-
fiant.

Israeli Arab protest gained in spleen for lack of alternative outlets. It was
a community that numbered approximately 700,000 in 1977, over a sixth of
Israel's population of 3.8 million. Yet its members owned no commercial or
financial institutions, no banks, no independent trade unions, no television
or radio stations, not even a newspaper unconnected to Israeli political parties (Rakach was not "officially" Arab, and Jews dominated its central committee). Since 1948 the Jews had devised an effective system of control, and
even their liberals intended to maintain it. Thus, Yigal Allon, a spokesman
for Labor's left wing, and a man widely regarded as a friend of the Arabs,
emphasized that the "Arabs have many states, the Jews have one state only.
. . . The fact that an Arab minority lives within [Israel] does not make ours
a multinational state. It only requires that the state grant equal citizenship
to every inhabitant . . . with no differences based on religion, race, or nationality." Altering American-style pluralism as a model for local Jewish-
Arab relations, the Israeli government revealed its distrust of the Arab minority in its choice of "Arab affairs" advisers to the prime minister. Five of
the six men who held this post were former secret service officials. The
Arabs possessed one of Israel's largest blocs of uncommitted votes. Yet hardly
ever were they invited to join the nation's established parties. Rather, as has
been seen, they were directed to Arab satellite factions of the mainstream
Jewish parties. Not until the establishment of Rakach in 1965 did Israeli
Arabs achieve a certain countrywide political structure. Even then, with its
anti-Zionist stance, Rakach was shunned as a coalition-partner by the Jewish
parties, and accordingly deprived of meaningful influence in the Knesset.

It was the impact of the Six-Day and Yom Kippur wars that undermined
the political and psychological status quo. In the aftermath of the 1967 con-

flict, branches of Arab families from both sides of the "Green Line," the prewar cartographic frontier, met for the first time in nineteen years, and some eye-opening experiences awaited them. Local Israeli Arabs found that a number of their West Bank relatives were high-ranking leaders in the PLO. West Bank newspapers, passionately nationalist, were distributed now in Israel's Arab villages, and whetted local thirst for an intrinsically Arab press. Then, following the 1973 war, Israel's reputation for invincibility was undermined on both sides of the former border. Initial signs of upheaval in the West Bank—burning tires, strikes, demonstrations, riots—for the first time began to evoke a parallel response among Israeli Arabs. So did Yasser Arafat's appearance at the United Nations, and the widening international recognition of the PLO. Israeli Arab students now stood before television cameras to declare themselves Palestinians. Indeed, by the mid-1970s, there was hardly an Arab in Israel who did not so define himself. Independent political organization was inevitable. In 1974 the National Council of Chairmen of Arab Local Authorities was founded to lead the struggle for equal rights and national identity. It was followed shortly by the National Council of Secondary School Students, then by the Rakach-sponsored National Council for the Protection of Arab Lands. Thus it was, almost in a single stroke, that passivity gave way to political activism.

Governmental obtuseness was surely a factor in exacerbating this minority unrest. Under Rabin, the practice was continued of occasionally expropriating Arab lands for security reasons. And it was under Rabin, too, that local Arabs suddenly revealed their unwillingness to accept these measures without protest. Early in 1976, when some 50 acres northeast of the village of Kafr Qasim were cordoned off by order of the Israel Lands Authority, the villagers announced that they would fight to the death. In turn, shocked and confused by this belligerency, the government ordered the operation "temporarily postponed." Other expropriations continued throughout the Galilee, however, a region whose substantial Arab majority posed frightening security implications to the Israelis. Whereupon, in response, the National Council for the Protection of Arab Lands took the drastic step of calling a general strike for "Land Day," March 30, 1976. It was a watershed event in the history of Arab-Jewish relations in Israel. Conceivably 10 percent of the nation's entire Arab labor force participated, including crowds of some 50,000 in the Arab towns of Nazareth and Shfar'am. When the oratory provoked violent demonstrations, troops had to be called in, and six Arabs were killed, scores of others wounded. From then on, moderation was a dead issue among Israeli Arabs. West Bank political leaders who visited Israeli Arab towns were received as heroes. Local Arab mayors now openly began to declare themselves supporters both of Rakach and the PLO.

As always, university students became the vanguard of the Arab nationalist camp. Numbering barely 2,000, these young undergraduates generated a virtual industry of pro-PLO literature on Israel's campuses, and many emblazoned PLO slogans on their T-shirts. Eventually the universities became hotbeds of Arab-Jewish confrontation. In 1975, when Arab students at the Hebrew University indignantly refused an administrative order to perform guard duty at the dormitories, the administration denied the Arab Students Committee use of campus facilities. In 1978, Jewish students at Haifa University sought to block Arab student demonstrations, even an Arab student conference, against the Sadat peace initiative, on the grounds that "this is an Israeli university, not a Fatah [i.e., Arab terrorist] training camp." Police harassment of Arab activists mounted. In 1977 Isam Mahoul, leader of the National Association of Arab Students, was arrested for incitement, as was Azimi Bishara, secretary of the Haifa University Arab Students Committee. Elsewhere, too, educated younger Arabs were intent upon asserting their leadership. In the 1973 elections, 23 percent of the candidates to local councils in Arab communities were under the age of forty. The proportion nearly doubled in the 1977 elections, and most of these newcomers were Rakach candidates. Gone by then were the days when the Arab vote was fixed by the heads of *chamoulahs*—extended families—and awarded (usually) to Labor lists. Through its younger candidates—teachers, writers, lawyers—Rakach disseminated its influence among the urban middle class, even among the peasantry.

The Rabin government hardly was more decisive in its approach to Arab militance than to Jewish militance. Following the unrest of Land Day on March 30, 1976, alarmed by the developing "palestinization" of the nation's Arabs, the cabinet made a show of launching an extensive review of its minorities policy. Soon afterward, a confidential government memorandum on the Arab question leaked to the newspaper *al-HaMishmar*, which published it in full. The document was written by Israel Koenig, district commissioner for the Galilee, and the senior arabist in the ministry of the interior. Without mincing words, Koenig advocated an intensification of Jewish settlement in the Galilee; a selective application of carrot and stick to Arab villages and elites; a coordinated campaign of defamation against Rakach activists; and the conscious harassment of "all negative [Arab] personalities at all levels and at all institutions." Various techniques also were proposed for reducing liquid savings (available for political contributions) in the Arab sector, for encouraging an emigration of Arab intellectuals, downgrading the effectiveness of Arab student organizations. Altogether, the memorandum was so shocking in its plainspokenness, and aroused such vigorous controversy among Jews and Arabs alike, that the government felt obliged to repudiate it. Yet Koenig was allowed to remain at his post. Indeed, Rabin subsequently chose Zvi Aldoraty, a co-author of the memorandum, as director of the Labor party's Arab affairs department.

Jewish popular suspicion of the Arab minority, then, far from diminishing after the 1967 victory, remained as widespread in the 1970s as in earlier decades. Those misgivings often were honestly and incisively described in Israeli fiction. Yitzchak Tischler's *The Last to Hold the Ridge*, an account of the Six-Day War, portrayed Jews recoiling at the sight of Arab refugees. Hanoch Bartov's vivid account of life in a moshav, as seen through the eyes of a child, identified Arabs mainly with heat, dirt, and flies. Amos Oz and A. B. Yehoshua wrote sensitively of Israel's unconscious guilt toward Arab refugees. But an even more common feature of Israeli literature was its sheer nonreference to Arabs altogether. If most Jews were unfamiliar with Arabs, they also were uninterested in enlarging their knowledge. Refugees and refugee camps seemed to belong to another world, although Gaza was within view of Israeli settlements and only 50 miles from Tel Aviv. Indifference doubtless was an unconscious protection against fear. Thus, whenever the citizens of a beleaguered nation gave attention to the swelling Arab minority in their midst, it was more comforting for them simply to allow the government to "take care of things."

THE INSTITUTIONALIZATION OF SUSPICION

During the early years of his tenure, preoccupied with foreign and military affairs, Menachem Begin tended largely to ignore Israel's non-Jewish communities. At the least, the prime minister was uninterested in providing development aid to the Arab sector. Shmuel Toledano, Arab affairs adviser under Rabin, had long favored the use of both rewards and punishments in dealing with local Arabs, the distribution of white-collar jobs and development funds to rival Arab cliques for the purpose of diverting them from Rakach and other radical movements. Under Begin, however, the new adviser for Arab affairs, Moshe Sharon, rejected this approach for good and always. The policy adopted now was brutally straightforward, and envisaged heavier taxation of Arab villages and progressive reductions in development aid to the Arab sector altogether. Cherut's decision makers in fact had long regarded a tough stance on the Arab minority as an article of political faith. The principal architect of their strategy was Amnon Linn, once a director of the Arab affairs department of the Labor party, and more recently a defector to Cherut. Under Linn's—Cherut's—new guidelines, Israel's Arabs should be obliged actively to prove their loyalty to the state by giving positive expression of their commitment to Israel and its Jewish-Zionist mission. Those unwilling to do so would be subjected to a "strong hand." For Linn, strength meant the economic blacklisting of "unreliables," the expulsion of Arabs suspected of involvement in guerrilla organizations, the nonadmittance of suspected radicals to Israeli universities, the "administrative detention" of Arab teachers or of other government employees known to support Rakach, and the use of police and border patrol units to quell demonstrations in Arab villages.

More than Moshe Sharon or Amnon Linn, however, it was Ariel Sharon who put the government's new approach into practical operation. The renowned hero of the Yom Kippur War, Sharon was now minister of agriculture in the Begin government and chairman of the cabinet's interministerial committee on land settlement. These were pivotal positions for dealing with Arabs—supremely the people of the land—both in the occupied territories and in Israel proper. In his dual role, then, Sharon let it be known that "I have launched an energetic offensive to stem the hold of foreigners on state lands." His use of the term "foreigners" to describe the Arab citizens of Israel, and of the phrase "Yehud HaGalil" (judaization of the Galilee), evinced a distrust more cold-blooded than at any time since the abolition of the military administration in 1964. Sharon's "offensive" was devoted at first to the prohibition of "illegal" construction outside Arab villages. At a time when the growing Arab population was creating serious congestion, their villagers now were forbidden to purchase land or build new homes. Sharon made the point in his typically frontal manner. In November 1977 security forces killed one Arab, injured twelve, and arrested thirty after the demolition of an "illegal" house in the village of Majd al-Qrum. Similar measures, including mobile "Green Patrols," were used to disperse Bedouin encampments and confiscate livestock grazing on kibbutz land. In the early years of the state, the military government had concentrated some 13,000 Bedouin within a 300,000-acre reservation in the northeastern Negev. Later, following a drought, the Bedouin were transplanted to the Lower Galilee. And now, in 1977, with their numbers grown to 40,000, and their flocks occasionally intruding again into kibbutz property, these Arab nomads were to be impacted into an even smaller area. Control over their lands would be vested in the Israel Lands Authority, and portions of their abandoned grazing area would be transformed into nature preserves. The tragedy inflicted on an ancient race was a cruel one.

It was the Galilee, however, that Sharon regarded as the proving ground for his offensive. By 1977 the prolonged outmigration of Jews, most of them orientals who had been transported to outlying Galilee farm and development towns during the 1950s, together with the higher Arab birthrate, had revived Arab preponderance in this strategic northern region. Successive Labor governments had established Jewish industrial communities in the Galilee, had appealed repeatedly for Jewish volunteers to take up residence there. Yet their effort failed to achieve a Jewish majority or even an impressive Jewish presence in the region. Now the Likud cabinet embarked on a new tack, one that went further than a simple refusal of Arab building permits. By 1981 Sharon had constructed thirty miniature settlements on hills overlooking Arab villages. Although only small numbers of Jews inhabited these Galilee outposts, they established a Jewish foothold, in the manner of the pioneering Zionist settlements of mandatory times, and they blocked the growth of adjacent Arab communities.

Far from being intimidated, Israel's Arab population responded to Sharon's new measures as emphatically as it had in the mid-1970s, with demonstrations and clashes. Meanwhile, from Lebanon, the PLO leadership turned their attention to this vibrant minority population west of the Green Line. Israel's Arabs evidently no longer were to be regarded as a negligible factor in the Palestine equation. They comprised one-eighth of the Palestinian people, after all, and almost one-third of those still in Palestine itself. Belatedly, then, the PLO issued an appeal for Israeli Arabs to play a central role in the Palestinian struggle, for the Arab world at large to salute "the heroic struggle of our people in the Galilee, the Triangle [a dense Moslem area near the former Jordanian border], and the Negev." Appreciation was reciprocal. In January 1979, twenty-eight chairmen of Arab local councils—more than half the Arab mayors in Israel—together with nearly 100 leaders of the Rakach

party, endorsed a resolution "welcoming the struggle of the Palestinian people under the leadership of the PLO to establish its independent state." In
ensuing weeks, Israeli Arab editorials vigorously denounced the government's policies in the occupied territories. On university campuses, a number of Arab student organizations set about provoking new confrontations
with Jewish classmates and the police.

If Israel's Arab activists were pushing their luck even under the Rabin
government, in December 1980 they stretched it altogether too far. A major
congress of PLO- and Rakach-sponsored elements was planned that month
for Nazareth. Whereupon Prime Minister Begin vetoed the gathering, then
announced his intention to outlaw any organization that scheduled a similar
congress in the future. No such prohibition had been enacted by any Israeli
government since 1964, when the military administration for the nation's
Arab-inhabited areas was dismantled. But, in fact, no such man as Menachem Begin had ever occupied the prime minister's office. For him, the
ideological heir of Jabotinsky, the time had come to throw down Israel's
challenge on all Arab fronts. East or west of the former armistice line, there
was one land to be redeemed as the Jewish people's inalienable right. With
the levers of power in his hands at last, it seemed inconceivable to Begin
that any force in the Arab world could withstand the long-delayed fulfillment
of that burning Revisionist dream.

CHAPTER III

EGYPT'S QUEST
FOR PEACE

SADAT TURNS TO THE RIGHT

The shift of Israel's political and economic course in 1977 was prefigured by a striking, and by no means dissimilar, reorientation in Egypt. Anwar al-Sadat orchestrated the change personally. With his reputation at new heights following the Yom Kippur War of October 1973, the Egyptian president felt secure enough to embark on a far-reaching overhaul of Egyptian society, a "liberalization in all fields" that would reverse the stagnation of the Nasser years. To that end, Sadat emptied the nation's concentration camps, permitted the return of figures who had been exiled by Nasser, authorized a new freedom of expression in cultural life. Early in 1977 he approved a gradual transition to a multiparty system.

The new liberalization similarly characterized the economic field. Here the president moved with caution, never overtly repudiating Nasser's state socialism, with its free schools and universities, its subsidized food and rents, its guaranteed job-tenure for university graduates. After the October 1973 war, however, wearing the mantle of a national hero, Sadat felt the moment appropriate to endorse a market economy. Egypt's population of 42 million was growing at the rate of nearly 750,000 a year; 4 million tons of grains, meat, and sugar had to be imported annually; the housing shortage was desperate. Frustration with the inadequacy of food and shelter was boiling over by then into occasional public demonstrations. There were major protests by students and workers in 1975 and 1976. It was plain, then, that wider economic incentives were urgently needed, that foreign investment funds would have to be attracted. In the president's view, the salvage operation could be accomplished only by a "New Look," the introduction of Western-style capitalism. It could also be achieved only by peace.

To foster his vision of national revival, Sadat made the calculated decision in June 1975 to open the Suez Canal. Seven years of intermittent hostilities along the waterway had left the isthmus cities of Port Said, Ismailia, Suez City, and Port Tewfiq ghostly ruins. Nearly a million of their inhabitants had sought refuge in Cairo, all but paralyzing the capital's motor functions. The throttled waterway also had cost Egypt $13 billion in lost tolls and inflated shipping costs since 1967. Now the vicious circle had to be broken—and if need be, unilaterally, without assurance of further Israeli withdrawal in the Sinai. For that matter, Sadat insisted that economic viability no longer could be reconciled with the burden of remaining on a perpetual war footing, of keeping 700,000 men under arms.

The president spoke for his people. After the Yom Kippur War of October 1973, their desire for an end to bloodshed had become a visceral passion. Estimates of Egyptian dead and wounded in all the conflicts with Israel ranged from 70,000 to 100,000. In the War of Attrition alone, between 1968 and 1970, at least 30,000 casualties were suffered, and the cost of the 1973 fighting was listed officially as 7,700 dead and three times that many wounded. The figure almost certainly was higher. These were painful losses for a warm, family-oriented people. Sadat recalled of the 1973 conflict: "I lost my younger brother, who was like a son to me, five minutes after the start of the October War. I have seen the victims of that war—young men destined to spend the rest of their lives in wheelchairs." Ironically, the successful crossing of the canal appeared to fortify the nation's desire for peace. With their honor restored, the Egyptian people found less humiliation in contemplating a modus vivendi with the enemy. Neither was Soviet pressure capable any longer of thwarting that accommodation. The Russians had been challenged even earlier, in July 1972, when Sadat had abruptly terminated their military mission in Egypt. Now, in March 1976, intent upon achieving complete freedom of action, the Egyptian president formally canceled the 1971 Soviet-Egyptian Treaty of Friendship and Cooperation, and in the following month revoked Soviet naval access to Egyptian ports. For Sadat, the field was clear by then to explore an additional disengagement with Israel. He explained his approach succinctly in his autobiographical *In Search of Identity:*

> We have recovered our pride and self-confidence after the October 1973 battle, just as our armed forces did. We are no longer motivated by "complexes"—whether defeatist "inferiority" ones or those born out of suspicion and hate. And this is why the opposing sides met soon after the battle dust had settled to talk matters over. . . . With the fighting over, we harbored nothing but respect for one another.

AN ESTABLISHMENT OF CONTACTS

The inauguration of a new administration in Washington added a unique, and distinctly personalized, momentum to the search for Middle Eastern peace. By the time Jimmy Carter entered the White House in January 1977, he had been persuaded that the Kissinger step-by-step approach to an Arab-Israeli settlement had reached the limits of its effectiveness. A broader and more definitive agreement was required now, one that might be achieved within the format of a revived Geneva Conference. Carter's vision of that final settlement was influenced by a 1975 Brookings Institution report. Among the document's authors were Zbigniew Brzezinski and William Quandt, both soon to play important roles in the Carter government's National Security Council, and Alfred Atherton, Jr., the State Department's Middle East specialist. These men strongly favored PLO participation at Geneva, together with an advance Israeli commitment to withdraw from the occupied territories, and an Arab pledge to sign a formal peace treaty with the Jewish state.

Initially, Carter floated the Brookings proposal to Yitzchak Rabin, who visited the White House in March 1977 as Israel's caretaker prime minister. The latter's response was chilly. Israel would never approve a return to the pre-1967 armistice lines, Rabin warned. Should even a territorially modified sector of the West Bank be restored to Hashemite civil administration (assuredly not to an independent Palestinian state), Israel would insist on maintaining full security control. As for a revived Geneva Conference, Rabin went on, PLO members would be unacceptable, either within or outside any other Arab delegation. It was a sharp rebuff. The president and his advisers still were evaluating its implications when Sadat arrived in Washington a month later. Here again the response to Carter's proposal was less than fulsome. In principle, the Egyptian leader did not oppose the notion of a revived Geneva conclave. Yet Sadat made clear that he envisaged little progress at a gathering attended by Syrian and Soviet delegates. And when Carter then queried his visitor about the chances for normalized Egyptian relations with Israel, Sadat could only shake his head vigorously and insist: "Not in my lifetime."

The election of Begin, afterward, seemed to dash even colder water on the American president's formula for Middle East peace. Within days the word came from Jerusalem that Israel, far from returning to its pre-1967 borders, now would intensify Jewish settlement in "Judea and Samaria." Nor would an Israeli delegation attend any conference in which PLO representatives participated. In July 1977, Begin arrived in the United States for his first meeting with Carter. Like Rabin, the new prime minister emphasized that all participants in a revived Geneva conference first would have to commit themselves to UN Resolutions 242 and 338, the mutual recognition of each's national right to live in peace and security; no other basis for negotiations would be acceptable, least of all that proposed by the Brookings Institution report. There appeared little in either the Egyptian or Israeli reactions, then, to sustain Jimmy Carter's optimism. Nor did Secretary of State Cyrus Vance evoke enthusiasm for a Geneva Conference in his tour of Middle Eastern capitals. The Jordanians were reserved. The Syrians had boycotted the original conference in 1973, and discerned no reason now to abandon their insistence upon a single, undifferentiated Arab delegation. It was a format Sadat never would have accepted, no more than would the Israelis. A wall-to-wall collection of moderate and hard-line Arabs would have guaranteed an impasse in negotiations. Not less than Carter, the Egyptian leader was all but frantic to achieve swift, meaningful results.

Indeed, Sadat early on had intimated his wish to reach a de facto accommodation with the Israelis. As far back as 1971 his signals to Washington, to Jerusalem, and to UN Mediator Gunnar Jarring revealed his willingness to negotiate interim arrangements for the canal as a quid pro quo for Israeli withdrawal. In 1974–75, too, his consent to the disengagement agreements was tangible evidence of his seriousness of purpose. So was his decision unilaterally to open the waterway in June 1975. If he had come this far, he was hardly prepared to allow his efforts to founder in a procedural quagmire at Geneva. Worst yet, from Sadat's viewpoint, the Americans evidently cherished the notion that Moscow was prepared to exert a moderating influence on Arab hard-liners. In early autumn of 1977 the two Great Powers had entered into their own discussions on the question of Geneva, and on October 1 their simultaneous communiques pledged both governments to reconvene the suspended 1973 conference within two months, and to negotiate

> withdrawal of Israeli armed forces from territories occupied in the 1967 conflict; the resolution of the Palestine question, including insuring the legitimate rights of the Palestinian people; termination of the state of war and establishment of normal peaceful relations on the basis of mutual recognition of the principles of sovereignty, territorial integrity and political independence.

If Begin was outraged by the pronunciamento, Sadat was hardly less so. He at least entertained no illusion about the role the Soviets intended to play. Earlier they had been closed out of the Middle East by Kissinger and by Sadat himself. Now they were virtually being entreated to reenter as a sponsor of the Geneva Conference. Backing Syria to the limit, too, they would be positioned to maneuver Egypt into a corner, forcing it to dance to the tune of the rejectionist camp. Thus it was, by agreeing to this sure and certain recipe for failure, that the Carter administration unwittingly forced Sadat's hand. The Egyptian president saw no alternative now but to act on his own.

As he evaluated his future course, Sadat was presented with an unexpected opening by the Israelis. It happened that, upon forming his cabinet, Menachem Begin had astonished his colleagues and the nation at large by his choice for foreign minister. It was Moshe Dayan. The man was not even a member of the ruling Likud bloc (although he had departed Labor and now called himself an "independent"). Since early 1974, Dayan had been something of a political leper, discredited by his lack of preparation for the Yom Kippur War, by the avalanche of criticism that had descended on him in the subsequent national postmortem. Nevertheless, he enjoyed the confidence of Begin. Dayan had been the most prominent hawk within the Labor camp, after all, a forceful advocate of Israeli control over the West Bank. Moreover, he was an arch-pragmatist, a gruff farmer-soldier who attached no mystic significance whatever to the Sinai. Indeed, between 1967 and 1973, it was Dayan alone in the Labor cabinet who had opposed the notion of holding fast at the Suez Canal. In his view, an immobilist stance along the waterway was a gratuitous provocation both to Egypt and to the international maritime community. As minister of defense before and during the War of Attrition, Dayan repeatedly hinted of possible Israeli withdrawals in return for an Egyptian commitment to nonbelligerency. Some years later, the disengagement agreements of 1974–75, then Sadat's decision to open the canal, appeared to validate that moderation. Thus, during secret talks between Rabin and Kissinger in 1976, the Labor prime minister followed

Dayan's earlier lead by intimating his own willingness to pull back even further, some two-thirds the distance of the Sinai, in exchange for an end to the state of belligerency.

Upon assuming office, then, Begin similarly shared this interest in a tradeoff. But the new prime minister had in mind an exchange less for an Egyptian commitment to nonbelligerency, or even to peace, than for an Israeli free hand in Judea and Samaria, the ideological obsession of his lifetime. Dayan, with his record as a dove in Sinai and a hawk in the West Bank, would be indispensable to Begin in selling the idea of such a tradeoff—first to Egypt, then to Israel. The prime minister thereupon outlined his strategy to Dayan, and the latter accepted it. Dayan sought his own assurance, however, that Begin was prepared to forgo an "official" annexation of the West Bank. Such a move would torpedo any possibility even of limited negotiation with Egypt, he argued. Begin conceded the point. Like Dayan, he appreciated that international opinion could not be flouted altogether, that the "fact" to be established was Israel's irretrievable presence and its right of settlement in an undivided Land of Israel. It was on this basis that Begin submitted Dayan's nomination as foreign minister to his party, then to the Knesset. Over considerable misgivings from the Likud bloc and much outrage from Dayan's former Labor colleagues, the nomination was approved.

From the earliest weeks of his incumbency, then, Begin set about dispatching signals to Cairo of his willingness to strike a deal. One signal took the form of a trip to Bucharest on August 28, 1977, a five-day ceremonial visit that included extensive discussions with President Nicolae Ceausescu. Seeking visibility as a "nonaligned" statesman, the Romanian leader had repeatedly made known his interest in brokering a Middle East settlement. It was Ceausescu in turn who arranged a secret parallel meeting between Begin and Said Merei, a visiting representative of the Egyptian National Assembly. The two guests conversed earnestly for several hours, discussing the possibility of a future meeting with Sadat. To Merei, as to Ceausescu, Begin emphasized his willingness to offer "extensive satisfaction" on the Sinai, to negotiate joint Egyptian-Israeli intelligence and defense measures against Libyan and Soviet penetration, then to negotiate some form of autonomous Arab administration for Gaza and the West Bank. Merei promised to convey Begin's message to Sadat.

It was Dayan who transmitted the second signal. Shortly after assuming office, he instructed Meir Rosenne, the foreign ministry's legal adviser, to formulate a draft of a possible Egyptian-Israeli peace treaty. Rosenne and his colleagues subsequently produced a forty-six-point document that elaborated upon the principles outlined by Begin in Romania. Immediately the paper was sent off to Washington for examination; U.S. endorsement still was regarded as indispensable for leverage with the Arab enemy. And, as Dayan hoped, Secretary of State Cyrus Vance and his advisers were impressed enough to request the president's personal intervention with the Egyptians. Carter agreed. On September 10, 1977, the Rosenne document was telexed to Cairo, followed by a private letter from Carter to Sadat. In his communication, the president entreated Sadat to test Begin's sincerity by acceding to an early revival of the Geneva Conference (a meeting that plainly had become a fixation with Carter).

Sadat's reaction to these feelers was influenced by an intriguing twist of history. Notwithstanding his desire for an accommodation, until then he had rejected the notion of secret negotiations or even secret contacts with the Israelis. His public addresses still were tainted with koranic aspersions against the Jews, even as his private comments to visiting Americans occasionally were tinged with antisemitic slurs. But in July 1977, well in advance of Begin's visit to Romania, the issue of contacts was given renewed urgency from an unlikely source. It was Muammar Qaddafi. In recent years the Libyan dictator had been exhorting Sadat to reactivate the moribund Syrian-Egyptian-Libyan Confederation, to detach Egypt from the American connection, and to resume a decisive military confrontation with Israel. Sadat was distinctly uninterested. Rather, in 1974 and 1975, he negotiated his two disengagement agreements with Israel and drew even closer to the United States. Whereupon Qaddafi became increasingly splenetic. Ugly territorial issues were raised along the Libyan-Egyptian frontier in the Western Desert, and the Libyan ruler ordered a massing of troops there. At that point, in May 1977, General Yitzchak Chofi, director of the Mossad (Israel's CIA), uncovered information of a Libyan assassination plot against Sadat. The "hit" team evidently consisted of trained Palestinian killers, men who were under continual Israeli scrutiny. In the past, whenever these plots had not been aimed specifically against Jews but rather against such Arab conservatives as Hussein of Jordan or Feisal of Saudi Arabia, Israel had turned over its information to the American CIA, which then had warned the intended victims under its own byline. But this time, informed of the plot early in June and preoccupied with the formation of his cabinet, Begin came up with a different proposal. It was to convey the information "directly" to the Egyptians. "That may warm the atmosphere between us," he suggested.

For "direct" communication with Egypt, Morocco's King Hassan was regarded as a valuable conduit. Like his forebears of the Alawi dynasty, Hassan was a moderate. Personally, he had made a point of treating his Jews well, and often had appealed for a fusion of "Jewish genius and Arab might" in reviving the Maghreb, the vast littoral of Moslem North Africa. The Israelis in turn had secretly offered the king help in fighting the Algerian-sponsored Polisario guerrillas in the western Sahara. Hassan was grateful. From 1975 on, it was his pet scheme to bring Israel and Egypt together, even to bring Israel into the Arab League. Occasionally, the monarch invited eminent Israelis of North African background (among them the writer André Chouraqui and Rabbi Aharon Abuhatzeira) to Morocco as his personal guests and as sounding boards for his ideas. In October 1976, Prime Minister Rabin himself paid a secret one-day visit to Morocco, to seek the king's good offices in arranging a meeting with Sadat. Hassan welcomed Rabin cordially, but observed that Israel still was too weakened by internal divisions after the October 1973 war to engage in productive negotiations.

Now, however, in June 1977, convinced that the moment for peace was ripe, Hassan was prepared to serve as intermediary. He proved to be an effective one. Through the Moroccan ambassador in Cairo, the Egyptians were informed that Israel possessed vital security information. But inasmuch as the data could only be transmitted directly, the king arranged a meeting in Casablanca between General Chofi and Egypt's director of military intelligence, Lieutenant General Kamal Hassan Ali. Hassan Ali was stunned by the detailed evidence the Israeli brought with him, including names and addresses of the Palestinian assassins in Cairo. Acting on the information, moreover, the Egyptian secret police seized the conspirators, with extensive incriminating documents and weapons. And afterward, on July 21, 1977, Sadat launched a military strike against Libya's secret staging basis, 22 miles across the frontier. The onslaught lasted six days and inflicted heavy casualties on the Libyans. Large quantities of their equipment were destroyed. Even as the fighting continued, meanwhile, Begin informed the Knesset that Israeli forces in the Sinai would stay put while Egypt was busy with the "common enemy," Qaddafi.

Sadat was genuinely grateful. He had been appalled by Begin's recent electoral victory, and by the presence in the Begin cabinet of hawks like Generals Dayan, Sharon, and Weizman. But the Libyan episode, and the subsequent overtures communicated through Ceausescu and Said Merei, made their impact. The Egyptian president accordingly sent word through Morocco that he was prepared to listen to serious Israeli proposals. Thus, on September 4, 1977, Dayan flew off secretly to Fez, arriving (by way of Paris) in the Moroccan king's private jet. Hassan welcomed the Israeli foreign minister graciously and confirmed that Sadat was indeed interested in a possible bilateral agreement—if Israel was prepared to be flexible. Dayan's response was affirmative. In ensuing days, too, the Egyptian president received from Washington a copy of the Israeli draft treaty, together with Carter's appeal for a revived Geneva Conference. While the notion of a conference remained distasteful to Sadat, he did not reject it out of hand; for there was a chance now that a private Egyptian-Israeli understanding could be reached in advance. Accordingly, two weeks later, on September 16, Dayan flew back to Morocco. Awaiting him this time in Rabat was Egyptian Deputy Prime Minister Hassan al-Tohami. Tohami was a useful choice as go-between. Years before, he had served in Vienna as Egypt's delegate to the International Atomic Energy Commission, and had become friendly there with Chancellor Bruno Kreisky and with the latter's Jewish millionaire friend Dr. Karl Kahana. Kahana in turn had remained a liaison for the Israelis after Sadat came to power, staying in touch with Tohami when the latter subsequently was appointed deputy prime minister and coordinator of intelligence services. As contacts multiplied through the summer of 1977, Israel had sent word that Tohami was their preferred intermediary, and Cairo had agreed to dispatch him to Morocco. Earnestly now, he entered into conversations with Dayan.

To the Israeli foreign minister, Tohami outlined Sadat's willingness to anticipate Geneva by quietly working out a "private" understanding for Egyptian-Israeli nonbelligerency. But Israel first would have to commit itself to the removal of all its troops from Egyptian soil, to a return of the entire Sinai to Egyptian sovereignty, and to acceptance of a Palestinian "arrangement" that would link the West Bank and Gaza to Jordan. Once this assurance was forthcoming, Sadat for his part was prepared to offer Israel every security guarantee, including United Nations forces on both sides of the Sinai frontier, and an intelligence partnership (as Begin had suggested in Romania) for combating Libyan and Soviet incursion into the Middle East. Dayan heard out the proposal. His reaction then was straightforward. Israel was willing to restore full Egyptian sovereignty over the Sinai, he explained, but not to abandon its settlements in the northeastern corner of the peninsula or its key air bases near the Gulf of Aqaba. Neither would Israel withdraw its settlements from the West Bank. Even so, there was room for negotiation on these and other issues. The two men discussed them at some additional length. The conversation lasted seven hours, with King Hassan occasionally looking in to ensure that all was going well. All was. Dayan and Tohami discerned possibilities for future clarification, and both agreed to return to Morocco within a fortnight with more concrete proposals.

AN EPOCHAL JOURNEY

Begin was gratified by Dayan's account of the meeting. Now at last the prime minister had his weapon for aborting, or at least reorienting, the threatened Geneva Conference. This new self-assurance became evident on September 19, 1977, when Dayan flew to Washington to meet with Carter and Vance—ostensibly to discuss the conference. Unfamiliar yet with Dayan's flinty character, and uninformed on the emergent dialogue in Morocco, the president confronted his guest with a virtual ultimatum. The Palestinians must be represented at Geneva, he insisted; Israel eventually would have to accept a Palestinian "entity" on the West Bank; Israel's settlements on the West Bank were illegal and would have to be dismantled. At this, Dayan emphatically rejected each assertion, as well as the very notion of Soviet or PLO participation in a resuscitated peace conference. The foreign minister was risking little by adopting this tough stance, of course. The return signal from Cairo was audible enough to persuade him that new and better alternatives were developing.

It was an accurate appraisal. Sadat was indeed contemplating a new departure. He was intent now upon dissipating the Israelis' chronic suspicion of their Arab neighbors, to enable them to overcome their "legalistic preoccupation with technicalities." He wrote later: "It was then that I drew, almost unconciously, on the inner strength I had developed in Cell 54 of Cairo Central Prison [under the British]—a strength, call it a talent or capacity, for change." He would launch a fresh approach to elicit Israeli trust. On October 30, the Egyptian president flew to Romania for a private discussion of his own with Ceausescu. "I asked Ceausescu about his impressions [of Begin]. He said: 'Begin wants a solution.'" This was heartening. All the more so inasmuch as Begin, several months earlier, had visited the Yamit cluster of eighteen Jewish settlements in northeastern Sinai and promised that they would remain. Now, as Dayan had intimated to Tohami, Begin evidently was softening his position. As a rightist, too, the prime minister was better positioned than were the Laborites to sell the Knesset a deal. The following day, Sadat continued on from Romania to Iran, and then to Saudi Arabia, for conferences with state leaders there. In Tehran he was exposed again to conciliatory advice. The shah, long a moderate on the Arab-Israel issue, urged his guest to enter into direct negotiations with Begin. Sadat was listening. Apparently it was on this Middle Eastern trip that he made a historic decision. As early as its first leg, en route from Bucharest to Tehran, he read an open letter from Abie Nathan, a flamboyant Israeli peace activist. Nathan had proposed an exchange of journalists between Egypt and Israel. Much later, Sadat revealed to a friend that it was Nathan's suggestion that began to tip the scales. "Why only journalists?" he wondered. "Why can't I myself make a spectacular visit to Israel?"

If the notion was dramatic, such gestures were hardly foreign to Sadat. As a youth, he had given thought to a career in the theater after finishing high school, and briefly held a job with a Cairo acting troupe. After his graduation from the military academy, many of his actions—his wartime spying for the Germans, his assassination plots against British and Egyptian officials—revealed an instinct for the spectacular. So did his surprise move later in expelling the Soviets from Egypt, and so did his decision to open the Suez Canal in 1975. The gesture Sadat had in mind now, however, was authentically mind-boggling. It was an invitation to Arab leaders to join him at the Knesset in Jerusalem, "to make it absolutely clear to Premier Begin that we were determined to prepare seriously for Geneva." Presumably in Jerusalem the hard Arab-Israeli issues would be resolved before the Soviets or the Syrians could sabotage an agreement. But as he conversed with Saudi Arabia's King Khaled in Riyadh, Sadat apparently sensed the unlikelihood of other Arab leaders' joining him. This conclusion was fortified on November 5, 1977, in the immediate aftermath of his grand tour, when he was visited in Cairo by Hussein of Jordan. The little Hashemite king cautioned Sadat against negotiating a Palestine deal with the Israelis on his own. As it happened, Sadat was determined to do precisely that. Faced with the reservations of Hussein, the only Arab leader (except for Morocco's King Hassan) to have engaged in secret dialogue with the Israelis, the Egyptian president envisaged no choice except to act on his own initiative. On November 9, then, addressing the opening session of the People's Assembly, he turned to the Palestine question. After reviewing the stalemate posed by Israeli intransigence, he concluded his remarks with the laconic announcement: "I am ready to go to the Israeli parliament itself and discuss [Israeli withdrawal] with them."

The impact of the declaration was stunning. Foreign Minister Ismail Fahmi promptly resigned. So did a number of other senior officials in Egypt's diplomatic service. In Washington, Carter and the State Department were confused and silent. Several days passed before the White House issued a rather constipated endorsement of Sadat's initiative. In Israel, Begin was hardly less astounded. He had anticipated private talks. But a public, ceremonial visit by an Egyptian president to Jerusalem? The prospect seemed all but unimaginable. Even so, swiftly regaining his composure, the prime minister was on the state radio the next day, issuing an open invitation to Sadat to come to Israel. Five days after that, in a note communicated through the U.S. embassies in Tel Aviv and Cairo, Begin formally and officially invited the Egyptian leader to address the Knesset on November 20. Sadat accepted on the spot. In advance, on November 16, he flew off to Damascus in an effort to win the understanding, or at least the forbearance, of Hafez al-Assad. He got neither from the Syrian president. The four-hour meeting was acrimonious. Even as Sadat departed for home, Assad instructed Syrian newspapers to announce a day of national mourning. The reaction was widely shared elsewhere in the Arab world.

In Israel, meanwhile, preparations began amidst uncertainty and considerable disbelief. General Mordechai Gur, the military chief of staff, suggested in a press interview that Sadat's announced visit actually might be a diversion for a surprise Egyptian attack in the Sinai. As the day of the scheduled arrival neared, however, the Israeli people came to accept that the trip evidently would materialize. The Arabic-language program on Kol Yisrael began broadcasting popular Egyptian melodies. Dozens of Israeli Knesset members on trips abroad hurriedly booked flights home. With the help of an advance party of Egyptian protocol and security officials, elaborate precautions were devised. Ten thousand Israeli police were placed on alert. Hospital operating rooms were put on an emergency footing, with blood reserves of the correct type readied for Sadat and senior members of his party. In Jerusalem, the King David Hotel was hastily cleared of its guests. Known Arab "radicals" were hustled out of town. Workshops were mobilized to sew Egyptian flags. Rush copies of the Egyptian national anthem were distributed to the army band.

At last, on twilight of Saturday, November 19, 1977, Sadat emplaned for Israel. Four Israeli Kfir fighters escorted the presidential jet as it entered Israeli airspace (circling several times to ensure that the Jewish Sabbath had ended). As the plane landed at Ben-Gurion Airport and taxied to the awaiting red carpet, millions of television viewers throughout the world watched the Egyptian president and his entourage proceed, to the echoes of a twenty-one gun salute, down the line of awaiting ministers, diplomats, political and religious leaders, and other dignitaries. Afterward, Sadat and Israel's President Efraim Katzir were sped in a heavily armed convoy to Jerusalem. During a brief opening chat with Begin at the King David Hotel, Sadat made plain immediately that he had not come to Israel to negotiate a separate peace, that his concern for the Sinai and Palestine issues was inseparable. For his part, Begin attempted to steer the discussion exclusively to the Sinai question, and reiterated his willingness to acknowledge Egyptian sovereignty over the entire peninsula. Meanwhile, Dayan was issuing a cautionary note of his own to Minister of State Butros Butros-Ghali, suggesting that Sadat avoid any reference to the PLO in his address to the Knesset. The advice was heeded.

The next morning, adhering to his prepared itinerary, Sadat attended the al-Aqsa Mosque for prayers, where he was greeted warmly by a large congregation of Arab notables, then was conducted on a tour of the Holocaust Memorial, an established formality on the schedule of distinguished Gentile guests. Early in the afternoon, then, the Egyptian president was ceremonially ushered into the Knesset to the burst of trumpets and applause. His forty-minute address was delivered in Arabic and telecast (with accompanying translation) abroad. As in his brief chat earlier with Begin, Sadat laid his major emphasis on the breadth of his purpose. He had come, he insisted, not to embark on negotiations for a separate peace between Israel and Egypt, but rather between Israel and all its Arab neighbors, the Palestinians among them. Here Sadat underscored the importance of total Israeli withdrawal from the occupied territories, including East Jerusalem. The "heart of the struggle" remained the Palestine problem, and only recognition of the Palestinians' right to a national entity would launch the first step to Arab-Israeli peace. Having adverted to the—presumably minimal—desiderata of other Arab nations, Sadat then went on to offer Israel the inducement its people had waited thirty years to hear:

> In all sincerity I tell you that we welcome you among us with full security and safety. . . . We used to reject you. . . . We had our reasons and our fears, yes. . . . [But] I declare to the whole world that we accept living with you in permanent peace based on justice. . . . Today through my visit to you I ask you, why do we not stretch out our hands with faith and sincerity so that together we might remove all suspicion of fear, betrayal and bad intentions?

Following prolonged applause, Begin took the speaker's lectern to reply graciously, but then to emphasize that Jerusalem would never again be divided, or the West Bank transformed into a PLO state. Like Sadat, he affirmed his government's willingness to go to Geneva. He reminded his visitor, however, that he, Begin, could negotiate only on the basis of UN Resolutions 242 and 338, with their recognition of Israel's right to live in peace and security. The prime minister then invited all "legitimate" spokesmen of Arabs living in the "Land of Israel" to meet with him and Sadat for serious discussions. If the Egyptian president was less than overwhelmed by this response, he managed to contain his reservations at the joint press conference he and Begin conducted afterward. Both leaders sustained the facade of cordiality. Yet Sadat was perturbed. He had been informed of Begin's meeting earlier in the day with a group of Egyptian editors. The prime minister had been harsh and unconciliatory, rejecting any notion of Arab self-determination on the West Bank. At the formal banquet that night, the mood on both sides was subdued. Later, Begin and Sadat went on to private conversation. Here at least they reached agreement on several principles. War was "rejected" as a means of settling disputes. Egyptian sovereignty over the Sinai would be restored. If there were a revived Geneva Conference, Egyptian-Israeli understanding on substantive issues should be achieved beforehand. With mutual protestations of goodwill, then, and assurances of early future meetings, Sadat and Begin ended their discussion. The next afternoon, November 21, after forty-three hours in Israel, the Egyptian president departed for home.

THE AFTERMATH OF EUPHORIA

Sadat's reception in Cairo was tumultuous. Nearly a million people cheered him on his open-car journey from the airport. Exploiting the groundswell of national support, the president was determined now to move quickly to restore Egyptian leadership at least among the moderate Arab nations. He would offer to convene his own international meeting as a "preliminary" to Geneva. In fact, the invitations he now dispatched to other Arab nations, to Israel, to the Soviet Union and the United States, were intended both to shift the site of negotiations to Cairo and the format of negotiations itself away from the rejectionist camp. The Begin government understood, and immediately accepted. In Washington, President Carter hesitated. A "preliminary" conference under Egyptian auspices might provoke the Arab hardliners, he feared. Indeed, Carter was right. Denouncing the very notion of such a gathering, the Arab rejectionist states and the PLO convened instead in an emergency session of their own, in the Libyan capital of Tripoli. There they condemned Sadat's initiative and froze diplomatic relations with Cairo. Although the Saudis and Hashemites declined to participate at Tripoli, their reservations on the peace initiative already were known. Moscow then loosed its own propaganda blast against the Egyptian president. At this point, Sadat recognized that the likelihood of a wider Geneva conclave was moribund; that the "preliminary" Cairo conference would devolve simply into tripartite negotiations between Egypt, Israel, and—Sadat hoped—the United States. It was a legitimate, if more modest, expectation. Upon appraising the diminished scope of his cherished Middle Eastern peace conference, President Carter did in fact agree to American participation.

Meanwhile, on December 2 and 3, 1977, Dayan and Tohami were secretly conferring again in Morocco. There Dayan revealed his government's first detailed proposals for the Sinai. They envisaged an early return of the peninsula to Egyptian sovereignty, linked with a deliberately calibrated territorial withdrawal. In the first phase of evacuation, Israel would pull back to a line between the towns of al-Arish and Ras Muhammad, maintaining control of Sharm es-Sheikh and the Rafa (Yamit) salient of Jewish villages in the northeastern quadrant, together with a group of Israeli military airfields; and the entire peninsula east of the Mitla and Gidi passes would be demilitarized. The second stage, of final withdrawal, would not be completed until the year 2000. For the interim, special provisions also would have to be made for a continued Israeli lease of Sharm es-Sheikh. The Egyptian deputy premier listened carefully to the formula, and in pained silence. Once Dayan finished, he observed coldly that it was not for this scheme of protracted Israeli evacuation that Sadat had risked the ire of the Arab world. Tohami also disabused Dayan of the notion that agreement could be reached on the Sinai without a parallel resolution of the broader Palestine question.

On that sober note, both statesmen returned home. Several days later, an "official" Israeli delegation arrived in Cairo, led by Eliahu Ben-Elissar, director-general of the prime minister's office. The talks in the Egyptian capital were friendly, but largely meaningless. On issues ranging from Sinai to the West Bank, the gulf plainly was too wide to be resolved except at the heads-of-government level.

Begin and his advisers in Jerusalem accordingly set to work formulating a more cosmetic package. Yet, as crafted by Aharon Barak, the prime minister's legal consultant, the plan was even less forthcoming than the original Dayan-Rosenne blueprint, the version that had been submitted to Carter and Sadat the previous September. Its prospectus for the West Bank, based on the 1977 Likud electoral platform, would "guarantee to the Arab nation in the Land of Israel a cultural autonomy, a fostering of the nucleus of their national culture, their religion, and their heritage." In fact, the plan's origins could be traced back to tsarist Russia, and replicated Jewish demands submitted to the revolutionary Duma of 1905. At the time, Vladimir Jabotinsky had contemptuously rejected the autonomist solution in favor of Zionism in Palestine alone. Thirty-two years later, ironically, in the aftermath of the 1937 Peel Report (advocating the partition of Palestine), Jabotinsky had revived the autonomist scheme—but this time as the blueprint for the Palestine Arabs. Now, forty years later yet, Begin was essentially submitting the plan as a formula for the West Bank. In the Barak-Begin version, Arab autonomy would be limited to cultural and religious activities, to health, social welfare, commerce, and tourism. Security and public order—in effect, control of the land—would remain the responsibility of Israel. Resident Arabs would be permitted to opt for Israeli or Jordanian citizenship, but not for citizenship within their own autonomous community. Begin's major concession in this plan was his declared willingness to postpone the issue of sovereignty "for the time being." As for Sinai, Israel's proposal already had been outlined by Dayan in Morocco. It envisaged a two-stage evacuation, with the Rafa settlements and the three largest air bases to remain in Israeli hands, at least until the end of the century.

Once again, even before presenting his formula to Sadat, Begin hoped to sell it initially to Jimmy Carter. He flew to Washington on December 15. Meeting with the president and with Secretary Vance the next day, he gave the Americans their first insight into Israel's thinking after Sadat's historic Jerusalem visit. The proposal Begin submitted now was far from their conception—or the Brookings Report conception—of Palestinian autonomy. Even so, Carter and Vance reacted with caution, observing only that the document was "encouraging," that it offered a "fair basis for negotiations." That equable response may have been a mistake. With his tendency to fantasize, Begin interpreted lack of vigorous objection as approval. Ironically, the one harsh rebuff he encountered was not in Washington but in Jerusalem. Upon returning home, the prime minister discovered that many in his own cabinet regarded the autonomy plan as a near certain format for an eventual Palestinian state. Begin was taken aback. He and Barak then quickly set about modifying the blueprint to win cabinet approval. Their revisions ensured that Israel would assume responsibility not only for public law and order within the West Bank, but also for the "permanent security" of the region's borders. Refugees who had fled in the aftermath of the 1967 Six-Day War would be allowed to return only in "reasonable" numbers and by unanimous decision of a joint Israeli-Jordanian-Palestinian committee. An identical unanimity would be required for the delegation of authority to a West Bank council. Finally, even these various approved arrangements would not come into effect until the conclusion of a peace treaty between Egypt and Israel. This, finally, was the refined and re-refined plan Begin took with him for his scheduled one-day return visit with Sadat in Ismailia, on December 25.

Until then, contacts between Egyptians and Israelis had been friendly. The "Cairo" delegation led by Ben-Elissar had been cordially received by the Egyptian government and people, even if its negotiations were exclusively procedural. Visiting Cairo even more briefly, Defense Minister Ezer Weizman had managed to establish a warm personal rapport with Sadat. But the prime minister's Ismailia trip was a disaster. Doubtless it was foredoomed, once Sadat had the opportunity to read Israel's autonomy plan before Begin's arrival. In fact, it was the draft seen by Carter in Washington. The toughened version now submitted by Begin was even more of a shock; while the proposal for maintaining an Israeli military and civilian presence in Sinai exacerbated the Egyptian president's outrage. Frigidly, Sadat informed his guest that the Sinai plan was not suitable even for discussion. The Palestinian suggestions would be "studied." Begin was unfazed by this response. Flying home that night, he remained confident that an accommodation with the Egyptians still was within reach. To that end, he punctiliously observed the format devised by the Ben-Elissar delegation and its Egyptian counterpart during the first Cairo visit. It was to divide negotiations into two tracks, military and political. The former would be conducted by Israeli and Egyptian teams in Cairo; the latter, by parallel teams in Jerusalem.

Initially, Ezer Weizman himself led his government's military delegation, before turning over responsibility to his subordinates. With his flamboyant, ingratiating personality, the defense minister continued to hit it off well with Sadat. The president in turn ventilated to "my friend, Ezra [sic]" his dismay at Israel's hard stance. In Bucharest he, Sadat, had received assurance that Begin was prepared to return the Sinai to Egyptian sovereignty. What kind of sovereignty was this? he asked. Hereupon Weizman explained that the Sinai bases, and particularly the huge Etzion base near Eilat, offered Israel defense in depth not only against Egypt but against Saudi Arabia; that only the Sinai offered Israel's air force the necessary space for dispersing its combat planes. Several years afterward, however, in his memoirs, Weizman admitted that psychology also influenced his government's position:

> Many of us had grown accustomed to regarding the Sinai as an integral part of . . . Israel. We had toured the length and breath of the peninsula, the bathing beaches in . . . Sharm es-Sheikh . . . were regularly inundated by hordes of vacationing Israelis. Radio and television reported the weather forecast for southern Sinai and the Gulf of Aqaba in the same routine fashion as they quoted tem-

peratures for the Galilee and the coastal plain. Furthermore, there was a new generation that could hardly remember Israel within the pre-June 1967 borders. . . . Suddenly, the Egyptians were confronting us with the demand that we give up the peninsula, whose size is much bigger than the entire country before 1967—in exchange for something abstract and intangible.

No agreement seemed possible for the while. Rather, as a precaution, Israel's "Sinai committee," a small cabinet group led by Dayan and Sharon, came up with a scheme to establish "dummy settlements" in Sinai. These presumably could be used as a tradeoff later for Rafa and the air bases. The scenario was less than inspired. Sadat exploded upon learning of it. When Carter in Washington added his own protest, Begin hastily ordered the fake communities dismantled. Otherwise, no progress was achieved in the joint Egyptian-Israeli military committee. On January 13, 1978, its discussions were temporarily suspended.

The political committee began its meetings in Jerusalem four days later, in the presence of Secretary of State Vance. Foreign Minister Ibrahim Kamil led the Egyptian delegation. Both sides had agreed in advance to a moratorium on public accusations and recriminations. Upon arrival at Ben-Gurion Airport, however, Kamil issued an uncompromising statement of Egypt's maximalist position. Then, at the banquet that evening in honor of the Egyptian visitors, Begin offered a patronizing toast to Kamil that left the foreign minister visibly pale. Evolving into a lengthy polemic, the toast extolled the virtues of a united Jerusalem, of "Judea and Samaria" linked permanently to Israel. Later that evening Kamil telephoned Sadat, insisting that Begin had foreclosed meaningful negotiations. Sadat agreed. Despite Vance's harried intermediary efforts, the Egyptian delegation was recalled only forty-eight hours after its arrival. By then Sadat was in despair. On February 3 he flew off to the United States, where he appealed for support directly to Carter. He got it. The American president fully shared Sadat's exasperation at Israel's hard stand on Sinai and the West Bank. He made his attitude clear to Begin in person, when the prime minister arrived in Washington for yet another—third—visit to the White House on March 22. Only days earlier, the Israeli cabinet had authorized the establishment of new settlements in the West Bank. Carter and his staff had reacted harshly, denouncing Israel's "expansionism." Nevertheless, in Carter's presence now, Begin remained unflinching, defending his people's "natural right" to settle within its "historic homeland." Only later did the prime minister admit to his aides that his meeting with the president was "one of the worst moments of my life."

IMPASSE AND ANGUISH

Sadat took little consolation from Begin's discomfiture. Whatever sympathy he evoked from Carter, the Egyptian president failed to make a dent on the Israelis themselves—or even on Begin's political opposition. In February 1978, Sadat invited Shimon Peres to meet with him at his vacation spa in

GAZA AND SINAI SETTLEMENTS IN 1977

● Existing settlements ----- 1949 armistice line
◇ New settlements -·-·- International boundary
○ Refugee camps ═══ Roads

0   MILES   10

Ashkelon

Jabaliya
Beach (Shatta) ○
MEDITERRANEAN SEA

Netzarim ◇ ● Gaza

Nuseirat ○

Breij ○
GAZA STRIP

Deir al-Balah ○   ○ Mughazi
Katif (Netzer Hazani) ◇   ◇ Kfar Darom
Khan Yunis ○

Khan Yunis ●   I S R A E L

○ Rafa   ◇ Morag
Talmei Yosef ◇

Yamit ◇   ● Rafa
                 (Yamit cluster)
Sufa (Succot) ◇   ◇ Pri'el
Holit ◇
Dikla ◇   ◇ Merkaz Avsholom
                 ── Bedouin resettlement zone
Haruvit ◇   Sadot ◇   ◇ Netiv Ha'asara

Nir Avraham ◇   ◇ Ogda

E G Y P T

S I N A I

Salzburg, Austria. It was a friendly enough two-hour conversation, but Peres supported his government on the Sinai issue. "As far as the [Rafa] settlements go," the Labor chairman explained, "Begin represents a national consensus. You said you would agree to border changes. Why not apply this agreement to the Rafa settlements?" Sadat admitted then that he had not opposed minor border changes—but hardly in Sinai, only in the West Bank. A later Sadat-Peres meeting on March 30, in Cairo, proved equally inconclusive. Four months later, stopping off in Salzburg en route to Vienna for a conference, the Egyptian president invited Ezer Weizman—his favorite Israeli interlocutor—to confer with him. As always, their discussion was cordial. It was also very frank. If he could make no further progress toward peace by October, Sadat warned, he would resign. Again he insisted upon a full Israeli withdrawal from the Sinai. On the issue of the West Bank, he might accept a more limited plan for Arab quasi-autonomy. But, in the interim, he must extract some dramatic, unilateral gesture of good faith from Israel—possibly an Israeli turnover of al-Arish or Mount Sinai.

Weizman listened sympathetically. The next day he flew back to Israel to report to the cabinet. Sadat was asking for little more than a "fig leaf" on the West Bank, he suggested. Could the government not respond with a somewhat more forthcoming posture? Begin listened thoughtfully. Five days later, however, on July 19, 1978, before the prime minister and the cabinet could react to Sadat's plaintive appeal, Dayan met in Leeds Castle, England, with his Egyptian counterpart Ibrahim Kamil, and with Secretary of State Cyrus Vance. And here, much to Dayan's surprise, the Egyptian foreign minister evinced no willingness at all to settle for a "fig leaf," either on the Sinai or the West Bank. Rather, Kamil adhered strictly to his government's original line: the Palestinians must be allowed full self-determination; all Israeli settlements and bases in Sinai must be abandoned without qualification; neither would there be a separate Egyptian-Israeli peace.

At this point Begin and his advisers were confused. What was Egypt's true policy? Weizman believed Sadat. Dayan believed Kamil. For Begin, in any case, there could only be one response. "Nothing for nothing," the prime minister informed a press conference. Both the intransigence and inelegance of the reply enraged Sadat. There appeared no further point to direct negotiations with the Begin government. On July 27, he ordered Israel's military delegation in Cairo to depart. Henceforth all communications would go through Washington. American pressure on the Israelis appeared to be Sadat's last hope. The Egyptian president had gambled his nation's prestige in the Arab world, and his personal reputation among his own people, on his dramatic trip to Jerusalem. It had been his assumption that Israel, in return for a peace treaty with its largest Moslem neighbor, would relinquish its enclave in Sinai and agree to withdraw at least from the major part of the West Bank.

Sadat had drawn hope, too, from the rising peace movement among the Israelis themselves. The development first surfaced in March 1978, four months after Sadat's appearance in Jerusalem, at a time when the evident failure of Begin's return visit to Ismailia had created a bleak mood in Israel.

Some 350 military reserve officers signed a letter entreating the prime minister to change his priorities, to accept an exchange of territories for peace. From this open letter, the "Peace Now" movement gained momentum, and soon won the endorsement of thirty Knesset members from six parties, from six eminent reserve generals, among them Chaim Bar-Lev, a former chief of staff. It was also the generals who questioned the putative military advantage of retaining the West Bank. Most of the Jewish settlements there, they argued, were irrelevant to Israel's security. Whether larger or smaller outposts, they could offer no meaningful obstacle to a surprise Arab attack, as the evacuation of Golan settlements in the first hours of the 1973 Yom Kippur War surely had made clear. If anything, the sixty-odd villages were militarily counterproductive; the army would be fragmented by the need to run in sixty different directions to protect Jewish civilians from the hostile surrounding population. Worse yet, the establishment of Jewish enclaves in Arab-populated territory only increased the danger of Arab terrorism, thus adding to Israel's military burden, obliging the army to keep more troops in the area, to maintain the law, enforce the curfews, undertake security checks. Troop morale would be eroded. In truth, that erosion already was apparent in a growing unwillingness of young soldiers to serve as an army of occupation—padlocking shops, demolishing homes, fighting off rock-throwing schoolgirls. The political argument was equally basic to the supporters of Peace Now. In their view, Israel simply had no business ruling a million Palestinian Arabs.

Above all else, the movement was animated by a passion to end Israel's thirty-year cycle of death and violence. Its partisans, including army officers, university professors, kibbutz leaders, and other prominent figures among the educated (largely Ashkenazic) middle class, were hardly less than terrified that Begin, imprisoned by his Revisionist theology, would sacrifice an unprecedented chance to end that bloodshed for good and always. It was a fear that lent the Peace Now demonstrations—some of them drawing 30,000 people—their anguished, visceral quality. By the summer of 1978 the movement had established fifteen branches throughout Israel. It was also beginning to evoke resonance abroad. Thus, writers and academics of the caliber of Saul Bellow, Alfred Memmi, Seymour Martin Lipset, political figures and Jewish activists such as Pierre-Mendès-France, Philip Klutznick, and Nahum Goldmann, identified full-heartedly with Peace Now. Goldmann, a former president of the World Jewish Congress, was perhaps the best known of the movement's advocates within the Jewish community. In a widely quoted article written for *Foreign Affairs*, in autumn 1978, the Diaspora elder statesman scoffed at the notion that Begin's, and the Orthodox bloc's, territorial claims were Bible-sanctified. Throughout history, at least fifteen different borders existed between Israel and its neighbors, Goldmann observed; and, in any case, "one may legitimately wonder why the Arabs or the Americans should be committed to the promises of the Jewish God!" Goldmann then urged the United States to exert its leverage on Israel, to save the little Jewish republic in spite of itself.

Informed of these mounting pressures within Israel and the wider Jewish community, Sadat could not have been unmoved. Yet it was ultimately with Begin, and Begin's coalition majority, that he would have to reach an agreement. The Egyptian president understood well, too, that a majority of Israel's citizens remained deeply suspicious of the Arabs after years of isolation and threatened destruction. These elements were not to be easily won over. Neither had the United States registered even the smallest success in bridging the gap between the two sides. By the summer of 1978, Washington had presented numerous draft proposals for the Sinai and Palestine. Jimmy Carter was waging a vigorous personal campaign against the establishment of new Israeli settlements in the territories. His advisers had forwarded an imaginative autonomy blueprint for a five-year transitional regime in the West Bank and Gaza. Thus far, however, Begin and his ministerial colleagues were unresponsive to American appeals or proposals. Secretary Vance's shuttle diplomacy was hardly more productive. Spending August 5 and 6, 1978, in Jerusalem with Begin and Dayan, then meeting afterward with Egyptian officials in Alexandria, Vance was obliged to return home at mid-month without imminent prospect of a breakthrough.

It was this very lack of success, on the other hand, that now evoked a critical procedural agreement between the Israelis and the Egyptians, and virtually at the last moment. The announcement was made in Washington on August 8. Sadat and Begin had consented to discuss their outstanding differences jointly with Carter at the latter's presidential retreat of Camp David on September 5. It would be the first personal encounter between the Israeli and Egyptian leaders since their disastrous Christmas meeting in Ismailia, and it would take place at the urgent request of Jimmy Carter himself. Aware that the two Middle Eastern governments were reaching a possibly irreconcilable standoff, the American president was determined that the momentum for peace between them would now have to be restored by any diplomatic artifice, and if necessary by exercising the fullest political influence of the most powerful country in the world.

CHAPTER IV THE PRECARIOUS EMBRACE

A SUMMIT AT CAMP DAVID

Jimmy Carter, a devout Southern Baptist, well familiar with the place names and dramatis personae of the Bible, regarded himself as a committed friend of the Jewish people restored to their ancestral land. Yet the president's overriding commitment was to Arab-Israeli peace, and thereby to a just settlement of the Palestine issue. As he saw it, Menachem Begin was the "insurmountable obstacle to further progress" in that effort. The man would have to be dealt with personally and forcefully, then. In fact, Carter's sense of urgency was justified. The millennial hopes aroused by Sadat's visit had long since been dissipated. Recriminations between the two Middle Eastern statesmen were being loosed against a blizzard of press accusations and counteraccusations. During a terse press interview of May 14, 1978, Sadat even had intimated that the 1973 October War might not after all be the last conflict between his nation and Israel. The mood of despair and bitterness would have to be reversed.

The idea of a summit had been mooted by Sadat himself early in August, 1978, in a conversation with U.S. Ambassador Herman Eilts. There was precedent for detailed American involvement in the peace process. Kissinger had initiated this activist role. More recently, in the aftermath of Sadat's visit to Jerusalem, Vice-President Walter Mondale, Secretary Vance, and senior State Department and ambassadorial personnel all carried letters, messages, and other communications between the parties. Nor were the Americans simply messengers. They offered advice and counsel, formulated plans of their own, commented on the proposals of others, used their persuasive powers to seek out compromises. Transmitted to Israel via Washington, then, Sadat's notion of a summit conference was warmly received by Begin and his advisers. This time, they noted, the Egyptian president had not demanded prior Israeli commitment to full withdrawal from the territories. Moreover, if Egypt's professional diplomatic corps was endemically hostile to Israel, Sadat at least could be trusted to move imaginatively and decisively to a compromise solution. Even the State Department, with its built-in suspicion of summits, could discern a model for the tripartite "pressure cooker" approach. It was President Theodore Roosevelt's mediation in the 1905 Russo-Japanese Peace Conference at Portsmouth, New Hampshire. The pattern would be no less useful for Jimmy Carter and his staff now.

In anticipation of the Camp David conference, General Avraham Tamir, Weizman's deputy at the recently suspended military talks in Cairo, chaired a committee to formulate a new working document of Israel's position. The committee's eighty-page "Blue Paper," reflecting Weizman's view, argued that Israel's principal objective under all circumstances must remain a separate peace with Egypt. The government accordingly had to envisage special arrangements for the Sinai that would assure Egypt "meaningful" sovereignty over the peninsula; Sadat would accept nothing less. As for the West Bank and Gaza, the report went on, the key must be verbal flexibility, both to fulfill American expectations and to provide the Egyptian president with a "fig leaf." Yet, even on the Palestine issue, it was the committee's view that Israel must insist on three provisions: its right to maintain both its troops and its settlements on the West Bank; its right ultimately to assert its own claim to the area; and, at the least, its flat rejection of a Palestinian state. Dayan read this Blue Paper for the first time on the plane en route to the United States and was not impressed. He was certain the Egyptians would never accept a formula that offered the Palestinians so little; even as he and Begin were unprepared to withdraw Israeli settlements and bases from Sinai.

At the outset, the foreign minister's reservations appeared justified. On September 5, 1978, a warm late-summer afternoon, Sadat and Begin arrived separately with their entourages at the presidential retreat of Camp David in Maryland's Catoctin Mountains. Despite the carefully orchestrated atmosphere of informality (all participants were encouraged to wear sport clothes), it soon became evident that negotiations would be difficult. That same evening, the Egyptians made the final corrections on their own position paper. Sadat read the document aloud to Carter and Begin the next afternoon, in the presidential cabin. It included every tough demand the Arabs ever had made on the Jews, from return of the 1948 refugees and reparations for all previous Arab-Israeli wars to complete withdrawal to the 1967 Green Line and a renewed division of Jerusalem. Incensed, Begin threatened to walk out of the conference then and there. With some effort, Dayan finally calmed him down. But after two additional harsh meetings between the Middle Eastern leaders that same day, Israel's senior legal adviser, Aharon Barak, cautioned presidential adviser Zbignew Brzezinski not to bring Sadat and Begin together again for working sessions. They were not. Nevertheless, the ensuing four days of prolonged three-way staff discussions generated little progress. Begin's position remained seemingly as inflexible as Sadat's. There would be no abandonment of Israel's key airfields in Sinai, no evacuation of the Israeli enclave in the Rafa salient, no withdrawal of Israeli settlements from "Judea and Samaria," no restrictions on Jewish land purchase or home construction there.

Even as tensions between the two delegations mounted, Carter, Vance, and Brzezinski continued to negotiate quietly from cabin to cabin. The president recalled: "I . . . knew it was a good negotiating tactic by either Sadat or Begin first to reach agreement with me and then to have the two of us confront the third. Sadat had understood this strategy before he arrived. . . . Begin was just now beginning to realize the disadvantage of being odd man out." Long before Camp David, of course, Carter had argued that the Palestinians must have the "right to vote, the right to assemble and to debate issues that affected their lives, the right to own property without fear of it being confiscated, and the right to be free of military rule. To deny these rights was an indefensible position for a free and democratic society." Yet it was a plan Begin had resisted from the outset, and now with growing discomfiture. He rejected every allusion to the Palestinian issue, including the phrases "resolve the Palestine problem in all its aspects," and "legitimate rights." The latter, he insisted, was a "tautology and . . . there was no telling where it might lead." Provoked by this obduracy, Carter warned of possible "strained relations" between Washington and Jerusalem. "Please, Mr. President, no threats," Begin countered sternly.

Carter persisted. With his advisers, he submitted one version after another of a "compromise" draft. By the end of the conference twenty-three such documents would have been formulated. All potential areas of agreement were reviewed and refined in turn by Aharon Barak and by Dr. Osama al-Baz, Sadat's closest adviser. At last, on September 12, in his conversation with the Egyptian president, Dayan sensed a possible opening. "Concentrate on the Sinai issues," the foreign minister told his colleagues afterward. "I'm sure this is what Sadat really wants." Indeed, this had been Ezer Weizman's instinct all along, and it was reflected in the Blue Paper prepared by his deputy. Belatedly now, Dayan suspected that a deal might after all be reached on a "fig leaf" for the West Bank—provided Israel demonstrated flexibility on the Sinai. The Americans soon learned of this shift in Israeli emphasis. Although Brzezinski feared leaving the Palestine issue in limbo, Carter overruled him. "Let's get agreement on first things first," the president insisted. Thereafter, the Sinai and West Bank issues were separated, with only ambiguous terminology agreed upon as linkage. Carter's plan now envisaged two distinct agreements, one dealing with peace between Egypt and Israel; the second, with a broader-ranging settlement of the Middle East conflict.

For his part, Begin still intended to hold firm on the Sinai airfields and settlements. On the eve of his departure for the United States, the prime minister had visited the Yamit (Rafa) village of Sadot and pledged never to abandon its inhabitants. If he were pressed at Camp David to renounce the settlements, he declared, "I will pack my bags and go back home." But now Dayan at least was wavering. On the morning of the fourteenth, the Israeli foreign minister was summoned to Carter's cabin. The president's tone was stern. In the event war broke out on the issue of the Rafa salient, he warned, Israel would not be able to count on American support. Afterward, Dayan was given further pause by his meeting with Sadat in the latter's cabin. Exhausted and nerve-frazzled, the Egyptian leader announced his intention to depart Camp David. He had sent word for a helicopter to be made ready. Point-blank, he asked Dayan now: would Begin yield on the Sinai or not? When Dayan remained uncertain, Sadat exploded: "Kindly convey this from me to Begin. Settlements, never! Why are you coming to me with such ideas you know I'll never agree to? Why should we torture President Carter with us?"

Unexpectedly then, Sadat added a sweetener. He hinted at the possibility of full diplomatic relations only nine months after the signing of a peace treaty—provided Israel agreed to withdraw the Sinai settlements. No mention whatever was made of Palestine. Hereupon Dayan returned to his prime minister and added his own appeal for moderation. Although glum and silent, Begin was listening. On the next evening, the fifteenth, he met with Carter. In a tough, four-hour conversation, the president tightened the screw. Discussions became heated. "I stood up for [Begin] to leave," Carter wrote later, "and accused him of being willing to give up peace with his only formidable enemy, free trade and diplomatic recognition from Egypt, unimpeded access to international waters, Arab acceptance of an undivided Jerusalem, permanent security for Israel, and the approbation of the world—all this, just to keep a few illegal settlers on Egyptian land." If no agreement were reached by the next day, he, the president, would terminate the conference and present a full report to Congress—with the intimation that the onus would be Israel's for the collapse of negotiations. At this point, then, Carter linked the warning with an inducement. If Israel agreed to abandon the Sinai bases, the United States at its own expense would build Israel two military air bases of "important scope and with extensive facilities."

That same evening, contemplating the offer, the prime minister received an unexpected telephone call from Jerusalem. It was from Ariel Sharon. Unknown to Begin, the agriculture minister had himself been telephoned in advance by Ezer Weizman, who requested him to intercede with the prime minister. Sharon had agreed. Now therefore, much to Begin's surprise, the hawkish former general declared that he personally anticipated no unmanageable security risks in evacuating all Sinai bases and settlements. Sharon's intercession proved decisive. Begin decided to give in. He did so grudgingly, still fearing the political consequences at home. But he knew that Sadat had his own problems by then, that his own foreign minister, Ibrahim Kamil, had resigned several hours earlier upon learning of a proposed "fig leaf" for the West Bank. Begin accordingly softened his position both on the Sinai and (as shall be seen) on Palestine. Indeed, the prime minister was willing even to honor Sadat's request for signature of a Palestinian agreement in advance of the Sinai understanding, thereby avoiding the appearance of a separate Egyptian-Israeli accommodation. Although these concessions were largely procedural, they were of great importance to Sadat—even as the substance of the accord on the West Bank was decisive to Begin and his advisers.

THE FRAMEWORKS OF AGREEMENT

The two components of agreement were entitled, rather elaborately, a "Framework for Peace in the Middle East," and a "Framework for the Conclusion of a Peace Treaty between Egypt and Israel." Under this curious joint rubric, it was the second framework that related exclusively to Egyptian-Israeli matters, and that provided for recognition of Egyptian sovereignty up to the old international Sinai-Palestine border; evacuation of Israeli bases and settlements from the Sinai; the right of Israeli maritime passage through the Gulf of Suez, the Suez Canal, the Strait of Tiran, and the Gulf of Aqaba; and construction of a highway between Sinai and Jordan near Eilat, with guaranteed right of passage by Egypt and Jordan. More detailed features provided for a phased withdrawal of Israeli troops, to be completed between two and three years after ratification of the peace treaty; for limited Egyptian forces in certain key areas; for United Nations forces in other specified zones; and, following ratification of the treaty and completion of the first phase of Israeli withdrawal (after nine months), the establishment of diplomatic relations between Egypt and Israel. It was a good arrangement for both sides. The Egyptians were guaranteed return of all their land. The Israelis won assurance that, for three years after treaty ratification, 40 percent of the Sinai would remain in their hands. For at least two of those three years, moreover, normalized relations would be maintained between Egypt and Israel, including open borders and commercial and cultural interchange. As former Prime Minister Yitzchak Rabin noted with approval later: "I cannot overemphasize the importance of testing Egypt's intentions not merely by virtue of what the Egyptians say, but by what they do for more than two years while Israel continues to hold on to such a large proportion of the Sinai."

Meanwhile, preceding the Egyptian-Israeli accord was a distinctly more complex "framework" for dealing with the West Bank and Gaza. It envisaged nothing less than the resolution of "the Palestine problem in all its aspects." Here, three stages of negotiations were outlined. In the first, Egypt, Israel, and Jordan (or Egypt and Israel alone, if Jordan chose not to participate) would lay the ground rules for electing a "self-governing authority" in the territories, and for defining the authority's powers. In the second stage, a transitional five-year period would begin once the self-governing authority was established and functioning; and Israel would dismantle its military government in the territories, withdrawing its troops to specified security locations. In the third stage—and not later than one year after the onset of the transitional period—discussions would be launched between Israel, Egypt, Jordan, and elected representatives of Palestinians living in the West Bank and Gaza, to determine the final status of the territories. Thereafter a separate committee of Israelis, Jordanians, and elected West Bank and Gaza Arabs would negotiate a formal peace treaty between Israel and Jordan. In the discussions, appropriate attention would be given "the legitimate rights of the Palestinian people."

Altogether, the Camp David accords represented an impressive diplomatic breakthrough for each of the participants. Sadat later was able to interpret the West Bank and Gaza "framework" as a major Israeli commitment to the Palestinians. Did it not promise "full autonomy" and respect for the "legitimate rights of the Palestinian people," phrases Israel had always resisted, and assurance of early Israeli troop withdrawal? The framework declared, too, that peace treaties should similarly be negotiated between Israel and its other Arab enemies, thereby apparently refuting any suspicions that Sadat was retrieving the Sinai without so much as a backward glance at Jordan and Syria. Yet, in perspective, it seemed unlikely that Sadat had achieved (or, as Weizman had insisted all along, had expected to achieve) more than a "fig leaf" on the Palestine issue. He had agreed to a convoluted, amorphous formula—essentially Israel's—that postponed the entire question of final Israeli withdrawal for future negotiations. Addressing Congress several days after Camp David, Jimmy Carter asserted that "the Israeli military governments over these areas will be withdrawn and will be replaced with a self-government with full autonomy." But it was precisely the degree of Palestinian autonomy that remained to be negotiated between Egypt, Jordan, and Israel. Each of these countries in effect retained a veto over the powers to be arrogated the "self-governing" authority, and it seemed unlikely that Begin would permit those responsibilities to be extensive. By the same token, Israel retained the veto over any Palestinians nominated to join in the autonomy negotiations, a right that also surely would be exercised. Nor was there reason to assume that other parties, notably Jordan, would hesitate to exercise their veto. By its very anticipation of Hashemite participation in the discussions, after all, the "framework" all but foreclosed the West Bankers' widely proclaimed hope for a Palestinian state.

The agreement meanwhile endorsed Israel's right to take "all necessary measures" to ensure "the security of Israel . . . during the transitional period and beyond." Beyond? Beyond the five-year period? Begin almost certainly would insist on the right to station troops at critical security points in and around the West Bank, and on a permanent or near-permanent basis. Their presence in turn was likely to give the coup de grâce to any scheme for genuine Palestinian self-government. Finally, the sensitive matter of Jerusalem was not so much as touched on in the accord. Carter had expected initially to finesse the issue through a private written assurance to Sadat that the United States recognized East Jerusalem as "occupied" territory. But when the Israeli delegation got wind of this proposal, it threatened to return home immediately Carter backed down. The entire question was relegated instead to separate letters from Sadat and Begin to Carter, with the Israeli prime minister reasserting his government's claim to a united Jerusalem as the capital of Israel, and the Egyptian president favoring Arab rule over the eastern part of the city.

Yet if Sadat had indeed given approval essentially to a separate Egyptian-Israeli peace agreement, to a scheme for the West Bank that differed little in spirit from the plan first submitted by Begin in December 1977, the "frameworks" hardly were without important—immediate, tangible—advantages to his own country. As a result of Camp David, Egypt would be able not merely to regain the long-coveted Sinai, but to shift its troops from the Israeli frontier to the Libyan border. This strategic redeployment in turn was but one of several verbal understandings reached between Sadat, Begin, and Carter. Expanded cooperation between the Egyptian and Israeli intelligence services was another, notably against Soviet penetration into Africa and the Middle East. The standing Egyptian army could be reduced now by at least half, to about 300,000 men; and, at that, it would be restructured and streamlined with modern American equipment. The United States also was prepared to organize a Western economic consortium to help replace Arab funds withheld from Egypt in retaliation for Camp David. Here, then, was the ultimate benediction that Sadat and the Egyptian people anticipated from the Camp David "frameworks." It was peace, and the assurance of desperately needed capital infusions to rescue the nation from an impending "indianization" of its already marginal, Middle Eastern, standard of living.

After twelve days of touch-and-go negotiations, a bone-weary Carter summoned Begin and Sadat to his lodge for a festive toast. That night, September 17, 1978, a televised ceremony of signing took place at the White House, in the presence of selected members of Congress. The president himself outlined the substance of agreement, observed that a formal peace treaty between Egypt and Israel was to be negotiated and completed within the next three months, and emphasized his government's commitment to "participate fully in all subsequent negotiations relating to the future" of the West Bank and Gaza. The following evening, the three leaders appeared again before a special joint session of the full Congress, this time with Carter the principal speaker, congratulating Begin and Sadat, and basking in the triumph of his diplomatic tour de force. Television viewers then were treated to the spectacle of Begin and Sadat embracing before an audience of cheering legislators. The road to peace now seemed dramatically shorter.

Indeed, Begin appeared to take the initiative on that road by honoring a promise he had made in one of the Camp David "side letters," namely, to place the new accords before the Knesset. On the other hand, the prime minister was unwilling to make the vote a matter of party discipline, and there was concern that several even of his closest political allies would oppose the concessions. None, surely, had to be reminded of the strategic role fulfilled by the Rafa salient, a function so palpable that the Labor members actually had taken a firmer stand on the Sinai than had the Likud bloc. For them, the Yamit cluster of settlements provided a security guarantee more legitimate to Israel's self-interest than did the Right's mystical obsession with Judea and Samaria. Nevertheless, in the Knesset debate itself, Shimon Peres asked his fellow Laborites to support the accords as the best current hope for peace. Most did. It was Likud that proved far more intransigent—and specifically on the issue of the West Bank. Notwithstanding Dayan's assurances of built-in procedural guarantees, many of the Right discerned in the Camp David formula a prescription for the "unthinkable," Palestinian statehood. After seventeen hours of debate, the legislators endorsed the "frameworks of agreement" by a margin of 84 to 19. Yet, of the 84 affirmative votes, only 46 came from Begin's parliamentary coalition, and only 29 from Likud's 43 representatives. Numbered among Begin's opponents were several of his oldest allies, including Geula Cohen, a fiery nationalist of Yemenite extraction, and Dr. Yuval Ne'eman, the former scientific adviser to the ministry of defense. These two Cherut veterans eventually would break with their party to establish a new, even more implacably right-wing faction, to be known as Techiya. In the short term, meanwhile, their criticism induced Begin to move with renewed caution toward the final peace treaty with Egypt.

At the same time, Sadat was not without grave concerns of his own. Chief among these was the anticipated hostile reaction of the Arab world. It was to defuse that expected outburst that the Egyptian president flew directly from the United States to Morocco. There he counted on the support of King Hassan, his most faithful intermediary. And there, to his astonishment, the Moroccan sovereign this time was reserved and noncommital. So was Hussein of Jordan, who was fearful even of meeting with Sadat, much less of joining the anticipated Palestinian negotiations without broader Arab support. In Saudi Arabia, too, King Khaled remained ominously silent. Sadat was obliged eventually to conduct his Rabat press conference in isolation. Afterward, he flew on to Cairo. Here, at least, on his native soil, he entertained no concern whatever about the public response. Nor was he wrong. As in the aftermath of his Jerusalem trip, the president was greeted rapturously, escorted to his home by nearly a million cheering citizens. Evidently they shared their leader's conviction that peace was in sight, that the lost Sinai would be restored in short order, the standing army reduced, Western investment capital provided, and that, with all these bounties, a quantum leap was imminent in the nation's standard of living. The hope appeared further vindicated in late November when the 1978 Nobel Peace Prize was awarded jointly to Sadat and Begin.

One of Sadat's initial priorities was to halt the propaganda campaign against Israel and the Jews that had been fostered in Egypt over nearly four decades. He had moved in that direction immediately following his 1977 Jerusalem visit; but with the lost momentum toward peace afterward, the campaign had revived. Now Sadat ordered it closed off decisively. In the ministry of education, a program was launched to dediabolize the Jews and Israel in textbooks and social science courses. In the media, denunciations of Israel and Zionism, and appeals for armed struggle, were eliminated. The faculty of al-Azhar University, long the repository of theocratic fundamentalism in Egypt, was instructed—and agreed—to mute its habitual appeal for *jihad* against Israel. Yet the most vivid evidence of the change in national temper was the welcome accorded visiting Israelis. Among these were journalists, professors, scientists, and other professionals who were selectively granted admittance in the aftermath of Camp David. The newcomers were greeted cordially by their Egyptian counterparts, and with visible emotion by the Egyptian man in the street.

In Israel, meanwhile, the vista of peace created a tense ambivalence between hallucination and disbelief. Only half-jestingly, travel agents spoke of possible two-way excursions between the former enemies, of Middle East package tours that no longer would be circumscribed by complex, indirect dogleg stopovers in Greece or Cyprus. For the average citizen the vision of escape from more than thirty years of claustrophobia, from the brooding peril of war and endless human and economic sacrifice, was overwhelming. Even so, well accustomed by then to earlier disappointments in the aftermath of the Six-Day and Yom Kippur wars, the Israelis were cautious about articulating that vision in detail. They waited, their hopes and dreams still barely at the threshold of consciousness. Fate dared not be tempted.

SADAT ALTERS HIS TERMS

Under the language of the Camp David "framework," Egypt and Israel were obliged within three months to complete the remaining details of their impending peace treaty. To that end, their respective negotiating teams gathered at Washington's Blair House on October 12, 1978. The Egyptian team was led by Defense Minister Kamal Hassan Ali and Acting Foreign Minister Butros Butros-Ghali; the Israeli team, by Dayan and Weizman. As always, Secretary of State Vance was an active participant. The atmosphere was hopeful. Both sets of negotiators anticipated signing and sealing the treaty well before the three-month deadline, and possibly as early as November 19, the first anniversary of Sadat's visit to Jerusalem. Yet, the days went by, and draft after draft failed to win acceptance. It soon developed that one of the stumbling blocks related not to the Sinai but to the West Bank, and particularly to Israeli settlements there. At Camp David no language on the settlements question had appeared in the "frameworks of agreement." Nevertheless, when the matter was raised between Carter and Begin shortly before the original White House signing ceremony, on September 17, it was Carter's understanding that Israel would call a halt to new settlements until the Palestinians elected their self-governing council. The "issue of additional settlements would be resolved by the parties during the negotiations," Carter recalled. "My notes are clear that the settlement freeze would continue until all negotiations were completed."

Within days after Camp David, however, chivied to remain honest by the religionists and other hard-liners in his cabinet, Begin declared that the settlement freeze would apply only to the three months (or less) needed to complete and sign the Egyptian-Israeli peace treaty; and even within that limited period, no restrictions of any kind would be accepted on the "thickening" of existing settlements. It was a painful affront to Sadat—and to Carter. Privately, then, Carter wrote the prime minister, imploring him not to raise impediments to the peace process. Begin held his ground. Indeed, his position was stiffened by an impolitic remark dropped by Assistant Secretary of State Harold Saunders, during the latter's trip to Jordan and the West Bank in late October 1978. Meeting with a group of Palestinian mayors and newspaper editors, Saunders assured his Arab hosts that Israel eventually would dismantle its settlements in the occupied territories altogether. Outraged by this palpable misreading of its intention, the Israeli cabinet in turn announced plans to enlarge existing Jewish settlements in Judea and Samaria without further delay. The challenge left Carter in an acutely uncomfortable position.

Neither did the president win consolation in the Arab camp. Immediately following the White House signing ceremony, he had dispatched Vance on a tour of Middle Eastern capitals to sell the "frameworks of agreement." The effort did not go well. In Amman, Hussein and his cabinet cross-examined their visitor intensely, then explained the risks they courted in defying the Arab hard-liners—that is, by endorsing an agreement that failed to provide for total Israeli withdrawal from all occupied territories, including Jerusalem and the Golan Heights. In Riyadh, the secretary encountered an equally brittle response from the Saudi government. And under the influence of the PLO (Chapter V), virtually all West Bank mayors denounced the Camp David accord, declaring their intention to boycott any planned elections for a "puppet," "powerless" administrative council.

If the Americans were given pause by this frigid reception, Sadat could only have been more gravely concerned. The Egyptian leader was quite prepared to withstand the execration of the Arab world. But the restiveness now beginning to surface within Egypt itself was more ominous. Evidently the initial public display of acclaim had concealed serious divisions over the Camp David frameworks. Leftists, Nasserists, members of the religious Right, including remnants of the outlawed Moslem Brotherhood, bitterly opposed the agreements. It is recalled that, within the diplomatic corps, Foreign Minister Ibrahim Kamil—himself a successor to Ismail Fahmi—had resigned in protest. Sadat's closest adviser, Osama al-Baz, a hard-liner, was known to be privately skeptical of the Camp David concessions. Thus, before the People's Assembly on November 2, 1978, Sadat reaffirmed the primacy of a West Bank solution over an exclusively Egyptian-Israeli pact. The agreements did not relate to Egypt alone, he insisted, or even to the Palestinians, but equally to Jordan, Syria, and Lebanon, "and that is the [only] accessible road to the liberation of the Arab land after 1967."

The appeal evoked no response whatever in the Arab world. Even as Sadat addressed the People's Assembly, emissaries from the Arab states were meeting again in Baghdad. This time the Saudis participated. For them, the volatile Palestinian guerrilla movements represented by far the most critical threat to Middle Eastern stability. These were elements that could be assuaged only by a tighter assurance of self-determination on the West Bank. With Saudi endorsement, therefore, the conference detailed elaborate sanctions to be imposed on Egypt in the event of a signed peace treaty with Israel. Sadat was gravely shaken. He sensed now that he had let Begin off too easily by not insisting upon a formal connection between the peace treaty and elections to an autonomous West Bank entity. Admittedly, the implication of a political, or at least a moral, linkage had been understood and quietly accepted by both sides. But in the immediate aftermath of Camp David the Israelis had emphasized, and the Egyptians had agreed, that each "framework" was designed to stand alone. After all, no one could guarantee that Jordan and the Palestinians would consent to participate in negotiations; and an Egyptian-Israeli peace treaty should not remain hostage to Arab indecision elsewhere. Nevertheless, Sadat felt obliged now to alter that understanding. On November 3, even as peace talks were continuing at Washington's Blair House, Acting Foreign Minister Butros-Ghali suddenly announced a shift in his government's position. Egypt would have to insist on a definite timetable for Palestinian autonomy before committing itself to a formal peace treaty with Israel. Moreover, as an inducement for Israel to fulfill its pledges on the occupied territories, formal diplomatic relations between Egypt and Israel should be delayed until one month after the establishment of an autonomous authority in the West Bank and Gaza.

The Israelis were stunned. Their shock soon became outrage when Vance then pressed them to accept the new linkage. After an urgent meeting four days later, Begin and his cabinet rejected the new proposal out of hand. Indeed, the linkage between a peace treaty and West Bank negotiations already was too close for several of the ministers' tastes. When yet additional requested changes for the draft treaty were floated by Cairo, they were given even shorter shrift. One of these called for a review of Sinai security arrangements after five years, including all demilitarized and limited force zones. Another related to Israel's scheduled interim withdrawal to the al-Arish-Ras Muhammad line. To save face among his critics, Sadat now asked Jerusalem to make the "gesture" of withdrawal six months, rather than nine months, following the treaty signature. Still another demand was for an Egyptian "presence" in Gaza during elections there for the envisaged administrative council. In each instance, the Israeli response was a peremptory no.

In a countermove, rather, that evinced their revived distrust of the former enemy, Begin and his associates submitted two new desiderata of their own. One raised the issue of priority of treaties. It had been understood at Camp David that the Egyptian-Israeli accord would be signed without regard to Egypt's treaty obligations to other Arab nations—that is, obligations for mutual defense in the event of war with Israel. But now, under mounting Arab pressure, particularly from Syria, Egyptian diplomats were quietly suggesting that their original Arab treaty commitments would indeed take priority. The Israelis got wind of this tactic. They riposted immediately by demanding a formal provision in the draft treaty affirming the document's precedence over all other Egyptian commitments. Otherwise, they warned, the signature of peace would be meaningless. The second proposed amendment related to Sinai oil. The Begin government wanted assurance of its right to purchase supplies from the Sinai, from wells its own engineers had drilled, enlarged, and managed over eleven years of Israeli occupation. The demand was not academic. Political unrest was mounting in Iran, a nation that provided 40 percent of Israel's oil. In any case, Cairo rejected both proposed Israeli amendments.

By December 1978, the likelihood of peace was occluded again. The three-month deadline for completion of the treaty was approaching. Beginning on December 11, therefore, Cyrus Vance spent a frenetic week commuting between Jerusalem and Cairo. Neither side would budge. The deadline was not reached. Afterward, the task of shuttle diplomacy was left to Assistant Secretary of State Alfred Atherton. Yet, following two weeks of strenuous negotiations in Israel and Egypt, and ten American draft versions of a treaty, Atherton returned to Washington as empty-handed as had Vance. All the stumbling blocks remained: linkage of the Egyptian-Israeli accord to civil autonomy in the West Bank; priority of Egypt's mutual defense pacts over an Egyptian-Israeli treaty; a five-year review of Sinai security arrangements; an Egyptian "presence" in Gaza; timing of the exchange of ambassadors; and the availability of Sinai oil to Israel. In a brief meeting between Dayan and Egyptian Prime Minister Mustafa Khalil in Brussels on December 22, the one gesture Israel appeared ready to "consider" was an advanced timetable of withdrawal from al-Arish. Otherwise, the deadlock remained. Nor was it resolved in a second, foreign ministers, meeting at Camp David, in late February 1979. Conferring with Dayan and Butros-Ghali, President Carter barely managed to suppress his frustration. "The occasional smile with which he tempered his words was thin and fleeting," Dayan recalled afterward, "never extending beyond lips and teeth. His expression was grave, his look harsh."

For that matter, press editorials in Egypt and Israel both expressed a fatigued sense of déjà vu. Was Sadat's historic Jerusalem visit to be regarded as nothing more than a brief, spectacular interlude in the long succession of diplomatic failures: the Rhodes Armistice of 1949, the Palestine Conciliation Commission negotiations at Lausanne in 1949–50, the intermittent Mixed Armistice Commission contacts of the 1950s, UN Security Council Resolution 242, the disengagement agreements of 1974–75? Was Camp David, too, then, to be regarded as still another fading landmark on an endless and chimerical quest for peace?

CARTER GAMBLES ON A BREAKTHROUGH

The last week of February 1979 began in an atmosphere of dense pessimism. Sadat had rejected Carter's invitation to attend a second summit. Instead, he proposed coming to the United States to denounce Begin before Congress. Begin in turn was unwilling to meet with Prime Minister Mustafa Khalil, and eventually agreed only to fly to Washington for personal discussions with Carter. It was during the Israeli leader's visit of March 1, 1979, however, and in the ensuing four days of give-and-take at the White House, that the president finally won Begin's approval for several imaginative new proposals. A telephone call to Sadat in Cairo followed. Immediately afterward, Carter announced his plans for a new initiative—"an act of desperation," he admitted later. In a calculated move to expose the dignity of his office and his person to the high-risk buzz saw of Middle East negotiations, the president had decided to travel personally to Egypt and Israel in an effort to achieve a final breakthrough to peace.

As it developed, Carter had come up with ideas on two seemingly intractable issues. On the question of "linkage" between an Egyptian-Israeli treaty and Palestinian autonomy, he now proposed a "side letter" to the treaty in which Israel would agree simply to complete negotiations for West Bank-Gaza elections within one year of the treaty signing. The elections themselves would not actually have to meet this deadline; and in the event they were delayed as a consequence of Jordanian or Palestinian noncooperation, Israel would be left blameless. As for the potential conflict between the treaty and Egypt's obligations to other Arab nations, Carter suggested that the Egyptian-Israeli pact be regarded as binding, without specifying that it necessarily took precedence over Egypt's other treaties. Begin reacted well to these formulas. Indeed, he cabled Jerusalem afterward, declaring that he had achieved a "great victory."

On March 7, therefore, the president and his staff departed for the Middle East, arriving in Cairo the next day. The Egyptian reception was warm, and Carter was touched. More important, he found Sadat entirely accommodating. "Over the opposition of some of his closest advisers," Carter wrote later, "Sadat accepted the troublesome texts, and within an hour he and I resolved all the [remaining] questions." Thirty-six hours later, the Americans traveled on to Israel in high spirits, confident that the most difficult part of the trip was over. It was not. The welcome extended to Carter in Jerusalem was less than enthusiastic. Heckling crowds greeted him, and he recalled a sign in English saying "Welcome, Billy's Brother!" During a tense, seven-hour bargaining session with Begin and Dayan, the president and his advisers learned that important difficulties still remained. These no longer related to the unspecified timetable for Palestinian elections or to Egypt's prior treaty commitments. Both sides had accepted Carter's terminology on these issues. Still unresolved, rather, were the questions of guaranteed right to Sinai oil, Egypt's claim to a "presence" during elections in Gaza, the timing of an exchange of ambassadors—with the Israelis still holding fast to the original schedule, midway through their army's Sinai withdrawal. Begin and his advisers were unyielding on their position.

Neither did the prime minister improve the atmosphere by his refusal to initial any agreement with the Americans without first submitting it to the cabinet, then to the Knesset, for debate and approval. Exasperated by this tactic, Carter accused Begin of indifference to peace. "[E]verything he could do to obstruct it, he did with apparent relish," the president insisted. The mood remained somber on the afternoon of March 12, 1979, when Carter addressed the Knesset and obliquely lectured Begin on the need for flexibility. In turn, the American president was obliged to sit in pained embarrassment as Begin's speech of response was interrupted by the tasteless hectoring of several right-wing members. It was a bleak moment for Carter, who had gambled more than his diplomatic credibility on the dramatic trip to Egypt and Israel. His very political future in the United States was on the line.

The prospects for a breakthrough were not yet frozen. Dayan and Vance were urgently experimenting with compromise solutions of their own. Dayan had outlined for Carter the economic and military significance of the Sinai evacuation. Egypt would receive "an ordered and well organized Sinai," the foreign minister explained, with infrastructure and installations constructed by Israeli forces over a decade. For Israel, conversely, the expense of replacing this massive plant would be back-breaking; and even Carter's earlier promise of two American-constructed air bases in the Negev would not begin to lift the burden. Vance was listening. He discerned some room for negotiation here—provided money alone were involved. The president concurred. At this point, he authorized the secretary of defense, Harold Brown, to dangle a tantalizing list of ordnance before the Israelis, together with additional promises of American military and diplomatic support. Finally, on March 13, a breakfast meeting between Carter and Begin produced a tentative understanding on a wide panoply of supplies. Hereupon the president took off immediately for an unscheduled stopover in Cairo. Deplaning in the Egyptian capital, he met with Sadat in the airport VIP lounge and reviewed the latest Israeli concessions. Would the Egyptian president find them acceptable? When several of the latter's advisers demurred, Sadat dismissed their reservations with a magisterial wave of his hand, endorsing the agreement on the spot. Carter's eyes visibly misted. In Sadat's presence, he jubilantly telephoned the news to Begin in Jerusalem. The Israeli prime minister was gratified.

The agreements negotiated during these hectic five days in Jerusalem and Cairo included two revised accords, three annexes, an appendix, agreed minutes, and six letters. As at Camp David, the second accord was the peace treaty itself, to be signed directly between Egypt and Israel. The first accord outlined a comprehensive peace for Palestine; although in this case, at the Israelis' request, the plan was defined in "side letters" from Sadat and Begin to Carter. By its terms, the two leaders agreed to begin negotiations on Palestine self-rule one month after ratification of the Egyptian-Israeli peace treaty, to make a "good faith" effort (but no formal commitment) to reach agreement on autonomy within one year. In the event agreement actually were reached, and elections for a self-governing entity actually took place, then, one month later yet, Israel would terminate its military administration in the occupied areas and withdraw its troops to specified zones. Thereupon, as agreed at Camp David, a five-year transition period would begin. No further details were cited in these brief "side letters" to Carter, but subsequent negotiations on the fate of the West Bank and Gaza presumably would adhere to the Camp David format. This time the question of Jerusalem was not so much as mentioned.

Yet, inasmuch as Israel had agreed to a "good faith" effort in its impending negotiations on the West Bank, and Egypt in turn had forgone its demand for a timetabled "linkage" between Palestinian elections and ratification of a peace treaty, the way was open for other compromises, specifically on the Egyptian-Israeli pact. This latter was a very formal and legalistic document. Under Article VI, the new wording read: "This treaty does not affect . . . the rights and obligations of the parties under the Charter of the United Nations." It was a formula that enabled Sadat to proclaim his commitment to the United Nations principle of regional defense—should other Arab nations be directly attacked by Israel. Conversely, Israel received the assurance it sought in subsection 5 of this same provision: "Subject to Article 103 of the United Nations Charter, in the event of a conflict between the obligations of the parties under the present treaty and any of their obligations, the obligations of this treaty will be binding and implemented." The minutes appended to the document stated that "there is no assertion that this Treaty prevails over other Treaties or agreements, or that other Treaties or agreements prevail over this Treaty." In short, Egypt and Israel simply would not argue in public about treaty priorities. To paraphrase Mahmud Fawzi's 1957 formula on the Gaza Strip, if Israel would keep its mouth shut, Egypt would keep its eyes shut.

Otherwise, Camp David's three-year schedule of phased Israeli withdrawal from the Sinai was reconfirmed. In a carefully delimited chronological sequence of evacuation, to be protected each step of the way by the temporary interposition of United Nations Emergency Force (UNEF) troops, Israeli forces would pull back to the al-Arish-Ras Muhammad line within nine months of ratification of the treaty. Unofficially, Begin was prepared to be even more forthcoming. During his Jerusalem negotiations with Carter, he had agreed to shorten the initial withdrawal to six months. Then, two weeks after this commitment, Begin and his military advisers decided to embellish the gesture even further. The army would complete its retreat to the interim line within two months—and fully seven months ahead of schedule. Responding as a "man of honor" to the Israeli concession (in a face-saving maneuver proposed by Carter), Sadat privately assured the American president that both sides would exchange ambassadors one month after Israeli forces had moved to the new line. In this fashion, Israel shared in the advantages of the foreshortened schedule. Sadat additionally relinquished his demand for an Egyptian "presence" at the Gaza elections. Neither would he insist any longer on a review of Sinai security agreements after five years.

The oil question was resolved literally on the eve of the signature ceremony, following intense bargaining. In minutes attached to Annex III of the treaty, the Egyptians agreed to permit "normal commercial sales" to Israel "on the same basis and terms as apply to other bidders for such oil." This commitment in turn was quietly reinforced by an American "Memorandum of Agreement" promising to maintain Israel's oil supplies (in the event of a boycott) for not less than fifteen years—that is, for ten years beyond the earlier Ford commitment to Rabin—as part of the 1975 "Disengagement II." The "Memorandum of Agreement" included an even wider-ranging American commitment. Worked out between Dayan, Vance, and Brown, it confirmed the United States' willingness

> to provide support it deems appropriate for proper actions taken by Israel in response to such demonstrated violations of the Treaty of Peace. In particular, if a violation of the Treaty of Peace is deemed to threaten the security of Israel, including . . . a blockade of Israel's use of international waterways, a violation of the provisions of the Treaty of Peace concerning limitation of forces or an armed attack against Israel, the United States will be prepared to consider, on an urgent basis, such measures as the strengthening of the United States presence in the area, the providing of emergency supplies to Israel, and the exercise of maritime rights in order to put an end to the violation.

The memorandum further defined Washington's obligation to veto any United Nations action that could threaten the peace treaty, to "be responsive to the military and economic assistance requirements of Israel," and to impose restrictions on "weapons supplied any country which might transfer them without authorization to a third party for use in an armed attack against Israel."

(Saudi Arabia and Jordan clearly were implied here.) It was a major American undertaking. Indeed, subject to congressional approval, it was an all but unilateral treaty of defense for Israel, and plainly a critical factor in winning Begin's final acceptance of the peace document.

The American-Israeli "Memorandum of Understanding" initially evoked sharp criticism from Egypt's Prime Minister Khalil, who resented its evident assumption that Egypt alone was liable to violate the impending peace treaty. But here the State Department reminded Khalil that Washington had offered a similar commitment to the Egyptian government, and the latter had refused it. The Egyptians in any case had little reason for complaint. Under a separate understanding, the Americans had agreed to provide them with some $2 billion in airplanes, tanks, and antiaircraft weapons. Israel, to be sure, would receive fully $3 billion in military and financial help. These aid pledges supplemented the "normal" assistance Washington earlier had provided the two Middle Eastern countries, a package that in 1979 totaled $750 million to Egypt and $1.8 billion to Israel—the bulk of this latter to be applied to the construction of the two promised air bases in the Negev Desert. Altogether, there were inducements aplenty for both sides, and they opened the way for professional diplomats to tie the few remaining loose ends. The treaty was scheduled for signing in Washington on March 26, 1979.

THE FORMALIZATION OF COMPROMISE

In the period between Carter's dramatic breakthrough and the ceremony of signing, the Arab governments were predictably enraged. By then not a shred of doubt remained that Sadat had abandoned their cause, that his treaty was a separate peace, a historic disaster, the lineal descendant of such earlier calamities as the Sykes-Picot Agreement, the Balfour Declaration, and the 1947 United Nations Partition Resolution. Sadat had been served warning at the Baghdad summit in November 1978. Now the Arab delegates reassembled in the Iraqi capital on March 28, 1979, to make good on their threat to sever all remaining political and economic ties with Egypt. Saudi Arabia and Kuwait formally canceled their promised large-scale economic aid. No reference was made to the scores of thousands of Egyptians working in these Persian Gulf nations, but their employment future manifestly remained uncertain.

These announced punitive measures did not appear to faze Sadat. He had committed himself too extensively by then, his gamble would have to be pursued to the end. Begin's reasoning followed the same line. So did the Knesset's. On March 22, by a margin of 95 to 18, Israel's parliament voted to ratify the peace pact. Four days later, in a White House ceremony attended by the Israeli and Egyptian delegations, by United Nations Secretary-General Kurt Waldheim, and by high officials of the American government and other dignitaries, Sadat and Begin affixed their signatures to a formal treaty of peace between their nations.

The preamble dutifully invoked the "framework" signed at Camp David, as well as Security Council Resolutions 242 and 338, and emphasized again the intention of both nations to achieve a wider regional peace among all Middle Eastern nations. The document then went on to terminate the state of war between Egypt and Israel. Israel undertook to withdraw all its armed forces "and civilians" from the Sinai. Egypt pledged itself to use Israel's evacuated airfields in the peninsula exclusively for civilian purposes. Each party agreed to respect the other's sovereignty, territorial integrity, political independence, its right to live in peace within "secure and recognized boundaries" (another Israeli victory, implying boundaries to be negotiated in the east). Each similarly undertook to conduct normal diplomatic, economic, and cultural relations with the other, to remove all discriminatory barriers to the free movement of people and goods.

The document proceeded to outline four permanent limited-force zones in the Sinai and in a narrow strip of Negev territory adjacent to the Sinai (a concession by Israel), and to authorize the emplacement of United Nations troops and observers—their national components to be approved by Egypt and Israel—in the last, eastern, two of these zones. Nor were these forces to be evacuated from their assigned zones unless specifically authorized to do so by the Security Council, acting on a unanimous vote of its five permanent members. The fiasco of 1967 would not be repeated. Israeli ships and cargoes were to be assured free access through the Strait of Tiran and the Suez Canal. The mutual determination to avoid potentially troublesome incidents was apparent even in the smallest details. Thus, an Egyptian liaison office in al-Arish and an Israeli liaison office in Beersheba would be connected by a direct telephone "hot line." Israeli war memorials would be erected in the Sinai and Egyptian war memorials in Israel, each side agreeing to maintain these tributes to the other's fallen soldiers.

Not later than six months after Israel's interim withdrawal, both countries would enter into negotiations on trade and commerce, on the free, two-way movement of nationals into each other's territory, the establishment of normal postal, telephone, cable, radio, and other communications, and the construction of a highway near Eilat between Egypt, Israel, and—with Hussein's consent—Jordan, for mutual use. Finally, in accompanying letters, Begin and Sadat affirmed to Carter their understanding on the issue of the West Bank and Gaza, a meeting of the minds that ensured signature of the treaty itself, and their expectation that the United States would participate fully in all stages of Palestine autonomy negotiations as well as all phases of enforcing the Egyptian-Israeli treaty. For his part, Carter reaffirmed his government's intention (subject to congressional approval) to "take any appropriate" actions to ensure compliance with the treaty. The United States would conduct aerial monitoring over the Sinai during the three-year period of Israeli withdrawal. Should the Security Council refuse to authorize the permanent stationing of UN personnel in the withdrawal zones, the United States would seek to establish an "acceptable alternative multinational force." The implication was plain here of possible American involvement, as in the 1975 disengagement agreement. These commitments, in addition to other

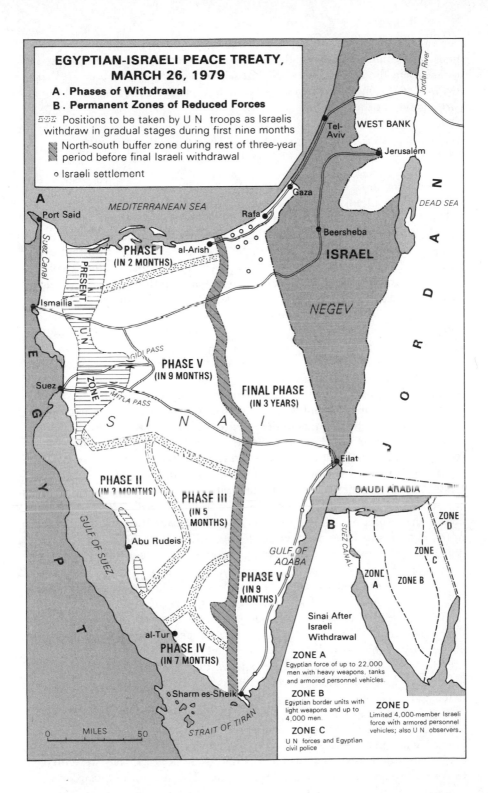

EGYPTIAN-ISRAELI PEACE TREATY, MARCH 26, 1979

A . Phases of Withdrawal

B . Permanent Zones of Reduced Forces

Positions to be taken by U N troops as Israelis withdraw in gradual stages during first nine months

North-south buffer zone during rest of three-year period before final Israeli withdrawal

○ Israeli settlement

A

MEDITERRANEAN SEA

Port Said

Suez Canal

PRESENT U N ZONE

Ismailia

GIDI PASS

Suez

MITLA PASS

E G Y P T

GULF OF SUEZ

al-Arish

PHASE I (IN 2 MONTHS)

PHASE V (IN 9 MONTHS)

S I N A I

PHASE II (IN 3 MONTHS)

Abu Rudeis

PHASE III (IN 5 MONTHS)

al-Tur

PHASE IV (IN 7 MONTHS)

GULF OF AQABA

PHASE V (IN 9 MONTHS)

○Sharm es-Sheik

STRAIT OF TIRAN

Rafa

Gaza

Tel-Aviv

WEST BANK

Jerusalem

Jordan River

Beersheba

ISRAEL

DEAD SEA

NEGEV

J O R D A N

FINAL PHASE (IN 3 YEARS)

Eilat

SAUDI ARABIA

B

SUEZ CANAL

ZONE A

ZONE B

ZONE C

ZONE D

Sinai After Israeli Withdrawal

ZONE A
Egyptian force of up to 22,000 men with heavy weapons, tanks and armored personnel vehicles.

ZONE B
Egyptian border units with light weapons and up to 4,000 men.

ZONE C
U N forces and Egyptian civil police

ZONE D
Limited 4,000-member Israeli force with armored personnel vehicles; also U N observers.

0   MILES   50

American promises of military aid to Egypt and Israel, and Carter's agreement to become involved in the forthcoming negotiations on Palestine, revealed again the importance both Jerusalem and Cairo attributed to Washington's role, and the extent to which both nations rejected any Soviet participation in their affairs.

THE CHALLENGES OF PEACE

In the aftermath of the treaty ceremony, none of the signatories underestimated the problems that remained, whether in implementing peace between Egypt and Israel themselves or in negotiating an autonomy solution for the West Bank and Gaza. Yet their solemnity was informed by an acute sense of urgency. For Sadat, it was the need to rescue his throttled economy. For Begin, it was the need to protect the territorial integrity of the "reunited Land of Israel." Neither was it possible to minimize the sacrifices each leader knowingly had incurred by accepting compromise. In the case of Sadat, abandonment of guaranteed linkage between the peace treaty and Palestinian autonomy ensured the imprecations of the Arab world, the termination of political relations with nineteen Arab states, the withdrawal of Saudi and Kuwaiti financial support, even the repudiation of Cairo as the site for Arab League headquarters.

Nor was opposition to the peace treaty lacking even within Egypt. Although the People's Assembly overwhelmingly approved it, a not insignificant minority of thirteen deputies, representing various political persuasions, voted against the document. For them, as for their constituents outside parliament, the treaty not only abandoned the Palestinians but deprived Egypt itself of effective sovereignty over the Sinai. The impending presence of an Israeli embassy in Cairo before the full withdrawal of enemy troops evoked memories of the British high commissioner in Egypt. Sadat was not indifferent to this criticism. Thus, after steering the treaty through the People's Assembly, he sought to overwhelm the opposition by conducting a referendum. On April 19, 1979, the Egyptian people were asked to say yes or no to "peace." By official count, 99 percent said yes. In the same referendum, the voters were asked whether the People's Assembly itself should be dissolved—two years ahead of schedule. Again 99 percent said yes. In the election campaign that followed, then, debate over the treaty was forbidden on the ground that all but 5,000 out of 40 million people had approved it. The very frenzy of this overkill may have revealed Sadat's awareness that trouble was brewing. If his critics had been silenced for the time being, he did not doubt that they were lying in wait. It was a measure of the man's courage that he steeled himself to confront their hostility.

Begin paid almost as heavy a price for agreeing to evacuate the Sinai. During their twelve years of occupation, the Israelis had invested in this crucial buffer peninsula $10 billion in military installations and $2 billion in roads and settlements. Now, within a period of three years, they were obliged to remove the totality of their infrastructure. The statistics of transfer were awesome. By initial calculations, 60,000 transport-days were needed to evacuate some 75,000 tons of equipment. In addition to 10 air bases, 2 of them comparable to the largest military fields available to NATO, the Israelis were obliged to dismantle and relocate 70 other military camps and installations encompassing 2,700 buildings. More daunting yet, the transplantation of this mass of arms and equipment required the construction of approximately 50 new camps in the Negev Desert, and 450 miles of new roads. Water lines and electric cables would have to be laid, sewage treatment plants and thousands of family housing units built. In the end, the cost of the military deployment to integral Israel was estimated at $4.4 billion. At Camp David, to be sure, the United States had agreed to contribute a quarter of this sum by constructing two major air bases in the Negev. But the Israelis would have to find the rest. Indeed, they would have to come up with the funds notwithstanding the crippled circumstances of their economy (by 1980 the annual inflation rate had climbed to 130 percent), and the fact that imported, South Korean, labor would build the new air bases—thus siphoning off money that otherwise would have remained in Israel. It was conceivable that the balance somehow could be generated over a period of twenty years, but surely not within the three years of phased evacuation required by the peace treaty. Worse yet, Israel would carry the back-breaking task of withdrawal just at the moment when it was abandoning its last Sinai oil supplies, resources that it had developed at the cost of $5 billion and that had met half the nation's energy requirements in the past decade. There could be little doubt, then, that Begin, like Sadat, was taking a prodigious gamble.

Were the economic prospects of peace sufficient to compensate both nations for the risk? For the threat of an Arab quarantine, in the case of Egypt? For the abandonment of defense in depth, of energy reserves, and of vastly expensive installations and settlements, in the case of Israel? Over the longer term, elimination of the threat of war seemingly justified almost any immediate economic sacrifice. For Sadat and his advisers, the opportunity to reduce the size of Egypt's standing army, to assure uninterrupted traffic in a revived Suez Canal, to gain access to a reopened Sinai and to Western capital—all were decisive inducements. The Sinai alone was envisaged as a potential cornucopia of natural resources, of coal, manganese, phosphates, and, supremely, of oil. Yet, in the short run, without major Western investment, neither the reduction of military expenditures nor the availability of oil and canal income alone was certain to compensate Egypt for the loss of Persian Gulf subsidies—a development Sadat had not envisaged upon launching his peace offensive. Considerably more time would be needed now for the dividends to come in.

For Israel, the pertinent question was whether much was to be gained in commerce with Egypt. During the 1930s, Egypt had indeed served as Palestine's main trading partner. But a half-century later the prospects for renewed two-way commerce were less than impressive. Egypt's main exports were cotton and cotton goods. Israel produced its own. Within the—still limited—Egyptian market, Israeli industrial technology would have to compete against European, American, Japanese products. Chances for large-scale tourism between Egypt and Israel were equally uncertain. The Israelis themselves were more than willing to pour into Egypt with all their hearts, souls, and pocketbooks. Yet the likelihood of transcending short-term, curiosity-impelled visitation depended on mutually strengthened economic infrastructures. For both countries, then, the economic advantages of peace were to be found less in trade, or even in reduced military expenditures, than in enlarged opportunities to attract foreign capital. Here it was that each nation pinned its expectations on the willingness of American and European companies to invest in Israeli or Egyptian projects, once the threat of war finally was lifted from the Near East.

A week after the treaty signing, Menachem Begin departed on his first official visit to the capital of his former enemy. He was met at Cairo airport by Vice-President Hosni Mubarak. The welcoming ceremony lasted a bare ten minutes, and the crowds Begin encountered as he drove through Cairo's thoroughfares were far smaller than those that had greeted Sadat in little Israel seventeen months before. Much acrimony and many frustrations had flowed under the bridge since then, of course. The prime minister appeared unperturbed. He spent most of April 3, 1979, on a guided tour of the principal tourist sites. Visiting the Sharei Shamayim Synagogue, he was tearfully welcomed by the less than 200 remaining members of Cairo's aging Jewish community. In the evening Begin attended a state dinner hosted by Sadat. Again the prime minister was correctly, if coolly, received by the assembled dignitaries. This time Begin struck a conciliatory note. He praised Sadat, the Egyptian people, the courage and honor of Egypt's fighting men. The remarks drew applause. For his part, Sadat replied in a vein more cordial than any he had employed toward Israel (and Begin) since his trip to Jerusalem. At the end of the twenty-eight-hour visit, the two leaders jointly announced that Israel would relinquish its control over al-Arish on May 26, 1979, seven months ahead of schedule. Here at last was the goodwill gesture Begin once had scorned ("nothing for nothing"), but which finally had been tendered, first through Carter in Jerusalem, then to Sadat in Washington on March 25, as an inducement for the Egyptian president to accept the oil concessions of the treaty. The joint communique further declared that Sadat and Begin would meet in al-Arish on May 27, and from there would travel to Beersheba, Israel. During these visits, they would ceremonially proclaim the Egyptian-Israeli border open, then would inaugurate an air corridor between Cairo and Tel Aviv.

Both men, too, had agreed to make of the al-Arish ceremony an occasion for the display of *sulh*, a peace far deeper than the perfunctory Arab *salaam*, denoting a more authentic reconciliation. To that end, Defense Ministers Weizman and Hassan Ali worked out the arrangement for a humanizing feature, a symbolic meeting of veterans from both sides. A selected group of 100 Egyptian and Israeli war wounded thereupon was brought to the Sinai coastal town on May 27. Their meeting was cordial, even emotional. Presently they were joined by Sadat, Begin, and Secretary of State Vance. When the excitement subsided, Sadat addressed the audience: "Premier Begin and

I have decided to give maximum impetus to the movement for peace, and to continue together the common effort to end the suffering of the countries in this area. . . . Your sacrifices will not be in vain. The road is open to peace. Together, we shall light the candle of life, consolation and redemption." Begin echoed these sentiments, pledging his nation to peace, then intoning the Hebrew prayer, Shehechiyanu: "Let us pray to God and bless Him for allowing us to live and to reach this moment."

Was this indeed the advent of a new era of communication and mutual forbearance? It was much to expect from perhaps the most notoriously strife-torn region in the world. There had been other formal treaties, after all, among the Arab nations themselves, and none had offered a guarantee of indefinite quietude or peaceful accommodation. At best, the threshold that had been reached at Camp David, at Washington, and at al-Arish was a precedent. In their future relations, diplomats and international lawyers on both sides would find it possible to hark back not simply to a tradition of unrelieved hatred, but to an impressive model for discourse, even for understanding and tentative cooperation.

CHAPTER V

FADING HOPES

ON PALESTINE

THE DYNAMICS OF PALESTINIAN IDENTITY

On April 30, 1979, an Israeli freighter passed through the Suez Canal. It was the first vessel flying the Star of David to make the passage. Egyptian officials and crowds of private citizens watched the event with mild interest. Their mood had been softened by the peace treaty of the previous March, and by Israel's gesture of goodwill, two months after that, in returning the Sinai "capital" of al-Arish well ahead of schedule. Neither did the problem of a multinational buffer force for the al-Arish-Ras Muhammad corridor impede the momentum toward peace. It had been anticipated initially that these troops would be taken from the United Nations Emergency Force—UNEF— established by the Security Council in the last days of the Yom Kippur War. The Soviet Union vetoed this possibility, however, and the United States was unable to persuade other nations to send contingents. But in September 1979, after intensive American-Israeli discussions, agreement eventually was reached for the United States Field Mission that had been dispatched under "Sinai II," the second Egyptian-Israeli disengagement agreement of September 1975, to monitor the new interim Israeli withdrawal. The mission's civilian personnel would be supplemented by troops from the United Nations Truce Supervisory Organization and by joint Egyptian-Israeli patrols in the buffer zone. A compromise was reached, too, on the thorny issue of Sinai oil, to be supplied by the Alma and Abu Rudeis wells Israel was evacuating under its treaty obligations. At first, the Egyptians demanded the premium rates obtainable at the Rotterdam "spot" market; but finally, almost at the last minute, they consented to accept the prevailing Organization of Petroleum Exporting Countries (OPEC) price.

Even then, the prospects were less than dazzling for wholly normalized relations between the two countries. Israel's name still did not appear on Egyptian maps of the Middle East. Jerusalem's first ambassador to Cairo, Eliahu Ben-Elissar, who took up his assignment one month after the Israeli withdrawal from al-Arish, found himself in near quarantine. The entire summer of 1979 passed before a landlord would agree to rent premises to the Israeli embassy, or personal lodgings to Ben-Elissar and his family. No newspaper was willing to accept a paid advertisement announcing the Israeli consulate's office hours. Egyptian Acting Foreign Minister Butros-Ghali complained to a friend that he spent most of his time receiving distinguished

Israeli visitors "because no other Egyptian official wants to deal with them." Even the foreign diplomatic corps in Cairo sensed the chill. Only two ambassadors ventured to invite the Ben-Elissars to their homes. Sadat himself unquestionably was serious about diffusing his nation's lingering suspicions. During a goodwill visit to Haifa in September 1979, he invited residents of the port city to visit Alexandria, a community with which they had been "twinned." Otherwise, the members of Egypt's political and economic establishment waited cautiously. They needed evidence of Israel's intention to fulfill that single feature of the peace package capable of redeeming Egypt from pariahdom in the Arab world. This was meaningful progress toward Palestinian autonomy.

As it turned out, the Palestine Arabs themselves would have something to say about their role in the anticipated Camp David formula. Numbering approximately 750,000 in the West Bank and Jerusalem (with another 400,000 in Gaza) by the summer of 1979, they no longer fitted their Western stereotype of hapless rustics and refugees. Despite the constraints of Israeli rule, their standard of living had been rising more impressively in the West Bank than anywhere else in the Arab Middle East, except for the Gulf oil kingdoms. Their circumstances could be measured not only in an annual 12 percent GNP growth rate since 1968, but in the marked enhancement of their village and town services, the decline of their infant mortality, the improvement of their nutrition, the enlargement of their homes. Economically integrated into the Israel-Palestine common market, commuting to jobs in Israel or working on subcontract from Israeli industrialists, the Palestine Arabs were rapidly being transformed into blue-collar workers, acquiring better take-home pay, new skills, new status and dignity. A growing minority among them thrived as merchants and technicians. Education both matched and stimulated this economic growth. Even under Hashemite rule, 95 percent of the children in West Bank refugee camps were attending United Nations Refugee Works Administration (UNRWA) primary schools; 70 percent, UNRWA high schools. Under Israeli occupation, government schools picked up the slack for nonrefugee children. Between 1967 and 1980, the number of pupils increased from 250,000 to 400,000. By then too some 70,000 Palestinians were attending either Arab or foreign universities. It was the highest per capita university enrollment in the Arab world.

These economic and educational changes produced widening ramifications. Numbers of Arab women found employment (although often in their own homes) as finishers of clothing or shoes for Israeli companies. Most of them came from refugee towns, from a male-dominated community. Earning wages now, they were prepared increasingly to question their social inferiority. So were Palestinian youths, also better educated and better paid than their predecessors. Until the mid-1970s, the informal Israeli-Jordanian administration in the West Bank permitted the region's mayors to serve as liaison between the two countries—and to benefit from the financial assistance of both. With few exceptions, these were older Palestinian notables, heads of conservative landed families, and essentially peaceful in their accommodation to the status quo. In recognizing and rewarding their leadership since 1967, Israel and Jordan precluded the development of a radical PLO alternative. Six or seven years later, however, it was the emerging younger, depeasantized, educated group that increasingly rejected the spokesmanship of the notables, and the latter's willingness to cooperate with the Israeli military government.

Curiously enough, the new generation's restiveness was kindled initially by Israel's very military prowess and staying power. It was significant that the literature of Palestinian "resistance" until the early 1970s was a phenomenon first and preeminently of the Palestinian diaspora, beyond the Holy Land itself; and, at that, a writing that tended to focus on the nostalgia of the "return," without linking this goal to any specific entity other than Israel itself. But the Jews' subsequent conquest of the West Bank eradicated forever the chimera of Israel's disappearance, and enabled the Palestinians for the first time to concentrate on a more tangible, and a politically more acceptable and negotiable, objective. It was the elimination of Israeli rule and the establishment of a Palestinian state. The fusion of this seemingly more rational objective with the emergent Palestinian younger, educated group in the early 1970s, proved tailor-made for PLO incitement.

Thereafter, it was the Yom Kippur War that provided the supreme catalyst for this fusion. "Within a few brief minutes and hours," recalled Raymonda Tawil, a brilliant Palestinian activist, "we were regaining the pride and self-confidence we had lost in the course of six years of defeat and humiliation." The PLO accordingly emerged from the 1973 conflict as the winner by default among the West Bank Arabs. Hussein, who had disgraced himself in the eyes of the Palestinians for ordering the expulsion and slaughter of PLO members and other refugees during "Black September" 1970, committed his second fatal mistake in not joining the Yom Kippur attack on Israel. The initial success of the Arab offensive was regarded as a feat entirely independent of Jordan, and the latter as a result lost all credibility to the emergent nationalist leadership. Ironically, the Israeli government itself fortified that reaction. When Rabin declined to enter into negotiations with Hussein in the aftermath of the 1975 Sinai disengagement agreement with Egypt, he also rendered the Jordanian option increasingly irrelevant. Earlier, in October of 1974, the Rabat summit conference of Arab nations for the first time had granted the PLO recognition as the "sole legitimate representative" of the Palestinian people. Now, after Sinai II, even Hussein had to pay grudging lip service to this formula.

Within the next three years, therefore, a string of diplomatic successes further enhanced the prestige of Arafat and his confederation of guerilla factions. Equally declaring its objective to be the establishment of a "democratic, secular Palestinian state," the PLO soon won recognition in one form or another by more than 100 governments. In November 1974 it achieved a spectacular diplomatic coup when Arafat was invited to address the UN General Assembly. The following year the General Assembly passed a resolution branding Zionism as "a form of racialism and racial discrimination." And then, in the summer of 1975, the world body came within a hairs-

breadth of depriving Israel of its membership in the United Nations altogether (only the sternest of American warnings averted this step). PLO officials subsequently were invited to participate, as representatives of the
Palestinian people, in UN-sponsored deliberations and conferences on the
Middle East. After the Sadat peace initiative, the UN Palestine Committee
cabled Arafat in January 1978 to express its "solidarity" with the PLO, and
added that it "could not support any decision which goes against the . . .
participation of the Palestine Liberation Organization."

Against this backdrop of international approbation, it was not surprising
that the PLO should have become the focus of West Bank Arab nationalism,
that younger, educated partisans of the PLO should increasingly have directed the efforts to contest Israeli land expropriations, oppose Arab land
sales to Israelis, and refuse tax payments to the Israeli-controlled local authorities. Shortly after Yasser Arafat's appearance before the United Nations
in November 1974, pro-PLO student demonstrations were mounted in the West
Bank. These were followed later in the month by strikes of business and
professional men in East Jerusalem and Ramallah. During the winter, civil
disobedience extended to smaller towns and refugee camps. The Israeli authorities struck back with additional curfews and deportations. By 1975 relations between the military government and the Palestine Arabs had deteriorated ominously. It was in the midst of this rising agitation, too, that the
Labor cabinet made a severe miscalculation. It authorized municipal elections in the occupied territories for April 1976. Evidently Prime Minister
Rabin and Defense Minister Peres anticipated that, by liberalizing the former Jordanian electoral regulations, by allowing women and men of the poorer
classes to cast ballots for the first time, the West Bank population would be
gratified—and placated. Instead, a debate raged for months among the Palestinians, almost to the last week before voting. Should legitimacy be given
to an Israeli sponsored election? To Israel's unilateral abrogation of the old
Hashemite voting restrictions?

In the end, the Arabs' decision was influenced by the opportunity at least
for a free choice. In almost every village and town, radical-nationalist slates
of young progressives had announced their candidacy, openly proclaiming
their opposition to the old, Jordanian-sponsored notables, and their enthusiastic support for the PLO. Thereupon, the overwhelming majority of West
Bankers decided to cast their votes. The April 12 elections accordingly produced a landslide victory for the radical nationalists, a massive repudiation
of the pro-Jordanian moderates. In town after town, from Ramallah to Tulkarem to Nablus to Hebron, PLO supporters took over the municipal councils. Altogether, nearly two-thirds of the elected councillors were new faces,
precisely the younger, educated vanguard of the West Bank's social revolution. During the entire campaign it was significant that none of the candidates had said a word about the need to improve municipal services. Their
election propaganda had dealt exclusively with the broader political issue of
Palestinian national identity.

When the impact of April 12 wore off, Israeli authorities made every effort to develop functional relations with the new West Bank officials. Some of the latter in fact were prepared to cooperate on a day-to-day basis. But most of them also openly castigated the presence of the military government, and particularly the establishment of new Jewish settlements. After Sadat's visit to Jerusalem, they soon ventilated their concern that an Egyptian-Israeli peace was in the offing at their expense. The Camp David "frameworks" appeared only to confirm their fears. Almost unanimously, then, the new West Bank leadership condemned the so-called autonomy plan. The Palestine blueprint made no provision for occupied Jerusalem, for the dismantlement of Israeli settlements, for the withdrawal of Israeli troops—above all, for Arab sovereignty in Arab-inhabited territory. "Autonomy is a sham," insisted Raymonda Tawil. "What kind of freedom will the West Bank and the Gaza Strip enjoy under the 'autonomy' plan? . . . What rights will we enjoy? The right to collect municipal taxes and supervise our own sewage works?" An émigré Palestinian scholar, Fayez Sayegh, underscored this reaction:

> A fraction of the Palestinian people . . . is promised a fraction of its rights (not including the national right to self-determination and statehood) in a fraction of its homeland (less than one-fifth of the area of the whole); and this promise is to be fulfilled several years from now, through a step-by-step process in which Israel is to exercise a decisive veto power over any agreement.

With hardly an exception, the Palestinian leaders resolved not to cooperate with the Camp David formula. As they saw it, their participation in the election of a "self-governing" authority—a council subject to Israeli supervision and veto—would be tantamount to acquiescence in a veiled prolongation of Israeli rule. Thus, after an initial conference in the village of Beit Hanina, the West Bank mayors called a series of meetings at Bir Zeit University, at the Catholic College of Bethlehem, and at al-Naqach College in Nablus. Speakers at these gatherings represented almost all factions of the Palestinian resistance. Unanimously, they rejected any alternative to "self-determination," a code word many Israelis equated with eventual Palestinian statehood. It was probable, to be sure, that the assembled notables did not yet speak for the majority of West Bank Arabs—for farmers, workers, and their families, most of whom still were concerned essentially with routine issues of peace and financial security. Yet even those West Bankers who discerned in the Camp David formula a gradualistic step toward eventual autonomy, men like Anwar al-Khatib, a governor of Jerusalem under the former Hashemite administration, were intimidated into silence. They risked stigmatization as traitors. By then, the nationalists dominated public expression in the territories.

For Begin, the maximalism of these new leaders, their evident commitment to PLO ideology, was all the evidence he needed—or needed to cite—in defining his government's own intractable position. There were concomitant security factors to stiffen that posture, after all. The prime minister and other (far more moderate) Israelis understood precisely the goal the guerrilla organizations had in mind. During the upsurge of Arab confidence following the Yom Kippur War, Israel lived with the fedayun—terrorist-infiltrator—program on a revived and continual basis. Thus, on April 11, 1974, in an effort to disrupt Kissinger's shuttle negotiations, 3 members of the PFLP "General Command" crossed the Lebanese border, reaching the Israeli development town of Kiryat Shmonah. There they burst into an apartment house and indiscriminately machine-gunned to death 18 men, women, and children before themselves dying in an Israeli army fusillade. A few weeks later, 3 armed guerrillas of the PFLP again infiltrated from Lebanon, this time entering the northern Israeli village of Ma'alot. Breaking into a school, the fedayeen took some 120 children hostage, then threatened to dynamite the building unless Israel released 26 imprisoned Arab guerrillas. Troops were ordered to charge the school. In the ensuing melee, which took the lives of the guerrillas and of an Israeli soldier, 20 teenage schoolchildren died, most of them girls. On June 19, 1974, a PFLP squad attacked Kibbutz Shamir in the Upper Galilee, killing 3 women. A week later, a trio of guerrillas slipped down the coast by boat, entered an apartment house in Naharia and killed a man and his 2 children, an Israeli soldier, and wounded 8 other Israelis before finally being shot themselves. Four months later, 3 other guerrillas shot their way into an apartment building in the Jordan Valley development town of Beit Sh'an, killing 4 civilians and wounding 23 others.

In March 1975, 8 members of Fatah, the largest of the PLO fedayun organizations, again sailed down the Israeli coast, this time disembarking in a rubber dinghy off the Tel Aviv shoreline. Entering a seafront hotel, the raiders murdered 8 hostages and wounded 8 others. Eventually they were killed in a shootout with Israeli troops. In mid-June, 3 members of a family living in the northern Israeli town of Kfar Yuval were slain by Palestinian infiltrators, 6 others were wounded. On July 4, 1975, a fedayun bomb hidden in a refrigerator exploded in downtown (Jewish) Jerusalem, killing 14 people and injuring 75. Four months after that, on November 13, the first anniversary of Arafat's appearance before the United Nations, another bomb exploded in the same street, this time killing killing 6 teenagers and wounding 30 other citizens. Yet even these atrocities were but a prelude to the violence that flared in the aftermath of the Egyptian Israeli disengagement agreements, of Sadat's visit to Jerusalem, and the ensuing Camp David accords and peace treaty. Between January 1974 and September 1979, a total of 1,207 people were killed and 2,950 wounded in PLO operations that included arsons, bombings, kidnappings, assassinations, shootings, and hijackings. During 1978 alone, 250 fedayun attacks against civilians inside and outside Israel left 245 people dead, 334 wounded. It was significant that a majority of the victims were themselves Palestinian Arabs known to favor coexistence with the Jewish state. To Begin, then, the case for Israel's "security" control of the West Bank was ironclad.

A PARALYSIS OF NEGOTIATIONS

Yet the prime minister's approach clearly signified far more than a reaction to Arab guerrilla violence. Menachem Begin was still the devout territorialist, still the fervent apostle of Revisionist Zionism. Not for a moment had he accepted the notion, embraced by moderates within the Labor Alignment, that the West Bank and Gaza represented pawns in a future territorial compromise with Hashemite Jordan. If the prime minister's obsession with an undivided Land of Israel occasionally appeared to soften, it was only for tactical negotiating purposes. "Begin had repeatedly promised full autonomy for the West Bank Palestinians," Carter recalled, "and I pushed him on how much freedom they would have. He replied that the only powers they would not be able to exercise would be those relating to immigration of Palestinian refugees and the security of Israel. This sounded like good news to me." Little time elapsed before the president was enlightened.

Thus, within days of the Camp David accords in October 1978, Begin shocked both Carter and Sadat by his refusal to postpone the establishment of new Jewish settlements beyond a three-month period; then by his belligerent riposte to a speech of Egypt's Prime Minister Mustafa Khalil. Addressing the Egyptian People's Assembly, Khalil, not lacking in his own presumptuousness, assured his listeners that the Israelis eventually would evacuate East Jerusalem and "all" the occupied territories; would dismantle their "illegal" settlements, and the Palestinians then would establish a state in their place. "Dr. Khalil," Begin answered rhetorically the next day in a speech to the Knesset, "I inform you that Israel will never return to the borders of June 4, 1967. Dear and honorable Dr. Khalil, please note this: Jerusalem, the united, the one and only, is the capital of Israel. It will never be divided. Dr. Khalil, a state called Palestine will never be established in Judea, Samaria, and Gaza." To the cabinet, later, the prime minister insisted that "this government was formed in order to preserve the Land of Israel." If any decision were passed undermining that task, he, Begin, would resign. "That is an assertion of principle," he warned, his voice breaking. "My old age will not shame my youth. I was born an Eretz Yisraeli [a Land of Israelite] and will remain one to the end of my days."

The prime minister soon left no doubt that his stance was authentic. In the "side letters" from Sadat and Begin to Carter, the Middle Eastern leaders had agreed that autonomy negotiations for the West Bank and Gaza should begin within a month following ratification of the Egyptian-Israeli peace treaty, "with the goal of completing the negotiations within one year," that is, by May 26, 1980. It was in anticipation of those discussions that the Begin cabinet adopted a statement of principles. These were forthright: the military government would be "withdrawn from its main centers, but not abolished"; Israel would retain responsibility for law and order, for land and water resources; Jewish settlements would be subject to Israeli law and sovereignty. So undisguisedly territorialist were these guidelines that Dayan and Weizman declared them unworkable. In turn, vexed by the evident dovishness of the two former generals, Begin proceeded to name Dr. Yosef

Burg as chairman of Israel's six-man negotiating delegation. Dayan and Weizman nominally served as members of the team, but thereafter exerted little influence on its policy decisions. Minister of the Interior Burg, in turn, seventy years old, was leader of the National Religious party, whose claim to Judea and Samaria ostensibly was embedded in Scripture. Indeed, it was Burg and his NRP colleagues who had extracted Begin's written pledge that the territories would never be relinquished—as if the prime minister's fixation with the Land of Israel needed written validation.

Discussions between the Egyptians and Israelis began on May 26, 1979, initially in the northern Negev city of Beersheba. The Americans were represented by Robert Strauss, a veteran Democratic party leader and, although a Jew himself, a man committed to his president's goal of meaningful West Bank and Gaza autonomy. The Palestinians, meanwhile, those whose fate specifically was being negotiated, boycotted the talks. So did the Jordanians. Yet it was less the procedure than the substance of the discussions that all but ensured immobility. The purpose of the meetings was to formulate "modalities for establishing the selected self-governing authority in the territories." At the outset, General Kamal Hassan Ali, leader of the Egyptian team, reverted to the maximalist Arab position on the "inadmissibility of acquisition of territories by war," and Israel's obligation to withdraw its forces and settlements equally from the West Bank, the Golan Heights, and East Jerusalem. Hassan Ali then demanded an immediate transfer of Israel's military government headquarters to a site outside the territories, the prompt return of all Palestinians who fled in the wake of the Six-Day War. It was the Egyptian position, moreover, that the self-governing authority should be invested with legislative as well as administrative powers, with the right to enact laws, assess and collect taxes, maintain public order, hold title to public domain, to government property and water sources, and—finally—to participate in all stages of negotiations during the transitional period, for the purpose of resolving the Palestine problem "in all its aspects." East Jerusalem, of course, would be the appropriate locus of the self-governing authority.

This imposing Egyptian formula was soon matched by an equally uncompromising Israeli prospectus. As in the original version submitted to Carter in December 1977, following the Sadat visit to Jerusalem, the Israeli conception of autonomy was limited to the residents of the territories and did not apply to the soil itself, "which was sacred to every Jew." In its essence, the plan now submitted by Burg traced back to Jabotinsky, whose vision of Arab autonomy was restricted to cultural, religious, health, and commercial matters. The jurisdiction of the self-governing entity thus would be narrowly administrative, and would relate exclusively to matters within the Begin-Burg (Jabotinsky) framework. Security, land and water, the custodianship of abandoned property would remain within the Israeli ambit. In practical terms, the scheme's tightly attenuated format would add little to the current situation in the West Bank and Gaza, where some 7,500 Arab (Hashemite-appointed) civil servants, operating under the supervision of perhaps fifteen Israeli bureaucrats attached to the military government, already were dealing with the day-to-day affairs of local inhabitants in the realm of commerce, health, and education. As the Labor newspaper *Davar* observed,

> The Arab residents of the West Bank have been enjoying almost complete auton-
> omy for years. . . . What the local Arabs want in addition . . . is the manage-
> ment and supervision of the prisons service, of the customs branch, and the cus-
> todianship of abandoned and state property. Then, with the addition of a few
> symbols, they will have their own state.

It was specifically this authority, however, and its possible transformation into statehood, that the Begin government was intent on aborting.

Thus, under the Begin-Burg plan, East Jerusalem with its nearly 120,000 Arabs (the single largest population concentration in the territories) was re-affirmed as an integral part of Israel, and its inhabitants accordingly were not to be included within the tenuous jurisdiction of the self-governing entity altogether. At Camp David, it had been understood that the autonomous Arab council would replace the military government. Yet, under Israel's new proposal, the military government would not be replaced but merely "with-drawn to specified areas," from whence it would continue as the highest authority in the territories, delegating only a fraction of its powers to the elected council. In sum, by retaining the ultimate legal "source of author-ity," the Israelis would be in a position to dissolve the council should it overstep its powers—whatever these eventually were determined to be. Or, as Begin stated the matter unambiguously, if the members of the council proclaimed a state, they would be "clapped in jail." And if the talks with the Arabs over "permanent status" broke down, Israel as the "source of the au-thority" could either suspend autonomy indefinitely or even reinstate the military government throughout the territories.

As for the crucial issue of security, the Camp David framework had re-ferred to a "strong" police force to be established by the autonomous council and manned by Arab inhabitants. Now, in the Burg Committee proposals, "internal security and the fight against terror, subversion and violence" would remain the prerogative of Israeli defense forces. The Egyptians vigorously challenged this position, of course, observing that, without responsibility for public order and internal security, the self-governing authority would be a sham in the eyes of the local population. The argument was applied equally to Israel's demand for control over the West Bank's water, its uncultivated state lands—the very guts of the region's present and future economy. In Jewish hands, these resources would continue permanently hostage to Is-rael's own agricultural and settlement requirements, its own powers of "em-inent domain." Conceivably, so grudging and hollow a conception of auton-omy had not been envisaged at Camp David, not even by Begin. It was more likely that the cabinet's theological hard-liners—the religionists—had obliged the prime minister to eviscerate even his limited earlier concessions.

Until the end of 1979, seven plenary sessions took place between the Egyptian and Israeli negotiating teams. They were held variously in Beer-sheba, Tel Aviv, Alexandria, Herzlia, Haifa—even London. The delegations periodically split into subcommittees to deal with the various issues. Here and there agreement was reached on procedural matters, but never on questions of substance. It is recalled that one year was projected as the time-frame of negotiations. Yet, by the early spring of 1980, any hope of reaching the May deadline clearly was becoming a pipe dream. The talks limped along, staggered, fell, picked themselves up, but failed to make discernible progress. To every Egyptian proposal of substance—that the self-governing authority should have legislative powers, that the inhabitants of East Jerusalem should participate in elections to the authority, that Egypt should establish a liaison office in Gaza—Burg and his negotiators insisted that Camp David "did not provide for that." As early as August 1979, Prime Minister Khalil was complaining that, at the current rate of progress, 250 sessions would be needed to reach agreement.

Sadat privately shared his negotiators' alarm. Still, publicly, he expressed optimism. He would praise the Israelis as often as he gently admonished them, observing that they were "honestly and faithfully" carrying out their treaty obligations in Sinai. By March 1980, however, the Egyptian president intimated that a "new situation" might be created if the discussions were not completed on schedule. In early May, yet another round of talks proved so barren that Sol Linowitz, who recently had succeeded Robert Strauss as Carter's representative, echoed that veiled warning. Finally, on May 12, reacting to a particularly charmless claim by Begin on "the one and indivisible Land of Israel," Sadat suspended the talks. Five days later Carter persuaded him to rescind the cancellation. Yet, with the target date already passed, the Egyptian president sensed that criticism against him was mounting at home. To cope with it, he assumed the premiership for himself, and appointed the well-regarded Kamal Hassan Ali as foreign minister. The talks would resume in mid-August, he announced then. But on July 30 the Israeli Knesset solemnly voted a "fundamental law," declaring Jerusalem to be Israel's united and indivisible capital (Chapter VI). Arab outrage worldwide was all but uncontrollable. In an aggrieved personal letter to Begin, Sadat observed that the Knesset vote precluded further negotiations on Palestine. Again Carter interceded, pleading with the Egyptian leader to continue. And again, over the objections of Hassan Ali, Sadat concurred. His proviso this time, however, was that the American president convene another Camp David summit, once he was reelected to a second four-year term in November 1980. Carter agreed.

And was not reelected. Few times in his life did Sadat feel a loss as deeply as this one. Now he would have to start over, building a personal relationship with Ronald Reagan. Months passed, with the autonomy discussions frozen. Indeed, hardly a single outstanding issue had been resolved, neither the scope, structure, or jurisdiction of the self-governing authority, nor the problem of security, nor the status of Jerusalem. In a rather tortured effort to put the best light on the impasse, Butros-Ghali in Cairo noted in a foreign ministry statement of autumn 1981:

> [D]espite the emphasis put by Egypt on the importance of confidence-building measures and the necessity of opening contacts with the Palestinians in the West Bank and Gaza, Israel—all though the previous phase of the autonomy talks—had

constantly refused even to consider the discussion of these two issues. . . .
Nevertheless, through Egyptian persistence, at least a reference to these issues
was inserted for the first time in the last joint communique issued by Prime Minister Begin and President Sadat in Egypt.

It was meager consolation.

THE SETTLEMENTS OBSESSION

Security remained the Begin's government most obvious rationale. No one
was allowed to forget that, before June 1967, the irregular bulge of the West
Bank pinched Israel's waist only 9 miles from the Mediterranean, and within
this narrow strip lived 75 percent of the Jewish population. Jerusalem had
been surrounded on three sides by Hashemite artillery. Indeed, Jordanian
guns in the West Bank could have reached any factory in central Israel.
More than a few senior officers and political right-wingers regarded the Jordan River as Israel's "permanent security frontier." For them, the hills rising
from the Jordan Valley offered a logical secondary line of defense against
attack from the east, even as radar stations provided early-warning capability
against air strikes from the east and north. Publicly, then, it was security
that the Likud government first invoked in its changed settlement approach,
from the "Allon Plan" areas—that is, the sparsely inhabited, purely strategic
sites of the West Bank—to all sectors of the Land of Israel, including the
densely populated Arab regions. Nothing less than widespread Jewish settlement would give Israel's armed forces and installations the space they required, argued Chief of Staff Rafael Eytan. Indeed, this amplitude was all
the more critical in the aftermath of the Egyptian-Israeli peace treaty. It
provided an indispensable substitute for abandoned Sinai bases and training
areas.

So it was that the Begin cabinet now chose to interpret Camp David's
formula of "defined security areas" in its broadest contours. Immediately
after the treaty signing, the Israeli authorities launched into an intensified
construction program of camps and depots in the West Bank. Agriculture
Minister Ariel Sharon, chairman of the settlements planning committee, took
pleasure in explaining the government's policy to visitors. Holding a map,
he drew attention to large colored blocks of land on the West Bank surrounding the Jewish settlements. It was precisely this terrain, enveloping
some two-thirds of the region's surface and the bulk of its Arab population,
that the previous Labor governments had rejected as inappropriate for a
major Jewish presence. Sharon (and Begin) saw matters differently. Here,
they insisted, not less than 300,000 Jews eventually would be ensconced, a
permanent "security buffer" for Israel.

Yet if "broad" security and ideological considerations were important to
the Likud government, they were profoundly interpenetrated by political
factors. Like Charles de Gaulle, who grasped the psychological importance
of projecting French "grandeur" in Europe following the withdrawal from
Algeria, Begin similarly understood the need to compensate his nation's right

wing for the evacuation of the Sinai. Among those rightist elements, none were more influential than the religionists. As noted, the Orthodox earlier had established the first "unauthorized" settlements in the heart of Arab-inhabited territory. By the time of the 1977 election, five Gush Emunim outposts existed in the Samarian mountain massif, most of them inhabited by religionists, and they embodied not simply the theological commitment of Rabbi Moshe Levinger and his self-annointed millenarians, or of the ultra-Orthodox Aguda factions, but of the traditionally more moderate National Religious party. It was something of a volte-face for this formerly centrist group. In its earlier incarnation as the Mizrachi and Mizrachi Worker parties, we recall, the NRP's leaders as a rule had accommodated to the territorial restraint of Labor. Their goals had been defined essentially within the ambit of Israeli domestic life, and rarely had its members exhibited signs of isolationist chauvinism. But here as elsewhere, the trauma of the Six-Day War, and no doubt the challenge laid down afterward by the Gush Emunim (Chapter I), exerted a decisive influence.

For that matter, the NRP itself was increasingly being taken over by younger militants like Zevulon Hammer, chairman of the B'nai Akiva youth movement, and the American-born Yehuda Ben-Meir. By 1972, B'nai Akiva activists had won a quarter of the seats at the NRP convention and increasingly forced their views on the older leadership. One of the Young Guard's tenets rejected the very notion of withdrawal from any part of the Land of Israel. In the West Bank, after all, was to be found the cradle of the Jewish past and identity. Here the patriarchs had spoken with their God and dug their wells. Here the Jewish judges had judged, the Jewish prophets prophesied, the Jewish kings ruled. Normally, the presence of an Arab majority on the West Bank might have intimidated even the most committed group of settlers. Not the religionists. Indeed, they did not disguise the hope that many of the non-Jewish inhabitants would be induced to leave in future years. Nor were they given pause by the exigencies of international opinion. As the Orthodox saw it, Gentile animus was built into history. Nothing was to be gained by efforts to placate it. By 1973, then, the hawkishness of the religionists was an accepted feature of Israeli political life. Few eyebrows were raised any longer when the Sephardic chief rabbi issued a *p'sak halachah*, a kind of rabbinical bull, "commanding" the government and all Jews individually to refrain from "so much as contemplating" the return of any of the sacred land promised by God to His people. Even Prime Minister Golda Meir, reestablishing her government early that year, was obliged to make advance commitment to her NRP coalition partners never to yield an inch of West Bank territory without first conducting a national plebiscite. The promise was repeated by Prime Minister Rabin in the aftermath of the second disengagement with Egypt.

And now Begin, who in 1977 had formed a coalition agreement with the religionists, including the ultra-Orthodox Aguda factions, was obliged to be even more acutely sensitive to their views than were his Labor predecessors. Once charged with forming a new government, Begin went directly from the president's residence to the home of Rabbi Zvi Kook, spiritual father of the Gush Emunim, to solicit the rabbi's blessing and to kiss his hand. It is recalled that, in June 1974, some twenty Gush Emunim families had illegally occupied a site in the Samarian mountains, which they called Elon Moreh. After an eight-month confrontation with the Rabin government, a "compromise" allowed the settlers to remain temporarily at a nearby army camp outside the Arab village of Qadum. This represented only a moratorium to the Gush Emunim, of course, and afterward its members greeted Begin's election with rapture. Neither did the premier-designate disappoint them. Almost immediately he visited Qadum, bearing a Torah in his arms, declaring emotionally that "there will be many more Elon Morehs." There were. Upon assuming office, the new government accorded full legal status to Qadum and to Ofra, another fledgling settlement in Samaria, and to Ma'aleh Adumim, an embryonic industrial park on the Jerusalem-Jericho road.

Thereafter, the Gush Emunim program, calling for the establishment of sixty communities in the Samarian mountains, received formal cabinet endorsement. And following the peace treaty with Egypt in March 1979, Begin elevated the Gush Emunim into the nation's leading settlement movement, showered money on its projects, even authorized local militias in its villages. To be sure, during the negotiations leading to the Camp David accords and the peace treaty, Begin had found it necessary to mute the question of Jewish settlements. But once the Knesset ratified both sets of documents, and autonomy discussions with the Egyptians began in May 1979, the government held to the principle of settlements more tenaciously than to any other feature of its coalition program. In the words of the Burg negotiating team, "the right of Jews to settle in all parts of the Land of Israel is a matter of historical ordination."

To coordinate the settlements program, meanwhile, the government and the World Zionist Organization established a joint planning committee only days after the signature of the peace treaty with Egypt. Its director was Matityahu Drobles, a veteran Likud member and currently chairman of the WZO rural settlement division. With the active encouragement of Sharon, who served as ex officio chairman, Drobles laid out the committee's guidelines. These stated:

> Settlement throughout the entire Land of Israel is for security and by right. A strip of settlements at strategic sites enhances both internal and external security, as well as making concrete and realizing our right to the Land of Israel. . . . The disposition of the settlements must be carried out not only around the settlements of the minorities [indigenous Arabs] but also in between them. This is in accordance with the settlement program adopted in Galilee and in other parts of the country, with the objective of reducing to the minimum the possibility for the development of another Arab state in these regions.

The words were not those of the Gush Emunim, but of the World Zionist Organization and of the Israeli government, which spoke of the West Bank Arabs as "minorities" and of the West Bank itself as an integral part of Greater Israel. The Drobles proclamation went on to emphasize that state and uncultivated land should be seized immediately in order that "in the coming five years 12 to 15 rural and urban settlements should be built each year so that 60–75 additional settlements may be established and the Jewish population will reach between 100,000 and 120,000." With this document, the historic policy of territorial compromise, which under decades of Labor leadership had become all but synonymous with governmental restraint and moderation, now at last was openly and officially abandoned.

THE DYNAMICS OF SETTLEMENT

Various techniques were adopted to achieve settlement. One, following the Drobles guidelines, was envisaged as a countermeasure to "uncontrolled" Arab growth. Areas of extensive Arab habitation were to be encircled by Jewish outposts in the first stage, penetrated and fragmented in the second. Restrictions similarly were envisaged for Arab subsoil rights. It was of note that these limitations were related both to Israel's own water needs in integral Israel and in the West Bank. Thus, despite a series of brilliantly imaginative irrigation achievements in the 1950s and 1960s, including the overland transfer of Jordan River waters to the west and south, Israeli hydrologists recognized that existing water supplies were inadequate for the nation's future requirements. New sources were urgently needed, and these could be supplied exclusively from two main aquifers. Both originated in the West Bank. Both drained westward toward the Mediterranean and already were being tapped by an elaborate system of wells along the Israeli coastal area. Plainly, any Arab water development on the West Bank capable of interrupting this vital underground flow would have to be blocked.

The Drobles Committee regarded it as a useful byproduct, moreover, that regulations on Arab water usage on the West Bank not only protect Israel but also restrain "uncontrolled" Arab agricultural and industrial growth. This goal was accomplished by regulations against the drilling of artesian wells without prior authorization, a flat ban on the drilling of wells close to the old Green Line, limited permission even on authorized drilling, with meters placed on each well and heavy penalties imposed for overuse. As anticipated, the restrictions had the effect of slowing further Arab agricultural development. The manipulation of electricity usage was still another example of pragmatic Israeli control. Before the Six-Day War, eleven commercial, municipal, or cooperative electricity companies functioned on the West Bank. They did not function very well, as it happened. Their stations broke down frequently, depriving numerous villages, even entire neighborhoods in East Jerusalem, of access to electric power altogether. Over the years, Israel exerted various forms of pressure to link these firms to the highly efficient Israeli electricity grid. As a matter of principle, several Arab firms refused. Indeed, the Jerusalem and Nablus companies held out for years. But most eventually gave in, connecting their plants to the Israeli power system. That system, once unified, proved a boon to Arabs and Jews alike. For Israel, of course, political considerations were decisive. Linked electric grids would facilitate the establishment of Jewish settlements in the area.

If Begin and his colleagues were inhibited by the Camp David accord from de jure annexation, meanwhile, still other methods were available to fulfill that goal de facto. Some of these were even less oblique than restrictions on Arab water or linked electric facilities, or the transplantation of Jews well beyond the modest configuration of the old Allon Plan. To achieve the "incorporation of Judea and Samaria into the [Israeli] national system," in the plainspoken language of the Drobles Committee, land would have to be acquired for settlement and for the construction of urban centers in all parts of the West Bank. More specifically, "incorporation" was predicated on the governmental right to sequester vast tracts of Arab land ostensibly for security purposes. That "right" in fact was put to the test earlier than Begin would have preferred. His hand was forced, typically, by the Gush Emunim. As premier-designate, he had found it one matter to promise the squatters at Qadum "many Elon Morehs." But afterward, involved in sensitive diplomacy with the Egyptians, he learned that it was another matter to disrupt sensitive negotiations by provocative frontal challenges. When therefore the government appeared to be dragging its feet, the settlers decided to move on their own.

In early January of 1979, with a maneuver reminiscent of earlier tactics against the Rabin government, a convoy of vehicles set off for a tract the settlers had chosen for Elon Moreh, near the Arab village of Rujeib. The procession was stopped by an army roadblock. Arguments, then negotiations, followed. After five weeks of delay and consultations, the Gush Emunim activists finally received government assurance that they soon would be allowed to move into their permanent site. The promise was kept. In June 1979, the military commander of the West Bank signed an order expropriating some 700 dunams (175 acres) of private Arab land, much of it under cultivation by the villagers of Rujeib. The pretext was national security. At this point, however, the outraged villagers took the unprecedented step of seeking legal help. Their lawyers (most of them Jewish) promptly appealed to the Israeli supreme court, sitting as the high court of justice, which agreed to hear the case as a matter of national priority. The government was caught by surprise. Yet it defended its action before the high court on traditional security grounds, and cited earlier precedents of expropriation.

Hereupon the government received a further shock. A number of former senior military commanders, including Generals Chaim Bar-Lev and Matityahu Peled, testified for the plaintiffs, arguing that the expropriated area was of negligible security value for Israel. The court then wasted little time ruling in favor of the Arab villagers. The government's purpose manifestly was political, the judges observed, a desire simply to tighten Israeli control over the West Bank; and its broadened interpretation of "national security" thus was unacceptable under the Hague Conventions. The Gush Emunim settlers were ordered to depart the confiscated tract within thirty days. They did—once it appeared that Begin and his associates were not prepared to contest a high court ruling. Eventually an alternative site was found for the settlers on a nearby hilltop—Jebel al-Qabir—on land that was not privately owned. At best, it was an interim solution. The government's blueprint for widespread future requisitions was now in serious jeopardy. To be sure, an act of the Knesset might have been a solution. But Attorney General Yitzchak Zamir warned Begin that legislative changes could not be enacted in the West Bank's legal status without violating the Camp David framework; the issue of sovereignty would have to await the expiration of the five-year transition period. Sharon, Hammer, and Burg in fact were quite prepared to run the risk of legislation. Weizman, Dayan, and Yadin were not. Faced with a divided cabinet, then, Begin felt obliged to postpone the issue repeatedly.

As it happened, other alternatives existed for nailing down Israel's presence on the West Bank. The area encompassed less than 3,000 square miles, after all. Within this territory, some 32,000 dunams (approximately 8,000 acres) had been owned by Jews prior to 1948, and still were registered in Jewish names in the land registry. These were reclaimed. Under the Begin government, too, all restrictions against private Jewish purchase of Arab land were dropped. Although a Jordanian law forbade such transfers to Jews or other foreigners on pain of death, numerous local Arabs continued quietly (indeed, secretly) to sell their lands through intermediaries, usually for foreign currency in deals that were negotiated overseas. One of the largest Jewish buyers, the Heymenuta Company, affiliated with the World Zionist Organization, was the owner of 73,000 dunams by 1980. Growing numbers of private individuals shared in the buying campaign, offering blue-chip prices. The temptation for Arabs to sell was all but overwhelming. By 1982, some 200,000 dunams had been transferred to Jewish ownership.

Additionally, thousands of Arabs had fled the West Bank since 1967, and their property had passed into the hands of the Israeli commissioner for abandoned private property. The latter acted as custodian, and exercised the right to lease out the land until its owners returned. The technique had been well perfected in Israel itself after 1949, of course. By 1981 an additional 430,000 dunams of West Bank property similarly were leased to Jewish settlements. The military government also inherited from its Jordanian predecessor the right of eminent domain. Under the Hague Conventions, this right was not applicable for military purposes. On the other hand, it could apply legitimately to "infrastructure," to roads, parks, and sewer systems. The Israeli military government thereupon invoked eminent domain to undertake such improvements as the construction of highways and access roads to and around Jewish settlements. By 1984, yet another 300,000 dunams had been requisitioned for purposes of "infrastructure."

Notwithstanding these and other outflanking techniques, it was the loophole of "security" confiscations of private land that had offered the Begin government its largest-scale opportunity for extensive Jewish settlement, and that loophole apparently had been closed by the Elon Moreh decision. Then, in March 1980, the government appointed a ministerial committee under the chairmanship of Sharon to come up with an alternate, legally foolproof, solution. Helped by the attorney general's office, the Sharon Committee wasted little time producing such a formula. It was to exploit a provision in the Hague Conventions that differentiated between privately owned and state land. Under this distinction, the right of legal redress was limited to victims of private-land seizures. Suit could not be brought in Israeli courts for land seized illegally from the public, or state, domain; there existed no recognized West Bank authority to file the action. The answer for the Begin cabinet, then, was to redefine "state" land.

At this point the government launched into an accelerated survey of the ownership and registration status of all land in the West Bank. The project was completed in the spring of 1981. Its detailed results were not made public. Nor was information available on the criteria used by the survey teams. Interestingly enough, the Jordanian government had begun its own registration process as early as 1953, but the task was only half-completed at the time of the Six-Day War. Even in registered areas, only a small minority of the landowners had received their title deeds, either because of uncompleted paperwork, failure to pay the small registration fee, or the interrupted mail service caused by the war. Now therefore, in the spring of 1981, the Likud government decided it had all the leeway it required. It simply declared all land outside the original Jordanian survey, as well as all "untitled" tracts within that area, to be "state" land—and thereby available for Israel's settlement needs. The one legal remedy open to an Arab who believed he held a valid private claim was an appeal to a special committee of Israeli officers within twenty-one days. And here, to make his case, the protesting landowner was obliged to furnish documents, maps, and measures, or receipts of land-tax payments. It was a bit much for a simple villager. Thus, requisitioning at an accelerated pace, the Israeli government managed to take over more "state" land during the remainder of 1981 than in all the preceding twelve years of military occupation. Altogether by then Israel had sequestered 31 percent of the West Bank's total land area.

The terrain was intended for use. Until 1977, we recall, most of the Jewish settlements on the West Bank had been established either near the former Green Line, around Jerusalem, or in the Jordan Valley. But with the advent of the Begin government, the more densely populated hill areas of Samaria and (to a lesser extent) Judea became the focus of Jewish settlement. Some forty new communities were implanted there: twenty-four in 1977–78, four in 1979, three in 1980, nine in 1981. Within three-and-half years the Jewish population in the West Bank tripled, to 17,500. It was still a less than impressive figure. Most of the "heartland" settlements were essentially nuclear, supporting less than 100 families. But the skeletal outlines at least were in place now for the government's master plan.

The Drobles Committee envisaged three principal belts of settlements. The first and earliest, completed even before 1977 under the Allon format, encompassed the whole of the Jordan Valley, and separated the West Bank's populated areas from the Hashemite East Bank. The second belt, well launched by 1980, comprised a series of agricultural complexes on the highlands of the Jordan rift, starting at the Jerusalem-Jericho road and connecting with the first belt of settlements along the northern part of the Green

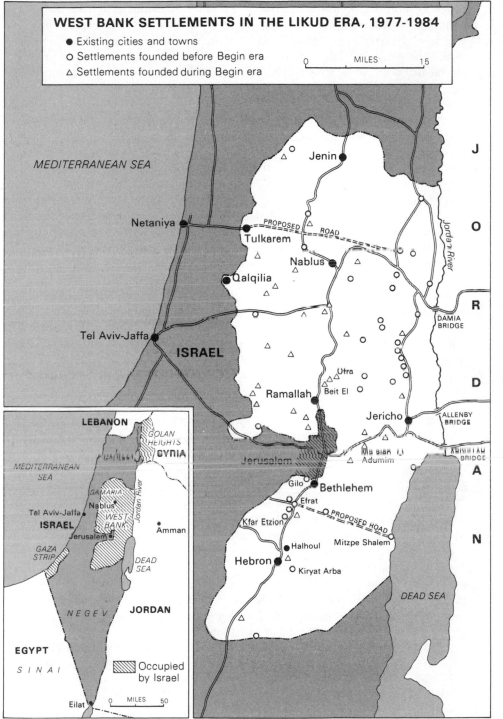

WEST BANK SETTLEMENTS IN THE LIKUD ERA, 1977-1984

- ● Existing cities and towns
- ○ Settlements founded before Begin era
- △ Settlements founded during Begin era

MILES 0 — 15

MEDITERRANEAN SEA

Jenin

Netaniya

PROPOSED ROAD

Tulkarem

Nablus

Qalqilia

Jordan River

DAMIA BRIDGE

Tel Aviv-Jaffa

ISRAEL

Utra

Ramallah   Beit El

Jericho   ALLENBY BRIDGE

Jerusalem   Ma'ale Adumim   BRIDGE

Gilo   Bethlehem

Efrat

Kfar Etzion   PROPOSED ROAD

Mitzpe Shalem

Halhoul

Hebron   Kiryat Arba

DEAD SEA

J O R D A N

Λ

LEBANON

GOLAN HEIGHTS

GALILEE   SYRIA

MEDITERRANEAN SEA

SAMARIA

Tel Aviv-Jaffa   Nablus

ISRAEL   WEST BANK

Jerusalem   Amman

Jordan River

GAZA STRIP

DEAD SEA

NEGEV   **JORDAN**

EGYPT

SINAI

Occupied by Israel

Eilat   MILES 0 — 50

Line. The third belt was projected as a chain of colonies along the entire length of the western highlands traversing the northern, central, and southern sectors of the West Bank, and moving the former Green Line deep inside the Arab populated areas. To fortify these settlements, a number of lateral highways were constructed or were under construction to link pre-1967 Israel with the three belts, and also to resect the areas of heaviest Arab concentration.

Nowhere was the Likud government's de facto annexationism more evident than in the extraterritorial rights Israel's settlers were claiming, and winning, for themselves on the West Bank. In March 1979, the military government signed an order establishing three regional councils there. Two more regional councils were added later, and still another after that for the Gaza Strip. In March 1981 five local (urban) councils similarly were established. In each instance, the regulations governing the councils' powers and functions matched those of their counterparts in Israel. These included the right to levy taxes and issue bylaws on all matters of local commerce, health, and security. But, additionally, the councils were involved in higher-level decisions on infrastructure, on legal, economic, security, land, and water issues affecting the West Bank as a whole. Indeed, emerging as a powerful lobbying bloc in Israel, their members swiftly won all but autonomous status for themselves. For regional security purposes, they were allowed to maintain their own home guard units, to perform their reserve duty in these militias. Enacting their own municipal and regional legislation, they also functioned extraterritorially under Israeli law, rather than under the Jordanian law that otherwise was operative on the West Bank. Thus, Israeli courts already had proclaimed their own competence to try Israelis living in the territories. The settlers paid Israeli taxes on income earned in the West Bank, even as they continued to enjoy Israel's insurance and unemployment benefits. In every respect, then, a dual legislative, judicial, and administrative system was emerging de facto in the territories, one for Jews and one for Arabs.

Here in essence, then, was Begin's "New Zionism." It divided the populated West Bank areas into smaller sectors; augmented the Israeli presence by purchase, eminent domain, exploitation of abandoned property, and redistribution of "state" land; pumped $100 million a year into a vast support program of infrastructure construction, linked electricity grids, controlled water supplies and the application of extraterritorial Israeli law. By all these measures, the government in 1981 was substantially embarked on the fulfillment of its pledge—to incorporate Judea and Samaria irretrievably into the Israeli ambit, to "redeem the Land of Israel."

No one grasped this purpose better than did the Palestinians themselves. As many as 350,000 of them responded to it simply by departing. Between 1968 and 1981 the combined Arab population of Jerusalem and the West Bank remained frozen at approximately 750,000. Had there been no prolonged occupation, had there been a return of Palestinians displaced in the early months of the Israeli conquest, and of their offspring, the natural increase of that population would have come to at least 1.2 million. As it was, the Israelis effected so profound a change in the territories that, in the words of one Arab observer, Hazem Nuseibeh, "it is no exaggeration that if a Palestinian [in 1981] were to visit his ancestral homeland in Palestine, he would find less difficulty recognizing Jaffa or Acre as he was forced to leave it in 1947–8 than he would Jerusalem and the heartland of the West Bank which was occupied in 1967."

A RISING BACKLASH

Without doubt, the momentum of the settlements program could not be ascribed exclusively to one party's, or even one religious bloc's, mystic vision. It is recalled that Zionism had always sustained a powerful current of territorial maximalism. If grave misgivings were expressed about the Begin program of de facto annexation, these were evoked not by doubts of Israel's legitimate claim to the Land of Israel, but rather (as in the past) by pragmatic realities. After 1977, one of those realities was the unprecedented chance for peace that had been opened by Sadat's dramatic initiative, and that now was in danger of being forfeited. Another, even more acute, reality was the debilitating social and political consequences for Israeli democracy of rule over hundreds of thousands of Palestinian Arabs; of reliance upon Arab labor for the menial work that formerly had been performed by Jews in their quest for "normalcy." Once again, the renowned historian Ya'akov Talmon defined the issue in a published letter to Begin of October 1980:

> Mr. Prime Minister . . . permit me to make an observation on the basis of decades of study of the history of nationalism. However ancient, special, noble, and unique our objective motives may be, the desire, at the end of the twentieth century, to dominate and govern a hostile foreign population that differs from us in language, history, culture, religion, consciousness, and national aspirations, as well as in its economic and social structure, is like an attempt to revive feudalism. The question is not a moral one. The idea is simply not feasible, nor is its realization worth the price . . . as France learned in Algeria. . . . It is hard to believe that any population would agree to this sort of "autonomy," without a legislative assembly of its own, or that such people would settle for a mere "administrative council" subordinate to the executive authority of an alien sovereignty.

Talmon's remarks, and those of others of Israel's best-known writers and academicians, might have been dismissed as the portentous moralizing of intellectuals had they not also been echoed by respected businessmen and labor leaders, by military officers, among them former generals of the caliber of Matityahu Peled and Yehoshafat Harkabi. The latter, formerly chief of military intelligence and afterward a professor of international relations at the Hebrew University, outlined the ultimate demographic and military dangers presented by a large, hostile Arab population whose endemic sabotage and terrorism surely would forestall Jewish immigration from abroad. Within the territories themselves, Harkabi added, there appeared no likelihood that Jewish settlement ever could modify the Arab character of the area; the Jewish population would have to be augmented at least thirty times over to match Arab numbers in the West Bank and Jerusalem.

If criticism was notably harsh from the Labor opposition, here the inter-penetration of politics and conviction might have been anticipated. Yet it was significant, in the aftermath of Camp David and the peace treaty, that few of the paladins of the Alignment were interested any longer in beating the Likud at its own game, that is, of engaging in a competition of nationalist sloganeering. Sadat's initiative had opened an era for the Middle East that dared not be ignored, whatever the legal case for Jewish settlement beyond the Green Line. "Mr. Begin is too precise and intelligent a man," insisted Abba Eban, the former Labor foreign minister, "for . . . [him] not [to] com-prehend how short a step separates this kind of 'self-government' from some form of eventual Arab sovereignty. Anyone who rules out the idea of ulti-mate Arab sovereignty in large areas west of the river ought not to have signed the Camp David accords." Adverting, too, to his nation's Jewish and democratic traditions, Eban warned that the "partition logic cannot be de-nied. . . . History has created such a duality of national identities that any unitary framework can only be coercive and morally fragile." Eban was no closet idealist, no more than were his Labor colleagues Peres, Rabin, Allon, and other leaders of the opposition. All of them rejected the notion of an independent Palestinian state, vulnerable to PLO domination. All of them appreciated that even the most flexible interpretation of territorial compro-mise required a significant extension of Israeli boundaries at Hashemite ex-pense; while for Jerusalem there could be no question of repartition alto-gether. Their desiderata still were based essentially on the original Allon Plan, on the expectation of retaining the most defensible terrain, while re-turning the largest part of the Arab population to Jordan. "We are prepared to relinquish parts of the West Bank [to the Hashemite Kingdom]," ex-plained Peres, "on condition that they remain demilitarized, that no foreign army ever again crosses the Jordan and menaces the gates of Jerusalem. Also we must have more defensible borders to deter a surprise attack." But the alternative solution, Begin's format of Arab "autonomy" under Israeli rule, was "as bad for us," in Rabin's words, "as an independent Palestinian state."

Begin was not the man to be discountenanced by jeremiads from the opposition or from the nation's intellectual elite. He was perturbed, how-ever, if not persuaded, by reservations expressed within his own cabinet. One source of these misgivings was the centrist Democratic Movement for Change, whose leader, Professor Yigael Yadin, held the position of deputy prime minister. The relationship with the DMC had been an arm's-length one from the beginning. Yadin had persuaded his colleagues to join the gov-ernment in the hope of influencing it from within. The hope proved vain. On domestic issues, none of the anticipated "constitutional" changes was ever effected; the party-list system remained intact. And on Middle Eastern issues, the DMC underwent a schism only a few weeks before Camp David, when seven of its Knesset members resumed the name of their original fac-tion, Shinui, and moved into opposition amid charges that Begin was stalling the peace process by his aggressive settlements policy. Within the cabinet,

Yadin and his vestigial DMC group meanwhile expressed their own reservations about Begin's hard line on West Bank autonomy. These were firmly rejected. In the autumn of 1981, with his lingering political influence eroded and his health increasingly uncertain, Yadin announced his retirement from political life. In the new election of that year the DMC no longer was entered on the ballot.

For Begin, however, criticism was voiced rather closer to his political home, in the pivotal foreign affairs and defense ministries. It was Moshe Dayan's pragmatism, we recall, that had played a key role in achieving the breakthroughs at Camp David and Jerusalem. Indeed, Dayan's original condition even for joining the Begin cabinet had been assurance that the annexation issue would be postponed "so long as there were prospects for peace negotiations." With his prime minister, Dayan rejected the notion of a Palestinian state or even of a Jordan-West Bank confederation. Yet he entertained no illusions that the Arabs would accept a truncated "administrative" council under Israeli rule. Neither would peace with Egypt get off the ground without a meaningful gesture of autonomy to the West Bank Arabs. Dayan expressed his views with discretion. Within the cabinet, he was listened to respectfully—and rebuffed with equal courtesy. Nor did he make headway with his colleagues as a member of the Burg negotiating team. His appeals for restraint on the settlements and autonomy issues were disregarded, and he himself was increasingly bypassed. In October 1979, after several months of this humiliation, he resigned as foreign minister.

He was joined in political exile not long afterward by his former brother-in-law Ezer Weizman. Once as militant as Begin, the charismatic ex-air force commander had been deeply sobered by the 1968–70 War of Attrition. It was in that nightmare of bombardment and counterbombardment along the Suez Canal that his son was gravely wounded, and Weizman later was given to reflection: "What did we do wrong that our children too must go on fighting?" His subsequent flexibility as defense minister, his willingness to accommodate to Egyptian sensibilities, were shrewdly appraised by Sadat. In 1977–78, the Egyptian president repeatedly turned to "my friend Ezra" as his favored confidant and conduit to the Begin cabinet. Vigorously opposing Sharon's plan to construct "dummy" settlements in the Sinai, Weizman collaborated effectively with Dayan in engineering the Camp David and Jerusalem compromises. Now, as a member of the Burg negotiating team on Palestine, he joined Dayan again, this time in a—vain—effort to win over his colleagues to moderation on West Bank autonomy. Seven months after Dayan's resignation, Weizman similarly departed the cabinet, issuing a scathing public denunciation of its "indifference" to peace. In a subsequent Knesset vote of confidence on government policy, Weizman raised his hand with the opposition. He was thereupon expelled from the Cherut party.

The prime minister took the discord in stride. Nor did he flinch before the mounting swell of criticism from beyond Israel. Egyptian impatience presumably was a factor worth considering. It was in March 1980 that Sadat quietly sent warning of a "new situation" if the autonomy talks stalled. Begin was not given pause. Sadat, he knew, had invested too much in Middle

Eastern peace to withdraw. In May, the Egyptian government did in fact sign an Agreement on Trade and Commerce with Israel. Each country accorded the other most-favored-nation status, and allowed the other to participate in all local fairs and exhibitions and to make payments in convertible currencies. Yet the agreements remained dead on the paper. "Not only is there no real common economic interest between Egypt and Israel," insisted Egyptian Minister of State Butros-Ghali in an interview of December 1979, "it is not Egypt's intention to establish any special relationship with the Hebrew State before a general and overall solution of the Middle East crisis has been achieved. . . . [N]o economic cooperation—Egyptian or Arab-Israeli—can develop without taking the Palestinian mission into account." During the next year and a half, Israel's exports to Egypt totaled barely $20 million, and Egypt's to Israel (except for oil) less than a fourth of that. While Israeli tourists increasingly descended on Egypt, the exchange was almost entirely one-way. If Sadat managed to contain his frustration when the autonomy target date of May 1980 was passed, his restraint was dictated by his unwillingness to jeopardize the final turnover of Sinai territory by Israel. Rather, it was Butros-Ghali who repeated his government's deeper misgivings in a Paris newspaper interview of September 1980. In the event of war between Syria and Israel, the minister of state intimated, Egypt would give priority to its collective security treaties with other Arab countries.

Meanwhile, the frozen autonomy negotiations did nothing to alleviate Israel's isolation in the Afro-Asian world. Diplomatic relations existed with a mere six Asian lands—Thailand, Singapore, Nepal, Burma, Philippines, and Japan—and only the latter three maintained full embassies in Israel. India permitted the Israelis to operate a consulate in Bombay, but remained consistently pro-Arab on Middle Eastern affairs. Even had these Asian nations so much as contemplated a diplomatic shift following the Egyptian-Israeli peace treaty, they held back as long as the Palestine question remained unresolved. An identical reserve characterized Black Africa, even among states with which Israel once had developed its most important nexus of diplomatic and economic relationships. Contacts, when they were restored, still tended at first to be unofficial, even clandestine (Chapter XI). It was in fact the Afro-Asian "third world" majority that had all but quarantined Israel in the United Nations. By 1980 the Jewish state no longer was represented on a single UN organ or appointive body. Since 1967, it had been subjected to ninety-three hostile Security Council and General Assembly resolutions. The most notorious of these was the "Zionism equals Racism" resolution of November 1975. The most threatening was a 1978 Iraqi proposal in the General Assembly calling for an arms embargo against Israel. It was adopted by a vote of 72 to 30, with 37 abstentions.

The abstentions were significant. They represented many European, and some Latin American, nations. Up to and through the Six-Day War, these democracies had largely supported Israel and had followed her development with much sympathy. The memory of the Holocaust and a concomitant sense of moral shame unquestionably influenced their attitude. So did Israel's image as a pioneering, socially progressive, militarily gallant nation. Until the late

1960s, too, few European governments made significant reference to Palestinian political rights. The West Bank continued to be regarded as an integral part of Jordan, the military activities of the PLO as condemnable terrorism rather than as honorable resistance. In the wake of the 1967 war, all members of the European Economic Community except France endorsed UN Resolution 242, with its implied acceptance of territorial modifications on Israel's behalf.

This was the high tide of the Jewish state's reputation. In ensuing years, West European nations began taking a more reserved view of Israel's continued occupation of the West Bank. The Yom Kippur War of 1973 and the subsequent Arab oil embargo appeared to confirm their misgivings. So now did Begin's foot dragging in the autonomy negotiations of 1979–80. Several of Israel's European critics expressed a genuine concern that, in an age of national liberation, the Palestinians apparently were being left out of the process. Otherwise, in the manner of realpolitik, the display of moral rectitude was a pretext. Europe was in urgent need of Arab oil, Arab markets, Arab investment funds. With almost mathematical exactitude, then, that dependence was paralleled by a rising tide of anti-Israel press criticism, of abstentions in the United Nations on resolutions of condemnation against Israel, and now, increasingly, of chill skepticism about the Egyptian-Israeli peace treaty itself as a job half-done.

A WANING OF AMERICAN PATIENCE

The best hope of international understanding remained with the United States. Commitment to Israel's basic security needs since the 1967 war had been accepted by every American president and by both political parties. From 1970 to 1977, Congress increased financial grants and loans to Israel by 30 percent, even beyond administration requests—and this at a time when Washington was slashing foreign aid to other nations by almost 25 percent. Pro-Israel strength in Congress was so extensive that various administrations utilized it to promote legislation on other issues. Thus, aid for the Greek military junta was advanced on the grounds that American support facilities were needed in Greece to help protect the security of Israel. Funds earmarked for Cambodia were attached to foreign assistance legislation for Israel. Menachem Begin counted on this tradition of American forbearance and generosity, even as he had depended on Washington's traditionally philanthropic approach to the Palestinians as refugees. Indeed, over the years, the United States had based its various proposals—the Clapp Plan, the Blandford Plan, the Johnston Mission—on the notion that Palestinian refugees could be employed outside the frontiers of Israel, and eventually resettled in other Arab lands.

Then came the Six-Day War, which eventually uprooted an additional 350,000 Arabs, most of these from the West Bank. In the aftermath of that conflict, it was Lyndon Johnson who first called for "justice for the refugees." This was a new approach for an American president, but it was subsequently embellished by Presidents Nixon and Ford, who envisaged "justice" within the framework of an all-inclusive Arab-Israeli settlement. Well before Jimmy Carter, then, Washington had shifted its Middle East priorities to the quest for a broader Palestine solution. And it was precisely that expectation of settlement, after Camp David and the Egyptian-Israeli treaty, that impelled Carter and his advisers to sustain the momentum of peace at all costs, to concentrate upon defusing the West Bank issue. There was little time to spare, either for the United States or Israel. Almost wistfully, William B. Quandt, Brzezinski's deputy at the National Security Council, warned Israel that time was not on its side, that Arab oil wealth and influence in world affairs could not be ignored, that the American people soon would face an agonizing choice of their best interests. The warning was echoed by Alexander Haig, Reagan's secretary of state until 1982 and a man who regarded himself as a warm friend of Israel. "The sympathy of world opinion," he wrote in 1984, "which had always before largely belonged to Israel, was in considerable measure transferred to . . . the Palestinian Arabs. Acts of terrorism against Jews and even against synagogues aroused less indignation than Israeli acts of reprisal." Haig too predicted that the situation would worsen.

Until the late 1980s, meanwhile, overseas Jewish loyalty to Israel remained one of the constants of the Middle Eastern equation. In the United States, the political influence of the Jewish community, while often exaggerated by non-Jews, remained a substantial force. Actively involved in public affairs, voting with greater regularity than their neighbors, Jews on a per capita basis contributed more generously to their favored candidates than did any other ethnic group in the United States. Stephen Isaacs, in his *Jews and American Politics*, estimated that Jews provided about 60 percent of all campaign funds for Democratic candidates in elections since World War II, and over 40 percent of Republican campaign funds. That financial support was taken seriously by both parties, and so was the depth of American Jewish commitment to Israel. Jimmy Carter rediscovered this fact of political life when he addressed a fund-raising dinner in Los Angeles, in the immediate aftermath of the joint Soviet-American statement of October 1977 (referring to the "legitimate rights of the Palestinian people"). So angered were normally dependable Jewish contributors that many refused to attend the dinner, an abstention that left half the $1,000-a-plate tickets unsold.

But for American Jews, too, much changed following the Sadat initiative, Camp David, and the Egyptian-Israeli peace treaty. Until these developments, even the numerous shortcomings and excesses of Israeli behavior had been overshadowed in the Diaspora by Israel's essentially defensive stance. No longer. With the autonomy talks launched, then paralyzed, Western Jews found it more difficult to reconcile their instinctive liberalism with Israel's evident determination to rule hundreds of thousands of Palestine Arabs. The tremolos of doubt, first evoked by Israel's Peace Now movement, grew with every month of impasse on the autonomy negotiations. Research conducted by an American academician, Steven M. Cohen, confirmed these misgivings. In response to Cohen's questionnaire, the leaders of American Jewish institutions and agencies concurred by a margin of 74 to 16 that Israel should offer the Arabs territorial compromise in Palestine in return for credible guarantees of peace, that a freeze on Jewish settlements was required now to improve the negotiating atmosphere. By a two-to-one margin, those American Jews who were queried agreed that "Palestinians have a right to a homeland on the West Bank and Gaza, so long as it does not threaten Israel." The reservations of the Diaspora, of Israel's single most disciplined and dependable ally, thus far hardly were expressed at all in public. But very soon, by the summer of 1982, concern would be transformed into anguish, and the traditionally compliant voice of the Jewish hinterland no longer would be muted.

CHAPTER VI **THE POLITICS**

OF MILITANCY

MALAISE ON THE DOMESTIC SCENE

If peace with the greater part of the Arab world still lay beyond reach in the summer of 1980, the Likud government continued to anticipate other compensations at home and abroad. Progress had been expected in immigration. In some degree, that hope was fulfilled. The latest source of newcomers was an unlikely one, however. It was Ethiopia. There, a small community of "Falashas," evidently descendants of converts of centuries earlier, continued to observe a form of crypto-Judaism. They were desperately poor, even by Ethiopian standards, eking our their livelihoods as craftsmen on the margins of the nation's agrarian economy. Traditionally despised by their neighbors as outsiders—the literal meaning of the term, Falashas—in recent years they had become victims of Ethiopia's debilitating civil war and crop failures, and of physical persecution encouraged by the Marxist regime in Adis Ababa.

The handful of Falashas who made their way to Israel in the 1960s and early 1970s sought to draw attention to the plight of their kinsmen. The response of the Israeli government and Jewish Agency was less than enthusiastic. Doubts raised by the chief rabbinate on the authenticity of the Falashas' Judaism may have been a factor in this hesitation. But so too, evidently, was the government's unwillingness to exacerbate Israel's racial and economic difficulties. Nevertheless, by mid-decade the circumstances of the Falashas had deteriorated so alarmingly, and appeals on their behalf by Jewish leaders worldwide had become so urgent, that Jerusalem agreed to give priority to their emigration. Thus, in 1980, the Mossad, Israel's foreign intelligence service, undertook a clandestine rescue operation on a limited scale. The effort continued for four years and brought over some 7,000 Falashas, before the venture became too dangerous to continue.

Yet, by 1984, the flight had achieved a momentum of its own. In a pattern reminiscent of the Yemenite emigration of the late 1940s, news that Jews were escaping to Israel sped through the close-knit Falasha community. With the worsening drought and famine and the opening of the western "Sudan route," thousands of Ethiopian Jews were no longer prepared to await Israeli direction. In a growing stream, they continued to leave their villages for the Sudan, trekking across a desolate landscape, running a gauntlet of torture and rape by bandits and imprisonment by the Ethiopian authorities. Even those who survived the ordeal and reached the Sudan achieved a dubious freedom. They were interned in refugee camps under subhuman conditions. In Um Requba, the largest of the concentration centers, some 2,000 Falashas perished of starvation and disease.

Then, in the autumn of 1984, through the quiet intercession of the United States government, Sudan's President Jaafar al-Numeiri was persuaded to allow the secret transfer of Falashas in exchange for American financial aid. Whereupon, in United States planes, a major airlift of Ethiopian Jews from Khartoum began on November 21, 1984. By late December, a planeload of Falashas was departing for Israel every twenty-four hours. Conceivably the entire remaining Falasha population of 17,000 would have been transferred to freedom had there not been a breach in the secrecy Numeiri required to protect his relations with the Arab world. The blunder characteristically was the result of American Jewish fund-raising publicity. When reports of the airlift appeared in American newspapers, the Israeli government was queried by reporters and admitted that the rescue was taking place. The Sudanese government then abruptly terminated the operation. Some 9,000 remaining Falashas were trapped either in the Sudan or in Ethiopia.

By then, to be sure, not less than 8,000 Jews had been carried to Israel via "Operation Moses"—the American airlift—where they joined the 7,000 who had been rescued in earlier years. It was a not inconsiderable number. Yet the Falashas were among the most backward of all Israeli immigrants. Although they were a gentle and law-abiding group, and exceptionally devout in their commitment to their beloved Holy Land, they would require years of social welfare, education, and vocational training before developing into a productive component of Israel's economy.

By contrast, the arrival of tens of thousands of Soviet Jews in the early and mid-1970s had been a source of particularly high hope. At the end of the decade, however, that initial exhilaration too was revealed as premature. The newcomers often were shocked by the shortcomings and inequities of Israeli life. Few of them had imagined that Israel suffered from class conflict, ethnic discrimination, serious pockets of poverty. Often married to Gentile spouses, Soviet newcomers were repelled also by the power of Israel's Orthodox parties, by the obstacles the chief rabbinate created for children of mixed marriages. Perhaps most fundamentally, they had expected to be treated not as mere immigrants but as heroes, entitled to the good apartments, modern appliances, and automobiles enjoyed by veteran citizens. Notwithstanding the government's strenuous exertions on their behalf, their Zionist commitment rarely matched their impatience. In letters to family members at home, they ventilated their frustrations. Applying for their own exit visas, the relatives in turn had opportunity to brood on the painful fate evidently awaiting them in Israel. Perhaps other alternatives were worth exploring, they speculated.

Those alternatives were not lacking. In recent years, Washington had quietly adopted the policy of admitting refugees from Communist nations on a priority basis. Once this fact became more widely known, thousands of Jewish visa-holders availed themselves of the eager cooperation of American Jewish philanthropies to bypass Israel for the United States. In 1977, the number of "defectors" was 8,483 out of 16,737; in 1978, 16,967 out of 28,868; in 1979, 34,056 out of 41,333; in 1980, 14,878 out of 21,478. The falloff was as unexpected as the original upsurge, but plainly far more unsettling. In any case, Moscow curtailed exit permits altogether after 1980. As the likelihood of détente with Washington faded, the Soviet regime discerned no further bargaining advantage even in selective Jewish departures. The gates of emigration were all but closed. It was a cruel blow to Soviet and Israeli Jewry alike.

By then, thousands of veteran Israelis similarly would have welcomed broader opportunities elsewhere. The nation's economy was fragile. No one was starving yet, but the average family in Israel in 1979 already was spending 45 percent of its income on food alone. Many thousands of citizens were finding it difficult to pay their bills at the grocer. Their plight was not exclusively the consequence of the Begin government's New Economic Policy. The peace treaty with Egypt effected no significant alleviation of the nation's defense burden. Rather, vast new military facilities had to be constructed in integral Israel to compensate for the abandoned Sinai bases. If the expense of calling up workers were added to the "normal" costs of weaponry and training, Israel's defense expenditures averaged 36 percent of the government's budget from 1978 through 1980—a quarter of the little country's GNP, or five times the comparable rate for the United States.

And yet, despite this prodigious outlay, large sectors of the population continued to live far beyond their means. In the boom years of the late 1960s and early 1970s, hundreds of thousands of Israelis had become accustomed to a high standard of living, and the Labor government lacked the political muscle to resist their expectations. Well before 1977, the nation was consuming a third more goods and services than it produced. Social welfare services and private consumption alike were devouring fully one-third of all domestic resources, compared to one-fifth or one-quarter in countries resembling Israel in their scale of development. Without adequate revenue to sustain this level of consumption, in turn, the government was obliged to borrow abroad at progressively higher rates of interest. A foreign debt of $2 billion in 1967 had reached $12 billion in 1978, and would reach $17 billion by the end of 1980. By then the cost merely of servicing the debt would reach $2.6 billion annually.

Far from resolving this problem, the Likud cabinet exacerbated it. Although the New Economic Policy initiated by Finance Minister Ehrlich intermittently abolished or reduced many of Labor's subsidies for basic food staples, the program also guaranteed a select elite immunity from serious taxation. As we recall, the tax structure created windfall profits for citizens— essentially those of European origin—with foreign currency holdings. With its attendant creeping devaluations, with an inflation rate exceeding 100 percent in 1980, Likud's domestic program succeeded only in further skewing Israel's already distorted class structure. If there were rich people, there developed a far greater number of genuinely poor than ever before, families whose higher absolute incomes often tended to disguise their shrinking purchasing power. Likud's promise that the New Economic Policy would ensure

a healthier society and a more widely diffused prosperity by 1980 was revealed as substantially hollow. The efforts of Yigal Hurevitz, Ehrlich's successor in the finance ministry, to hold the line on expenditures proved unavailing. Instead, Hurevitz shocked orthodox economists by sitting still for a
9 percent increase in defense expenditures, then resigning in January 1981
to protest the government's decision to grant salary increases to striking
teachers. At this point, Begin's muddled domestic policies no longer evoked
respect from either the Right or the Left.

Hurevitz's departure had been preceded by the resignations of other pil-
lars of the Begin cabinet. These included Ehrlich, Dayan, Weizman, and
Justice Minister Shmuel Tamir, whose defection in July 1980 precipitated
the final collapse of the cabinet's centrist faction, the Democratic Movement
for Change. Then, in December, Minister of Religious Affairs Aharon Abuhatzeira, of the National Religious party, was indicted on charges of bribe
taking. That same month, the NRP's Dr. Yosef Burg, minister of the interior, fired the national police chief, Herzl Shafir, for investigating charges of
corruption in the interior ministry itself, as well as accusations of reputed
wrongdoing by Burg's NRP colleague, Abuhatzeira. The crisis of governmental disarray by now left no doubt that the Likud cabinet was functioning
on borrowed time. Under public pressure, Begin reluctantly consented to
advance the date of the Knesset election from its original schedule of November 30 to June 30, 1981.

MOVEMENT TO THE POLITICAL OFFENSIVE:
JERUSALEM

The prime minister was not the man to forfeit the political initiative to his
critics. Better than any of his cabinet colleagues, he knew how to mobilize
the religio-nationalist passions of his constituency. The Holy City was a sure
and certain talisman. Moreover, for Begin the invocation of that symbol
manifestly transcended considerations of political tactics. Jerusalem's role as
undivided capital of an undivided Land of Israel was the very epicenter of
his romantic theology—so fundamental that, in his speech at the White House
peace-signing ceremony in March 1979, he gratuitously described the "liberation" of Jerusalem as one of happiest days of his life. Neither did any
issue more sharply divide Israelis and Egyptians in the subsequent autonomy negotiations. For Burg and his associates, it was unthinkable that the
Arabs of East Jerusalem should be regarded as inhabitants of the West Bank,
or entitled to participate in the election of a West Bank entity. Rather, they
should be counted as residents exclusively of the City of Jerusalem, which
itself, in its totality and fullest geographic contours, was "nonnegotiable" as
an integral part of the Jewish state.

For that matter, few Israelis of any background were impressed by Arab
claims to the city. They understood that Jerusalem exerted a certain religious significance for Moslems, although hardly comparable to the mystique
associated with Mecca and Medina. Yet, under Hashemite rule, the east-

ern—Arab—sector of Jerusalem had been permitted to atrophy. During the 1950s, all former government offices had been transferred to Amman, Jordan, and UNRWA headquarters to nearby Ramallah. The authority even of the municipal administration was systematically whittled away. East Jerusalem, with its 65,000 inhabitants (on the eve of the Six-Day War), was administered by a municipal council that had been elected on a narrow franchise. If the rights of local Moslems were largely ignored, those of Christians were flatly rejected. Churches were banned from acquiring land. The provisions of the 1949 Rhodes Armistice agreement, meanwhile, granting Jews access to their holy places in East Jerusalem, were given even shorter shrift by the Hashemite regime. Synagogues, graves, and other of Jerusalem's Jewish religious shrines were systematically desecrated under Arab rule.

During that same period before 1967, West (Jewish) Jerusalem only slowly recovered from its near strangulation in the original 1948–49 Palestine War. The "New City's" tentative revival was evident only from the mid-1950s, when Gershon Agron, the American-educated editor of the *Jerusalem Post*, was elected mayor. Under Agron, and Agron's successor Mordechai Ish-Shalom, Jewish Jerusalem enjoyed a decade of good government, of rationalized municipal services. Yet the city's most impressive takeoff dated from 1965, with the new mayoralty of Teddy Kollek. Vienna-born, Kollek had arrived in Palestine thirty years earlier, and in the next decade and a half fulfilled a variety of missions for the Labor movement, the Haganah—the pre-state Jewish underground—and the Jewish Agency. In 1952, after two years as counselor in Washington, he became director-general of the prime minister's office under Ben-Gurion and in that role swiftly won a reputation as a brilliant administrator. Then, suddenly, in 1965, Kollek took the risky step of joining Ben-Gurion in political defection, abandoning Labor for the breakaway Rafi party. In an even more dangerous move shortly afterward, he campaigned for the mayoralty of Jerusalem on the Rafi ticket. The gamble paid off. Widely admired for his personal qualities and achievements, Kollek gained support from Jerusalemites of all backgrounds. He was elected.

A chunky, energetic man, Kollek proved as much a go-getter as mayor as he had in his earlier national assignments. His work day typically started at dawn with a personal tour of the city, and often finished at midnight. In only two years, his record of dramatically improving the capital's public services, beautifying its parks and gardens, building museums, attracting tourists and benefactions from abroad, would alone have ensured his reputation. Yet Kollek's greatest moment came during the Six-Day War. When the Jordanians began shelling Jerusalem, the mayor traveled fearlessly in a car that was soon bullet-riddled, visiting the most exposed neighborhoods, encouraging people in the shelters and ensuring that emergency maintenance crews repaired disrupted facilities. Afterward, when the fighting ended, Kollek was equally vigorous in extending public services to the captured Arab parts of the city. Even before the cease-fire, municipal employees from the New City were crossing into East Jerusalem to repair broken water pipes and severed electrical wires, to dismantle barriers, roadblocks, and barbed wire, and to cart away debris. The electrical grid and telephone systems of the

Arab and Jewish municipalities were unified. Chronically water-short, Arab Jerusalem was linked to the plentiful West Jerusalem reservoirs. Soon the physical quality of life in East Jerusalem—until then a sprawling, underdeveloped Jordanian provincial community—was measurably improved beyond its prewar level.

Kollek and the national government plainly were inspired by more than humanitarian considerations. They were also making clear that they had come back to a reunited city to stay. In late June 1967, the Knesset passed the series of laws that applied Israeli "law, jurisdiction, and administration" to all corners of Jerusalem, Arab and Jewish. For years afterward, to be sure, the government's official position was that it had not formally annexed the incorporated areas, that these still were not juridically part of Israel. The argument was a sophistry. Israeli law henceforth was applied to the territory of East Jerusalem and to everyone—Arab and Jew—who lived there. It was an arrangement that differed significantly from that operating in the West Bank, where Israeli law was applied only to Israelis—Jews. The distinction hardly gratified East Jerusalem's Arabs, of course. Although they were given the option of choosing Jordanian or Israeli citizenship, they were simultaneously declared citizens of united Jerusalem. The new status carried with it the less than appreciated "right" to vote and to be elected to the (Jewish-dominated) city council. Despite Mayor Kollek's best persuasive efforts, only a few thousand Arabs chose to exercise that right in the elections of 1969, 1973, and 1978. At most, a few hundred others were prepared to accept employment in the municipal administration. The majority avoided any electoral or administrative commitment tending to legitimize the Israeli conquest.

Under Kollek's imaginative leadership, nevertheless, the ensuing years of Jewish control preserved Arab dignity and encouraged a certain functional cooperation between the two peoples. The mayor's principal liaison in this effort was his deputy, Meron Benvenisti, a brilliant, multilingual native Jerusalemite with a shrewd insight into the Arab mentality. It happened that, shortly after the Israeli occupation, under pressure from the Orthodox, the municipality had scrapped the Jordanian school curriculum and textbooks in the Arab sector and had substituted those used in Israeli Arab schools. Immediately the Arab teachers went on strike, and Arab parents kept their children in private schools. Although both teachers and pupils returned several months later, they did so grudgingly, under protest. Finally, in 1975, after nearly seven years of impasse, Benvenisti negotiated a compromise under which the Jordanian curriculum and texts would be retained, but expurgated of anti-Israel material. During their final three school years, moreover, Arab children might choose a curriculum specifically designed to prepare them for matriculation in Arab universities. The formula worked. Indeed, each year afterward, an official of the Jordanian ministry of education visited Jerusalem to supervise the university entrance examinations.

Initially, too, the government was determined to apply Israeli law and supervision to Jerusalem's Moslem religious institutions. As in Israel itself, the government would appoint all Moslem *qadis* (priests), administer *waqf* (charitable foundation) properties, give Israeli civil law priority over Islamic law in questions of personal status. Once again the Arabs of East Jerusalem mounted bitter opposition. And once more Kollek and Benvenisti developed a functional compromise. Under a system of "mutual nonrecognition," Israel would refrain from enforcing the decision of the Moslem courts, from subsidizing Moslem religious institutions (as it did in Israel itself). Yet, the authorities similarly would refrain from interfering with Moslem religious autonomy. In the case of the Mosque of Omar and the al-Aqsa Mosque, a self-appointed Moslem Council unofficially would be permitted to take responsibility for the two shrines, and Jews were forbidden to conduct prayers in their immediate vicinity. It was in fact through this Moslem Council that Kollek and Benvenisti kept in touch with the Arab community, learning of its needs, helping solve its problems. Contact also was maintained through the traditional (but equally unofficial) system of mukhtars, who were authorized to register births, deaths, and land ownership and to notarize documents for the courts; and through the Chamber of Commerce of East Jerusalem, whose local notables exercised power of attorney, authenticated high school diplomas for students applying to Arab universities, even transmitted Hashemite funds to subsidize former civil servants who declined to work in the Israeli administration.

There were still other examples of mutual accommodation. When East Jerusalem businessmen and professionals refused to accept Israel's municipal licensing system, Benvenisti negotiated a plan to accept the current—Hashemite—credentials; Israeli licenses were required only for Arab merchants and professionals who qualified after 1967. By the early 1970s, then, as Israel's authorities sought to introduce Israeli taxes into the East Jerusalem area, they had learned enough to proceed gradually, to assess Arab income and properties at rates far lower than their actual value, to augment the assessments and collections slowly, year by year. Even here, euphemisms proved useful. It was untenable for Arab Jerusalemites to acknowledge tax payments to the Israeli government. Rather, at Benvenisti's suggestion, local Arabs were permitted to make their payment as an "advance" on future taxes to be paid the "legitimate" government. For their part, East Jerusalemites were not unaware that Israel allowed them to sustain a virtually free press, in contrast to a single, government-controlled newspaper before 1967, to travel reasonably freely either to Israel or to Jordan, to welcome their visiting kinsmen from the West Bank and Jordan, and beyond.

The early 1970s were the best years of coexistence. Under Kollek's leadership, the combined Jewish and Arab population of Jerusalem grew from 260,000 in 1967 to 400,000 by 1980. The 70 percent increase enabled Jerusalem to overtake Tel Aviv as the largest city in Israel (although Tel Aviv remained the largest all-Jewish city). The increase represented government policy, of course, a determination to thicken the Jewish presence throughout the entire surrounding Judean mountain area. As early as the Rabin government, several of Jerusalem's newly constructed Jewish quarters encompassed more people than did a number of key development towns in integral

Israel. Even so, during this same period, the Arab population of Jerusalem, 65,000 in 1967, also nearly doubled, reaching 120,000 by 1980. Some of the growth was natural increase, but more of it represented an influx from the neighboring West Bank, essentially of Arabs who arrived to work in the city's booming contruction industry. In 1969, 38 percent of the total population of the Judean (southern) area of the West Bank resided in the Jerusalem metropolitan area. In 1980, the figure exceeded 50 percent. Thus, for the Begin government, it was critical to isolate—in effect, to disfranchise— the Jerusalem Arabs from participation in an autonomous West Bank entity. Within Jerusalem itself, meanwhile, this growing Arab presence was containable both politically and demographically. It was matched by the government-encouraged settlement of Jews, after all. Indeed, the ratio between the two peoples remained constant well into the 1980s, with Jews comprising three-quarters of the total population. Altogether, by then, Jewish domination of greater Jerusalem's economy and topography appeared irreversible.

So did the physical unity of the city. The Egyptians tacitly acknowledged the fact when the autonomy talks began in May 1979. Even the Jordanians sensed that relations between the city's Arabs and Jews were functional, if less than friendly. But Sadat (and Hussein) also understood that an undivided city was one thing, Israeli rule another. No element of the Arab population was inclined to accept the permanence of Israeli annexation, formal or otherwise. As its partisans wrested leadership from Arab moderates in the West Bank, the PLO also began to make inroads in Jerusalem. From 1974 on, the Jewish New City became a target for guerrilla violence. Bombs went off in crowded Jewish shopping centers, taking a mounting toll of civilian life. In Arab East Jerusalem, the quietude of pragmatic accommodation increasingly became the silence of festering resentment. "I feel less comfortable walking the streets of the Old City today," wrote Rafik Halabi, an Israeli Druze journalist, in 1981, "than I did in 1967." As Halabi saw it, the Israelis once had enjoyed an opportunity to make Jerusalem a showcase of mutual tolerance, provided Mayor Kollek's circumspect and forbearing approach had been followed consistently. Instead, they appeared to have forfeited their chance.

There were several reasons for this unexpected deterioration of relations. For one thing, the 1973 Yom Kippur War exerted its impact in East Jerusalem, as elsewhere in the Arab world. Conversely, Israel's posture in the autonomy negotiations disillusioned even moderate Arabs. A formula might have been devised, as Kollek was prepared to develop it, giving Arab East Jerusalem borough status, in the model of the boroughs of London. Arab Jerusalemites could have been offered free choice not only of Jordanian or Israeli citizenship (an option even Begin conceded), but of residence and voting rights either within Jerusalem or within an autonomous or quasi-autonomous West Bank. The prime minister was having none of the latter formula, however. Jerusalem must remain united under Israeli governance, he insisted, its Arab population cordoned off politically from the West Bank hinterland. To Sadat and Hussein, this stance was rendered all the more

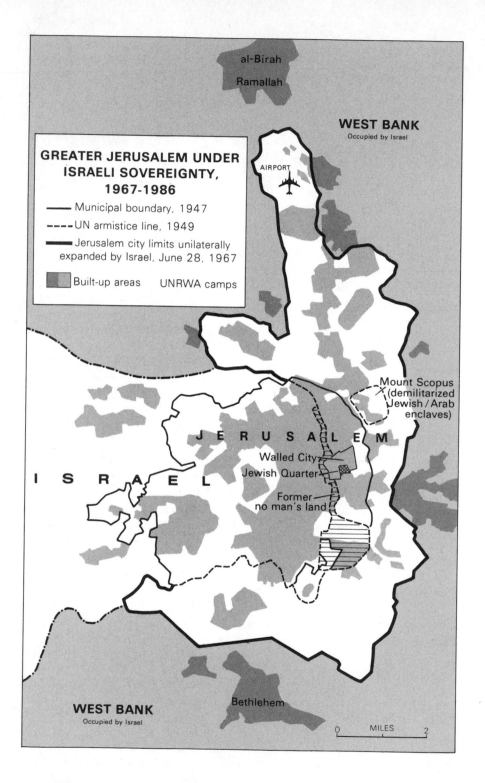

GREATER JERUSALEM UNDER ISRAELI SOVEREIGNTY, 1967-1986

——— Municipal boundary, 1947

----- UN armistice line, 1949

▬▬ Jerusalem city limits unilaterally expanded by Israel, June 28, 1967

Built-up areas   UNRWA camps

al-Birah

Ramallah

WEST BANK
Occupied by Israel

AIRPORT

Mount Scopus
(demilitarized
Jewish / Arab
enclaves)

J E R U S A L E M

Walled City

Jewish Quarter

Former
no man's land

I S R A E L

WEST BANK
Occupied by Israel

Bethlehem

0   MILES   2

galling by Begin's impassioned rhetoric, by warnings that "Jerusalem was and shall remain forever the undivided capital of the Jewish people," a pronunciamento guaranteed to send the prime minister's constituents into dithyrambs—and the Arabs into consternation.

Apparently even Begin's stridency was not commitment enough for the ultraright-wing Techiya faction. In July 1980, one of this party's three Knesset members, Geula Cohen, a renowned Jewish Pasionaria, submitted a draft law to establish united Jerusalem as the capital of Israel. From the practical standpoint, there was little new in the proposed legislation. For thirteen years Israeli administration and law had been functioning in East and West Jerusalem alike. But a statement of "united" Jerusalem as the capital, once adopted by the Knesset, moved perilously close to open, de jure, annexation. This was a step no Israeli prime minister, not even Begin, had quite yet dared take, and surely not in the aftermath of the Camp David framework. Even so, formulation of the bill now struck the Techiya group as astute politics. It would taunt the prime minister into putting his vote where his ideology was. For several days Begin remained inscrutable. As the date for the Knesset vote drew nearer, then, Washington quietly appealed to the Israelis not to disrupt the autonomy negotiations. Cairo also warned of possibly devastating consequences for the peace process. Begin continued to say nothing. The Americans and Egyptians began to breathe easier, assuming that the Jerusalem issue would be discreetly shelved.

Suddenly, in late July, the prime minister confounded the opposition by announcing that he would support the bill, although he would not impose his views on the government coalition, which presumably would vote for the measure only if it sensed a broad consensus. The move was a shrewd one. As Begin intended, the ball now passed to the Labor Alignment's court. And it was there, after some confusion and hesitation, that Peres and his colleagues felt obliged to match the Right in their commitment to the "integrity" of Jerusalem. Even such renowned doves as Abba Eban were outflanked into supporting the measure. On July 30, 1980, the bill passed its final reading and became law. Begin and his partisans were exultant. The fact that the prime minister had been maneuvered into the vote was conveniently forgotten. Whatever his domestic failures, apparently Begin had come fighting back, oblivious to all threats and pressures, to reassert his devotion to a united, sovereign, Jewish Jerusalem. This was the kind of proud Jewishness, and contempt for the perennial Arab enemy, that the nation's oriental majority appreciated. There were consequences elsewhere. All eleven of the foreign embassies in Jerusalem were immediately moved back to Tel Aviv, and the Turkish consulate-general in Jerusalem was closed. An outraged Sadat suspended the autonomy talks, and the Saudi government appealed to the Arab world for a *jihad* against Israel. Begin professed to be unfazed. His posture of bemused indifference merely fortified the pride of his nation's underprivileged.

PREFIGURATIONS OF A CONFRONTATION
IN LEBANON

The prime minister's uncompromising assertion of Israeli freedom of action took similar cognizance of authentic and long-standing military dangers. One of these was Arab guerrilla activity from the north—from Lebanon. Until 1970, some 180,000 Palestinian Arabs had subsisted in Lebanese camps as refugees from the original 1948 war. Then, following the 1970–71 Black September massacres of Palestinian camp-dwellers in Jordan, tens of thousands of additional fugitives swelled the Palestinian presence in Lebanon to approximately 300,000 by the mid-1970s. Their circumstances were bleaker than anywhere else in the Arab world. In neighboring Syria, over 190,000 Palestinians were allowed to travel, join the army, work as government employees. In Lebanon, by contrast, the refugees were placed in an indeterminate category, neither "foreigners" nor "nationals," and were barred from employment in any but marginal sectors of the economy. Their quarantine reflected Lebanese Christian fear that integration of the Palestinians (virtually all of them Sunni Moslems) would upset the nation's delicate sectarian balance. "The Palestinian is like a spring," observed Joseph Gaylani, chief of the Palestinian section of Lebanon's Deuxième Bureau, the government intelligence branch. "If you step on him, he stays quiet, but if you take your foot off, he'll hit you in the face." Thus, for years, the refugees were exploited at near starvation wages as agricultural workers or as an urban subproletariat. Their frustration in the camps was a compound of anomie, sheer economic misery, and growing bitterness at the elitist Christian establishment of Lebanon itself.

Thereafter, the PLO tightened its grip on Lebanon's refugee encampments through a shrewd mixture of revolutionary fervor and economic acumen. With $400 million in annual subsidies from Saudi Arabia and other Gulf oil states, Arafat and his colleagues purchased a massive storehouse of weaponry; organized in Lebanon a series of employment, welfare, and other services for the Palestinians; operated some thirty factories employing 6,500 full-time and 4,000 part-time refugee workers who produced religious articles, hand-embroidered dresses, kitchen utensils, safari suits, Louis XV chairs, high-heeled shoes, lace tableclothes, and foodstuffs ranging from butter and poultry to candy. In Lebanon, too, the guerrilla confederation maintained its own hospitals and clinics, its own news agency, research center, radio station, magazine and film center, its own bank and telex services. Altogether, with its bounty of Persian Gulf funds, of Soviet and Libyan weaponry, with its command of the archipelago of Palestinian shantytowns, the PLO won a substantial autonomy for itself on Lebanese soil, a functional independence that included the right literally to administer its own refugee camps under its own Palestinian flag. There were fifteen of these camps extending throughout the southern part of the country, from West Beirut down to the Israeli border. Ultimately they commanded so wide a supply route that they became the operative government in a large part of south Lebanon, as they had been several years earlier in Jordan. Their self-assurance in the end would exert a decisive impact on the nation's precarious Moslem-Christian equilibrium.

The PLO's key military bases were concentrated in the rocky wasteland around the foot of Mount Hermon on the borders of Syria and Israel's Upper Galilee. Here, in the Arqoub—called "Fatahland" by the Israelis—the guerrillas tunneled hundreds of caves to store their ammunition, and relied on a network of sympathetic Moslem villages to double as their supply centers. Here, too, they carried out military training, with 15,000 well-armed men, and planned their incursions into Israel. From their PLO headquarters in West Beirut, meanwhile, they laid even broader-ranging plans for attacks against Israelis and other Jews abroad. As noted (Chapter V), between January 1974 and September 1979, some 1,207 Jews and Arabs were killed and 2,950 wounded in all PLO operations.

Even under Israel's Labor government, stern reprisals for this mayhem were carried out by air and occasionally by land. Nor were casualties limited to PLO, or even Palestinian, targets. The Lebanese government learned early on that it too would be held responsible for PLO violence launched or even plotted from its territory. Thus, as early as December 1968, when a PLO attack against an El Al plane in Athens inflicted several Israeli casualties, Israeli commandos raided Beirut airport, blowing up thirteen commercial jet aircraft. In reprisal for the July 1972 Munich Olympics massacre, Israel mounted a brigade-sized crossing into "Fatahland" the following September; and during the two-day incursion, heavy casualties were inflicted both on the guerrillas and on Lebanese troops who attempted to intervene. In April 1973, an Israeli commando team was landed on the Lebanese coast, made its way to a selected apartment building in the heart of Beirut, and methodically assassinated three of Arafat's top aides. Several Lebanese policemen also were killed. Prime Minister Rabin made the warning explicit in a press interview of 1975. Under threat of retaliation, he pointed out, Jordan had managed to throttle guerrilla activity against Israel from its territory. So, now, Lebanon would be held responsible for similar anti-Israel violence. Rabin's analogy was misplaced, as it happened. Lebanon's undersized and ill-disciplined army was all but helpless to control the burgeoning PLO guerrilla confederation. Neither would the nation's Moslem majority have countenanced such an effort. Under the circumstances, the Lebanese government and people were obliged to watch helplessly as Israeli retaliation in forty-four separate attacks between 1968 and 1974 killed 890 Lebanese and Palestinian civilians, together with far many more PLO guerrillas.

At mid-decade, Lebanon's Christian communities, particularly its Maronite elite, recognized that forceful action soon would have to be taken against this swelling Palestinian presence, and the PLO militants who dominated West Beirut and southern Lebanon. Israeli retaliation forays were painful enough. But an even graver menace was the imminent disruption of the internal balance of power. Indeed, fedayun aggressiveness already had evoked the support of native Moslem radicals, those who had long awaited the opportunity to end Christian domination of the nation's army and economy, the constitutionally ordained six-to-five Christian numerical advantage in parliament. The Maronites accordingly were not prepared to wait until they were inundated by a vindictive Moslem avalanche. Among their leaders, the wealthy and powerful Chamoun and Gemayel families had been preparing a decisive confrontation for several years. They had never disguised their hatred of the guerrilla juntas, their desire to drive the volatile PLO from the country altogether.

On April 13, 1974, then, a Maronite "Phalangist" military unit ambushed and killed a busload of Palestinians outside Beirut. When PLO troops laid their own counterambush the next day, and killed several Christians, fighting spread rapidly to the capital itself. Phalangist gunners shelled the Tel Zaatar refugee camp in the western slums of the city. PLO forces counter-shelled Christian East Beirut. Other firefights broke out during the week. Soon groups of Lebanese Moslems joined the Palestinians against the Christians, and the bloodshed, widening in scope and savagery, approached the dimensions of a civil war. By the end of 1975, the better-disciplined Christian militias appeared to be winning. At this point the Syrians, who for thirty years had not so much as reconciled themselves to Lebanese independence, dispatched several Palestinian brigades that had been trained on Syrian soil. The tide of the fighting soon shifted. Indeed, Moslem strength grew so decisively that the Christian militias faced annihilation. But here again Damascus intervened—ironically, to save the Christians this time. President Hafez al-Assad's purpose was hardly philanthropic. If the Palestinians were allowed wholly to dominate the future of Lebanon, there was a danger that radical elements, emancipated from Syrian control, might be tempted to precipitate a major war with Israel, and Syria then almost certainly would be drawn in. The risk was too great. Exerting heavy military and diplomatic pressure, therefore, the Assad regime managed to negotiate an end to the civil strife, to carnage that by mid-1976 had taken at least 70,000 lives. Afterward, units of the Syrian army remained in West Beirut and in eastern and central Lebanon as "peacekeepers."

From Jerusalem, meanwhile, the Rabin government monitored these developments with much sympathy for Lebanon's Christian victims. Not a few of those victims had fled southward. There they were allowed entrance to Israel through the "good fence," the northern Israeli town of Metulla. Food and medical care were provided them, occasionally even employment. Rabin and his colleagues were hardly oblivious to the looming Syrian danger. Yet they had sent word to Damascus of their terms for refraining from intervention, and Assad had concurred. Henceforth the Syrians would not use their air force against Lebanese Christians or interfere with Israeli surveillance flights over southern Lebanon. Neither would Syrian forces move south in Lebanon beyond the area east of Tyre, a buffer zone for Israel's northern frontier that Syrians and Lebanese alike were prepared tacitly to acknowledge. Within this "Red Line" zone, moreover, a Christian militia under the command of Major Sa'ad Haddad, a renegade Lebanese officer, was encouraged to assume command of the territorial strip closest to Israel's border. There Haddad enjoyed access to Israeli weapons and training, and cooperated with Israeli intelligence in blocking PLO access in the region. For the time being, the arrangement was not intolerable for Israel—no more than for Syria.

Even the PLO, aware that the Palestinians had suffered 20,000 of their own casualties in the Lebanese civil war, seemed reasonably quiescent at first. Yet no one in Israel believed that the fedayun leaders would remain passive for long. Indeed, by the time of Begin's election in May 1977, the guerrilla confederation was fast recouping its strength, and its infiltrators were beginning to penetrate the Red Line area again. By then Syria was relaxing its control over the Palestinians, and soon would loosen it even further in an effort to disrupt Sadat's peace initiative. Thus, in early 1978, with Syrian acquiescence, the PLO took military control of the southern Lebanese town of Nabatiyeh and the neighboring Crusader citadel, Beaufort Castle, overlooking the northwestern Galilee. From Beaufort, the guerrillas fired increasing numbers of rocket and artillery rounds into Israel. Fedayun border crossings mounted again. Immediately, Israel's air force resumed its strikes against Palestinian targets. The tactic was reasonably effective in limiting PLO shellings and cross-border forays. It was by no means foolproof, however.

Israel's renewed vulnerability became apparent early in March 1978 when a small group of fedayeen embarked by landing craft from Tyre, and made their way down the Israeli coast to a point about 40 miles north of Tel Aviv. There they murdered a young American woman, who was photographing birds in a nature reserve, and went on to hijack a bus carrying passengers to Tel Aviv. Two miles north of the city they were stopped by Israeli troops. In the exchange of fire, thirty-four Israeli civilians were killed, including a number of women and children. Begin promptly authorized a decisive reprisal. The action, "Operation Litani," began on March 16. Some 8,000 troops battered through PLO positions south of Tyre, with the aim of destroying southern Lebanon's guerrilla enclaves. The expedition was not a brilliant success. The attack forces moved cautiously and unimaginatively. Tank commanders frequently lost their way. Squadrons were broken up. Very few guerrillas were trapped, although scores of Lebanese villagers were killed in the heavy advance bombardment. Operation Litani's single, rather limited, achievement was to induce the UN Security Council to assign an interim force—UNIFIL—to Lebanon, a 3,000-man unit that was emplaced in the Red Line zone east of Major Haddad's Christian militia, to block future Palestinian infiltration and Israeli reprisals. Yet even the UNIFIL failed to secure effective control over the buffer area. It proved incapable of stopping PLO rocket launchings from more withdrawn emplacements. There was also serious likelihood that Arafat soon would test his strength against the Jews. He enjoyed broader Syrian support as a consequence of Egyptian-Israeli peace negotiations.

In ensuing months, Begin authorized far-ranging air and land strikes against PLO targets in Lebanon, as well as the dispatch of additional equipment and advisers for Major Haddad's militia. Yet, in the end, it was Lebanon's shifting Christian-Moslem equilibrium that was destined to precipitate a renewed Arab confrontation with Israel. In March 1981, a Christian Phalangist unit moved against a strategic bridge in Zahle, an access way to the Syrian-dominated Beka'a Valley, and in the process inflicted heavy casualties on a Syrian army detachment. This was a mistake. Several months earlier President Assad had suppressed an internal uprising against his regime, and subsequently had reconsolidated his power at home. He was prepared now to reassert himself in Lebanon. Indeed, to emphasize the immovability of the Syrian presence, Assad decided to threaten the Phalangists' territorial base. He rushed in additional troops by helicopter. It was the use of airborne personnel, in turn, that violated the 1976 agreement with Israel. The question arose of an appropriate response.

Before this latest crisis, Dayan and Weizman had served as moderating influences in the Begin cabinet. Now, with both men gone, Begin was functioning as his own defense minister. Elections were scheduled for June. Even as he contemplated the possible usefulness of a riposte against the Syrians, moreover, Begin was exposed to the forceful influence of General Rafael Eytan, the military chief of staff. Native-born, raised on a cooperative farm, Eytan in the early 1950s had served under Sharon in the famous Commando Unit 101, then in the 1956 battle of the Mitla Pass. His subsequent performance in halting the Syrian offensive during the Yom Kippur War already was a legend. A bulldog, not quite as imaginative as Sharon, Eytan fully shared Sharon's passion for battle. His instinctive reaction, then, to the latest Syrian threat was for action. Begin needed little persuasion. The prime minister had since convinced himself that Israel bore a moral obligation to the Christians of Lebanon, whom he equated with the Jews of Europe in World War II. The policy of Israel's earlier, Labor, government had been essentially to help the Lebanese Christians help themselves. In contrast, Begin made a solemn commitment to take all necessary steps, including active intervention, to prevent a Christian "genocide." Both the approach and the analogy were challenged by Deputy Defense Minister Mordechai Zippori, who pointed out the risks of direct involvement in Lebanon's sectarian fighting. Zippori's misgivings were brushed aside.

Accordingly, when two Syrian helicopters later were detected carrying supplies to Syrian positions in southern Lebanon, they were shot down by Israeli fighters. The provocation infuriated Assad. This time the Syrian leader responded by moving SAM (antiaircraft) missiles into Lebanon's Beka'a Valley. Here was yet another violation of the Red Line understanding. Would Israel now bomb the missiles? In the event Begin had the idea in mind he was given pause, almost at the last moment, in May 1981 (and one month before the scheduled Knesset election) when President Ronald Reagan dispatched a special emissary to the Middle East, the veteran diplomatic troubleshooter Philip Habib. Shuttling between Jerusalem and Damascus, Habib exerted every effort to resolve the missile crisis. Evidently he failed. The Syrian government held firm. The Israeli government in turn continued to threaten air bombardment against the SAMs. Still, Begin refrained from overt action. The tactic in fact was an adroit mixture of bluff and diplomacy. On the one hand, it allowed the prime minister to continue through the June election, mobilizing national unity against the Syrian danger. On the other, it enabled him to play the disciplined statesman, willing to give peace a chance.

PREEMPTING THE NUCLEAR BALANCE

From the outset, Begin viewed the Reagan government's somewhat inactive diplomacy on the Middle East as a welcome contrast to the irascible pressures of the Carter administration. It was a benign neglect that tended to confirm the image of Reagan and Haig as friends of Israel. The president often had referred to the PLO as a terrorist organization, after all, even as Haig frequently had expressed deep suspicion of the Syrians as Soviet clients. Begin and his advisers accordingly were persuaded that a militant Israeli foreign policy would evoke Washington's understanding and sympathy. In fact, disillusion was rapid on both sides. As the prime minister took increasingly drastic measures to strike at the PLO in Lebanon, to suppress Arab opposition and thicken Israeli settlements on the West Bank, State Department criticism became sharper.

The tension became particularly severe in the summer and autumn of 1981, on the issue of a military aircraft sale to Saudi Arabia. As it happened, the decision to sell a squadron of advanced F-15 jets had been taken during the Carter presidency, and Reagan intended to do little more than fulfill the earlier agreement. Thus, Begin made no public protest at first. But soon it was learned that the Americans also had consented to include a squadron of radar-carrying, Advance Warning and Command Systems (AWACS) aircraft in the sale. Alarmed, the Israeli prime minister immediately expressed opposition to the entire package. The security of "moderate" Persian Gulf governments surely did not require this kind of sophisticated hardware, he argued. Afterward, visiting Washington in September 1981, Begin took the unprecedented step of going over the president's head to make his case directly before the Senate Foreign Relations Committee. Reagan was deeply offended. In a nationally televised attack on Israel's lobbying effort, he noted sternly that it "is not the business of other nations to make American foreign policy." At his request, too, former Presidents Nixon, Ford, and Carter declared their support of the package sale to Saudia Arabia, and added their criticism of Begin's "interference." Whereupon, exploiting the evident shift of national consensus on the Middle East, the oil lobby and other pro-Arab corporate groups applied pressure at the grass-roots level with the slogan "Reagan or Begin?" For the first time, a number of traditionally friendly senators and representatives turned their backs on Israel. The Senate confirmed the package sale to Saudi Arabia, 52 to 48.

The uncharacteristic eruption of anti-Israel feeling may well have been triggered earlier, on June 7, 1981, by Begin's single most forceful gesture of international defiance since assuming his nation's prime ministry. The challenge related to Iraq's nuclear reactor. Here it is must be noted that Israel's involvement in the nuclear game long antedated that of the Arabs. As far back as 1952, at a time when Israelis still were living on rations, the Ben-Gurion government set up its own atomic energy commission under the umbrella of the prime minister's office. A tiny 5-megawatt "swimming pool" research reactor subsequently was constructed at Nachal Soreq, near Rehovot, and in 1955 the United States agreed to sell Israel modest quantities of enriched uranium fuel. By then, too, Ben-Gurion and his young protégé, Shimon Peres, director-general of the defense ministry, had developed warm relations with France on the military level, and set about extending that cooperation to Israeli nuclear development.

They succeeded. In September 1956, a month and a half before the Sinai Campaign or even before France launched upon its own nuclear "force de frappe," the Guy Mollet government in Paris agreed to help build a major Israeli nuclear facility. The formal contract was signed a year later, and construction afterward was begun secretly at Dimona, in the northern Negev, on a 24-megawatt natural uranium reactor. From the outset, both governments understood that the facility would be devoted to the production of weapons-grade plutonium. Indeed, to that end, the French consented to fabricate a small adjacent plant in Dimona to reprocess plutonium from the spent nuclear fuel. Israel also was given unrestricted access to French nuclear test explosion data, thus sparing its scientists the need to test a bomb themselves.

Then, several years after Charles de Gaulle acceded to the presidency of France, the relationship with Israel began to cool. In 1964, Paris sent word that it was halting its supplies of natural uranium for the Dimona reactor. The blow was a grave one for the Israelis, for they were deprived suddenly of the wherewithal to produce weapons-grade plutonium from their reprocessing plant. Nor was any other country willing to pick up the slack; Israel, like France, had refused to sign the NPT—the nuclear nonproliferation treaty—(the implied threat of nuclear retaliation was the little nation's ultimate deterrent to its Arab enemies). Nevertheless, through circuitous and often illegal routes, the Israelis in ensuing years managed to lay their hands on some 200 pounds of newly enriched uranium. The quantity was just sufficient to make their Dimona processing plant operational. When additional supplies were needed, the Israelis again found ways.

"Operation Plumbat" was one example. A Belgian company, Société Générale de Belgique, stored large quantities of "yellow cake"—uranium oxide—in a silo near Antwerp. The material was not available to Israel. Euratom, the Common Market nuclear agency, did not permit sales to non-NPT countries. Whereupon, in 1968, the Israelis set up a front operation through a small West German chemical firm, Asmara Chemie, whose owner persuaded the Société and Euratom that he needed 200 tons of uranium oxide as a catalyst for the manufacture of petrochemicals. Arrangements subsequently were made for the yellow cake to be shipped to Genoa, then carried overland for "special processing" in Milan. In fact, the vessel to be used for the first leg of the trip had been purchased secretly by Israel, and its captain was a disguised Israeli. Loaded in Antwerp, the freighter sailed off—not to Genoa, but to the east Mediterranean. There its cargo was transferred at sea to an Israeli merchantman, protected by Israeli gunboats. Six months passed before Euratom began to suspect that the yellow cake had gone astray. By that time it was already at work in Israel's processing plant in Dimona.

By then, too, the American CIA had prepared a report suggesting that Israel might already have begun fabricating atomic weapons, the first Middle Eastern nation to possess that capacity. The report leaked to the press. Jerusalem refused to confirm or deny it. Some years later, however, during the Yom Kippur War, when their forces were hard-pressed, the Israelis themselves leaked a story to *Time* magazine that thirteen 20-kiloton (Hiroshima-strength) bombs had been assembled at a secret underground tunnel and rushed off to specially equipped Phantom and Kfir jets. Fortunately for Israel—and Egypt—the bombs were not needed. Whether or not the story was true, Sadat may have believed it. Conceivably, it was one of the factors that motivated his trip to Jerusalem in 1977. In the ensuing peace negotiations he pressed Israel to forego its nuclear capability. The Israelis equivocated, while admitting nothing.

If the little Jewish nation was indeed producing atomic weapons, it was a matter of some urgency that none of its Arab neighbors share that capability. Earlier fears of Egypt as an atomic threat had long since evaporated. The reactor in Einshas, built by the Soviets, was a tiny affair and used exclusively as a prototype for generating electricity. But in 1974 the Saddam Hussein government in Iraq began making plans to go nuclear. Saddam himself was a particularly dangerous character. He had seized power eight years earlier through ruthless political intrigue and assassination. Afterward, upon organizing and consolidating an ardently Ba'athist regime, he had set about promoting his own brand of pan-Arab "socialism" elsewhere in the Middle East through subversion and terror. Those close to the man regarded him as the logical heir of Gamal Abd al-Nasser. In fact, Saddam intended to widen his influence as Nasser never had by becoming the first Arab leader to command a nuclear arsenal. The ambition was not unrealistic. Like Israel in the 1950s, Iraq was able to rely on cooperation with France. The Giscard d'Estaing government was more than eager to secure guaranteed access to Iraqi oil supplies, to the vast Iraqi capital market for industrial and military equipment. A commitment to build two nuclear reactors for Iraq did not seem an extravagant price for that access. News of the Franco-Iraqi nuclear deal reached Washington in 1976. Concerned, the Ford administration interceded with France. "We cannot let our American and European allies continue their offensive against our nuclear industry," came the reply from Paris. Israeli appeals were spurned even more contemptuously.

By 1978, nevertheless, misgivings were developing even within the French foreign ministry. The contract with Iraq promised delivery not only of two nuclear plants but of nuclear fuel enriched with uranium 235. The fuel was to be shipped in six installments of 13 kilos each, with the entire consignment to be delivered by 1981—sufficient enriched uranium for the manufacture of nine weapons. As a compromise, the French foreign ministry now offered to sell Iraq an alternative "caramel" fuel, enriched with just enough uranium 235 to operate the first scheduled reactor, but not enough to manufacture the bomb. Baghdad flatly rejected the compromise. It would accept no "inferior" fuel. Nor did the French press the issue; they dared not endanger their commercial relationship with Iraq. Hereupon, the Israelis decided to adopt more vigorous measures on their own. In April 1979, saboteurs broke into a warehouse near Toulon and blew up the casings for Iraq's first scheduled "Osiraq" reactor, components that had been scheduled for shipment within the week. The setback was painful for Iraq, but not irretrievable. French Prime Minister Raymond Barre personally assured Baghdad that the damage soon would be made good, and it was. New casings were delivered in the autumn of 1979. The first installment of enriched uranium reached Iraq in June 1980. After three more shipments, the Iraqis would be able to irradiate enough fuel from their Osiraq reactor to produce weapons-grade plutonium. The reactor in turn was scheduled to become operational in Tuwaitah, outside Baghdad, early in 1981. By the mid-1980s, the Saddam Hussein regime could be expected to possess a stockpile of at least six atomic bombs.

The danger to the Jewish state was not remote. Israeli intelligence was aware that Iraqi Mirages, MiG-23s, and TU-22 "Blinder" bombers were nuclear-capable, that nuclear warheads could even be fitted on Iraq's Soviet-supplied SCUD missiles. Retaliatory weapons posed only a minor deterrent. The 300-mile range of Israel's Jericho missile reached little more than half the distance to Baghdad. Although the Israeli air force was unexcelled, the value of airplanes against warhead-carrying rockets clearly was limited to preemption. Thus, one way or another, Saddam Hussein's nuclear program would have to be aborted. If diplomatic intercession failed, then, as Israeli Deputy Defense Minister Mordechai Zippori tersely informed a press conference in January 1980, "Israel would have to revise its options." Meanwhile, other less than diplomatic steps were taken. In June 1980, a year after the initial Toulon sabotage, Professor Yahia al-Meshed, the Egyptian-born physicist who directed Iraq's nuclear research program, was murdered in his hotel room in Paris.

By then Menachem Begin was contemplating an even more decisive measure. This was aerial bombardment of the Osiraq reactor itself. In December 1980 he discussed the notion with Shimon Peres, the opposition Labor leader. Peres was not enthusiastic, fearing the possibility of an international diplomatic quarantine. For that matter, Begin's cabinet was split on the issue. Only Sharon and Foreign Minister Yitzchak Shamir unequivocally favored an air strike. But in the final vote of the ministerial defense committee, Begin won out. He and Chief of Staff Eytan then discussed the timing of the attack. The first choice, May 10, 1981, turned out to be the scheduled date of the French presidential election, and Begin was uninterested in creating a popular backlash in Giscard's favor. The raid was set then for June 7. Preparations began immediately and continued through the winter and spring of 1981. In the interval, François Mitterand was elected president of France. No political figure in Europe counted himself a better friend of the Jewish people and Israel. Several of Mitterand's closest advisers were Jews. His sister was married to a Jew. As a Socialist, Mitterand had developed close ties over the years with Israel's Labor party. During his election campaign, he had vigorously criticized Giscard's Middle East policy—above all, the sale of atomic materials to Iraq—and had committed himself to alter that policy, if elected. Now he intended to make good on his promise.

As one of his first presidential acts, Mitterand instructed the foreign ministry to come up with a tough new plan for guaranteeing that the Iraqi reactor would be used for peaceful purposes only. Foreign Minister Claude Cheysson thereupon turned the project over to a special ministerial task force. After ten days of intensive effort in late May, the committee produced a formula. It envisaged unrestricted access to the Osiraq reactor by French technicians; "spacing out" of deliveries to ensure that each shipment of bomb-grade material would be immediately irradiated—in effect, consumed—by the reactor; and limitation of all subsequent shipments to "caramel" uranium. The details of this plan were completed on June 1, and its final written version was brought to Cheysson's office four days later. The foreign minister did not have time to review its contents with Mitterand, however, before the Israelis decided the issue on their own.

On June 7, 1981, at 4:00 P.M., eight of Israel's fleet of F-16 jets took off from their Etzion base in the Sinai. Originally designed for aerial combat, the planes were provided with extra fuel tanks and special bomb racks that carried two 2,000-pound MK-84 "iron" bombs. Once airborne, the F-16s moved into formation with an escort of six F-14 fighter-interceptors. The Israelis also readied an unknown number of other planes, including F-15s equipped with oversized tanks to provide aerial refuelling. The trip would cover more than 1,000 miles to the target and back, a distance stretching the F-16s' range to the limit. Any deviation from plan, any battle with enemy planes, would require additional fuel, and refuelling would have to take place over Arab territory. No more room for error existed here than in the 1976 Entebbe rescue operation. The main force of eight F-16s and six F-15s accordingly headed east across the Gulf of Aqaba and along the northern width of Saudi Arabia, near the border with Jordan. Military intelligence had selected the route after identifying its occasional radar blind spots.

The attack group flew in tightly bunched formations, to project the radar "signature" of a large commercial aircraft. Crossing Saudi and Hashemite air space into Iraq, the planes sped on past Baghdad to Tuwaitah, and approached low to avoid detection. One of the F-15s initially overflew the Osiraq reactor to draw antiaircraft fire. Meanwhile, the F-16s had swept up into the sky to allow the pilots to aim. With the setting sun behind them, they dived at the target, sending the bombs hurtling against the reactor's concrete-and-lead protective dome. The tactic had been refined during months of practice against a mock-up in the Sinai. The first bombs were fitted with delayed-action fuses, allowing them to penetrate the dome before exploding, opening gaping holes in the side. The next bombs struck only moments later, finding the holes with "stupefying accuracy," in the words of a French observer, and destroying the reactor inside. One French technician was killed in the attack. Iraqi antiaircraft fire spattered harmlessly. All Israeli planes returned safely.

Was the raid justified? From all evidence, the Iraqis were indeed moving toward the production of atomic weaponry. Yet, they were at least two years away from developing their first bomb, and the Israelis unquestionably had time to wait, to give Mitterand a few months or even a year to close the loopholes in the Iraqi nuclear contract. Later, Begin observed that he could not have given the order to attack once the reactor became operational; the explosion would have disseminated a lethal radioactive cloud over Baghdad. The Iraqis then would have proceeded with impunity to build atomic bombs in their "hot" plant, and Israel would have faced another genocide. "Never again," the prime minister declared solemnly. "Never again." There could be no doubt that Begin's motive, of protecting his people, was entirely forthright. His argument was less so. The reactor would not have become critical until September at the earliest, and probably later. Even if Osiraq were "hot," its destruction by aerial bombs hardly was likely to have diffused a radioactive cloud over the reactor site, let alone over the city of Baghdad. Begin's one legitimate concern may have been the public relations consequences of bombing a "hot" reactor.

The air strike's diplomatic "fallout" was far-reaching, in any case, as Shimon Peres had warned. Obviously confused, Reagan at first condoned, then criticized the attack both in official statements and in his support of a UN Security Council resolution of condemnation. A scheduled shipment of F-16 aircraft to Israel was halted. Few of Israel's congressional supporters were prepared to countenance a unilateral military action against another sovereign state. When Begin arrived later to testify against the AWACS sale before the Senate Foreign Relations Committee, their cross-examination of the prime minister was harsh (and made all the more so by Begin's pugnacious manner). Israel's violation of Saudi air space merely fortified Riyadh's appeal for effective radar protection—against all potential threats. In Paris, meanwhile, Mitterand and his colleagues were offended and aggrieved. The French president quietly sent word to the Israelis that they might have placed greater confidence in his intentions. Although his friendship even now remained intact, his ability to plead Israel's case in Europe was seriously weakened. In Egypt (as shall be seen later in this chapter), the repercussions of the bombardment were more calamitous yet.

But if the air attack reaped only nettles on the international scene, there was one area at least in which Begin achieved substantial success. This was in Israel itself. Here news of the bombing evoked widespread admiration as a legitimate act of self-defense. The sheer audaciousness of the raid, no less than the brilliance of its execution, visibly enhanced the prime minister's political standing only days before the Knesset election.

IN SEARCH OF A RENEWED
POLITICAL MANDATE

From the moment of his electoral triumph in May 1977, Begin had navigated a roller coaster of popularity. Its initial peak was the Sadat visit. The ensuing diplomatic stalemate produced a decline. Afterward, Camp David and the peace treaty appeared to restore public confidence, although at a small cost in the defection of the Techiya party on the right. But from late 1979 through early 1981 yet another stalemate followed on the Palestinian negotiations. The impasse, together with the growing economic crisis and two serious bouts of illness for Begin himself, resulted in additional defections, by Dayan, Weizman, by half the DMC faction. It also accounted for a steep fall-off of Likud's standing in the polls. Under the circumstances, Begin's militance on the Jerusalem and Lebanon issues, on the sale of F-15s and AWACS to Saudi Arabia, and finally his electrifying air strike against the Iraqi nuclear reactor, proved rather more valuable as political tactics than as expressions of national policy.

So, on the domestic front, did a shift in economic policies following the resignation of Finance Minister Yigal Hurevitz in January 1981. Hurevitz's replacement, Yoram Aridor, was a young Cherut activist of commendable intellectual skills but of little practical business experience. He confronted a grim economic and political landscape. By then the prognosis for Begin's reelection unquestionably remained guarded. Opinion polls still classified his administration as the worst in Israel's history. He had failed to carry out the largest part of his 1977 domestic promises. The inflation had increased from 40 percent a year at the time of his accession to 100 percent, and appeared certain to go much higher. With food subsidies sharply cut, staples like meat were becoming luxuries for tens of thousands of Israelis. The number of citizens living below the poverty line had reached 100,000.

It was Aridor, therefore, with Begin's concurrence, who promptly adopted a disjointed version of supply-side economics. On the one hand, he slashed taxes and import duties on a wide range of consumer goods. On the other, he pumped the nation's modest dollar and Common Market currency reserves into Israel's banking system, thereby artificially shoring up the value of the Israeli shekel. Ignoring Labor (and some Liberal) warnings of economic suicide, the finance minister then genially presided over a frenzied squandermania, as shoppers filled the stores, snapping up freezers, television sets, video recorders, and other imported commercial items. Among the working classes, the government's reputation was partially salvaged, too, by awareness of its earlier series of populist measures, among them the extension of free secondary education, "Project Renewal" in slum neighborhoods, and David Levy's brawny interventionist efforts on behalf of subsidized housing. Then, later in the campaign, on the eve of Passover, prices on a number of recently unsubsidized basic foods were suddenly cut—that is, resubsidized.

Much to the pollsters' surprise, the government's economic failures in any case played only a minor role in the developing election campaign. For one thing, the largest numbers of citizens still did not consider themselves badly off. Notwithstanding the drastic rise in the cost of living, the widening gulf between rich and poor, thousands of oriental voters appeared to have achieved a certain lower-middle level of education and income (if not of purchasing power). Their disaffection remained entangled, rather, in unfulfilled status expectations. In 1981 as in 1977, the Europeans still dominated the upper strata of management, of politics and economic power; and the identification of the Ashkenazic elite with Labor remained unshakable in the eyes of the non-Europeans.

Even more fundamentally, economic issues altogether were of less importance to the orientals than those of foreign and military policy. And here, during the campaign, Begin emerged again as the dominant figure. With his aggressive rhetorical style, his uncompromising determination to retain the "whole Land of Israel," he appealed uniquely to the frustration-rooted chauvinism of the non-Europeans. In placard and newspaper advertisements, Likud election propaganda declared unabashedly that a "vote for Shimon Peres is a vote for a Palestinian state under Yasser Arafat's leadership." During May and June, we recall, Begin had adopted his belligerent position on Syrian missiles in Lebanon, continually threatening to destroy these batteries if they were not removed (they were neither removed nor destroyed). Then the bombardment of the Iraqi reactor was carried out only three weeks before the election.

Although ethnic resentments plainly had long been a fact of life in Israel, and a major, if unspoken, factor in the elections of 1977, those tensions became a focal issue in the 1981 campaign. The heat of the summer electioneering did not help. The dominant atmospherics were the emotionally charged issues of democracy versus fascism, political violence against political tolerance, the "growers" of tomatoes (ostensibly the Ashkenazim) against the "throwers" of tomatoes (presumably the orientals). Slogans, prejudices, and personalities, the frustrations of one ethnic group and the fears of another, now became more decisive than ever before in Israel's political history. Begin personally set the tone. Appearing before working-class audiences, where he was greeted rapturously as "Begin, King of Israel," the prime minister invoked the image of Israel's unyielding stand on Lebanon, of Syria's President Assad as a "chicken," of Peres as a "saboteur," of the Ashkenazic-dominated Alignment as "kibbutz millionaires, with their swimming pools and beauty parlors." When Peres in turn sought to address a large election rally in Jerusalem's Sacher Park, well-organized claques of Moroccan rowdies hooted and booed him, pelting him with tomatoes until he had to stand down, unable to utter a word. Other Alignment speakers often encountered the same reception. Their automobiles were overturned, their branch offices vandalized. Addressing rallies in "safe"—Ashkenazic—neighborhoods, Labor politicians responded with thinly veiled ethnic slurs of their own. The "nice people," they suggested, the "real fighters" and the "army officers" were with Labor, while the Likud camp was supported by the "chakchakim" (a derogatory term connoting Moroccan riffraff). Mordechai Gur, a former military chief of staff and now an important figure in Labor, warned a group of oriental hecklers that "we will beat you as we beat the Arabs." In his exasperation, Peres described the vandalizers of Labor property as "Khomeinistis," and suggested that they return to their home countries.

Ultimately, Labor's fundamental political error lay in its decision to concentrate on foreign policy rather than to strike frontally, even harshly, on painful social and economic issues. Peres and his associates could have made a stronger impact by dramatizing in plain language such bread-and-butter problems as inflation, declining real income, and bureaucratic mismanagement. With the departure of Dayan from the Labor camp, to be sure, the Alignment finally had managed to achieve a rough consensus on foreign policy. But it was hardly of the kind to set crowds dancing in the streets. In common with Likud, Peres and other Labor spokesmen rejected the notion of a Palestinian state, of a redivided Jerusalem, of anything less than total demilitarization of the West Bank. Yet they emphasized the need for territorial concessions within the framework of an authentic peace settlement, and warned that annexationism posed a danger to Israel's Jewish character. "A form of 'living together' that can only be sustained by day-and-night protection is socially worthless and politically explosive," Peres insisted. Jewish settlements in the West Bank should be established or retained only to ensure "genuine security," and even these might be expendable for the sake of a true peace settlement. In its moderation, the Labor Alignment's approach seemed to be little more than a pallid reversion to the Allon Plan, and thereby all but tailor-made for Begin's fiery sloganeering and withering invective.

Neither did Labor help its cause by internal dissensions. In 1980, even as its leaders appeared to have reached a meeting of minds on the territories, and as Peres managed gradually to rebuild his grass-roots strength, Yitzchak Rabin launched a campaign to regain the party chairmanship. In recent years, the former prime minister's resentment of Peres had grown increasingly open and bitter. His newly published memoirs were steeped in personal recriminations against his rival. Although Peres handily won reendorsement at the Labor convention in December 1980, Rabin accepted the vote gracelessly, and continued to cast aspersions on the chairman's integrity. For his part, Peres was too astute a politician to ignore Rabin's value as a vote-getter—the man had won the 1967 war, after all—and thus felt obliged to declare Rabin as his choice for defense minister in a future cabinet. Yet the desperation tactic merely appeared to confirm Peres's vague reputation for political opportunism. Meanwhile, the Begin government launched its program of renewed subsidies and import-duty cuts, of militance in Lebanon and bombing in Iraq. The combination was devastating. As recently as February 1981, the polls had showed Labor winning 58 seats to Likud's 20. Four months later, the June 30 election results snatched defeat from the jaws of victory.

In fact, the Alignment increased its strength rather impressively, from 32 to 47 Knesset seats. But Likud also gained, raising its total of seats from (the equivalent of) 43 to 48. If Peres and his associates had imagined that the 1977 election represented essentially an aberration, they now understood with perfect clarity that the norm of Labor dominance had ended. Thus, the difference between the 1977 and 1981 votes for the Alignment was a mere 75,000 votes. In the same period, Likud added some 375,000 votes to its total. The nation clearly had shifted its political axis. Even more than in

1977, the Right's political consolidation was rooted in ethnicity. Of those 375,000 new Likud votes, over 60 percent came from the oriental communities. Among the Ashkenazim, 60 percent of the vote went to the Alignment.

Still another noteworthy feature of the 1981 election was the polarization of strength between two main blocs. Three of the former thirteen Knesset parties were obliterated. Between the remainder, Likud and the Alignment received 75 percent of the vote. Dayan's venture to establish a centrist "Telem" group achieved only his own election. The Independent Liberals failed to elect a single candidate. The largest of the smaller factions was the National Religious party; and even the NRP won only 6 votes, half its total in the 1977 election. The more fundamentalist Aguda kept its earlier number, 5. And yet, ironically, the basic erosion in the religionists' numerical strength produced no weakening in their political bargaining position. The reason lay in Likud's wafer-thin, 61-seat Knesset majority. The right-wing bloc now depended for its one-seat edge precisely on the goodwill of the Orthodox— of the NRP, the Aguda, and "Tami" (this latter a breakaway splinter of oriental religionists from the NRP). That dependence was hardly a source of consolation to Israel's moderates. As shall be seen, fifty of the eighty-three clauses Begin was obliged to negotiate as his coalition "pact" were commitments on religious issues.

And on territorial issues, finally, there was no difference whatever between the Beginites and the Orthodox. With Dayan and Weizman gone by then, and the unregenerate Ariel Sharon soon to be appointed minister of defense, the new government was in a position to submit an exceptionally forthright statement of principles to the Knesset. In unambiguous language, the guidelines affirmed Jerusalem's status as "the eternal capital of Israel, indivisible, entirely under Israeli sovereignty," and the "right of the Jewish people to the Land of Israel, an eternal, unassailable right that is intertwined with the right to security and peace."

A TRAUMA OF CHANGE IN EGYPT

If Israeli doves were aghast at Begin's reelection, and the tactics that had achieved it, the malaise in Egypt was even more profound. Sadat himself was an increasingly isolated man by then, in his own country no less than elsewhere in the Arab world. His promises of "regional peace," of an era of "great prosperity," had not materialized. The United States was providing Egypt with $2 billion a year, but in relation to population the amount was proportionately fifty times smaller than the sum accorded Israel; and, at that, it still was channeled largely into the upkeep of the armed forces. To be sure, the nation had other sources of revenue, principally oil and remittances from émigrés working abroad. By 1980, however, 53 percent of Egypt's GNP was devoted to the purchase of imports, including basic food staples. American businessmen were not hurrying to the rescue of the floundering Egyptian economy with private investments. In some desperation, the government now solicited foreign capital on terms that would have been unthinkable in the earlier period of Sadat's *infitah*, his opening to the West. The editor Muhammad Heikal later observed that "not since the days of Khedive Ismail had Egypt been the scene of looting on such a massive and organized scale." Cairo became a city of middlemen and commission agents. It was an open secret that contracts were awarded wholesale to favorites and friends of the president. Sadat himself probably did not benefit from the profiteering, but his brother Esmat suddenly became an immensely wealthy owner of many companies.

Infitah's palpable lack of success now became a major factor in Egypt's growing unrest. A majority of the nation's political parties reproved Sadat for the widespread corruption of society, and they condemned his deference to Washington as both sterile and dangerous. Intent upon linking Egypt to the Western defense system, the president had shifted virtually all his arms purchases to American sources. By 1980 AWACS planes of the United States air force were flying missions out of Qena, 280 miles south of Cairo, and in April of that year Qena was a staging base for the abortive rescue attempt of American hostages in Tehran. In the summer, another base, Cairo West, was put at the disposal of the Americans. In November, 1,400 American troops, accompanied by tactical fighter-bombers, joined the Egyptian army in desert training exercises. Later Sadat informally offered Washington access to yet a third air base, Ras Banas. By then, however, Egyptians were asking what their president had gotten for this American connection. His talks with Reagan in August 1981 had been unproductive. Far from exerting pressure on Israel, or agreeing to enter a dialogue with the PLO, Reagan had expounded on the dangers of Soviet infiltration, and the necessity of a Middle Eastern "strategic consensus" in which Israel and Saudi Arabia would occupy key positions.

From the Israelis, meanwhile, even less was forthcoming. Superficially, relations between the two nations were correct. In October 1980 Israel's President Yitzchak Navon paid a highly successful visit to Cairo. Graciously received, Navon in turn charmed the Egyptians with his complimentary and conciliatory speech, delivered in fluent Arabic before the People's Assembly. Other exchange visits took place on the ministerial level. In December 1980, Egyptian Prime Minister Mustafa Khalil and Acting Foreign Minister Butros-Ghali were honored guests at Israel's Labor party convention in Tel Aviv. By 1981 Israeli tourists were arriving in Egypt at the rate of several hundred a day. Nevertheless, in the progress that mattered, that is, a fundamental accord on Palestinian autonomy, there was no movement whatever. On June 4, 1981, Sadat and Begin held a one-day summit meeting at Sharm es-Sheikh. There both men essentially marked time by denouncing the Syrian presence in Lebanon and reaffirming their commitment to the Camp David formula.

Only three days later, however, Israel carried out its bombardment of the Iraqi nuclear reactor. The timing of the attack was mortifying to Sadat. Although the Egyptian president condemned the air strike before his People's Assembly as an "act of international aggression," he failed to dispel suspicions that he may tacitly have approved the raid. Those misgivings were enhanced by Sadat's unwillingness to go beyond verbal reproaches. Instead, fearful of jeopardizing Israel's scheduled departure from Sinai in April 1982, he negotiated an agreement with Begin on a 3,000-man multinational force for the anticipated zones of evacuation, and the understanding was initialed six weeks later in an Egyptian-Israeli ceremony in London. Unquestionably, the intervening Israeli election was a grave blow for Sadat. But, even then, intent upon salvaging the peace process, the Egyptian leader met with Begin for still another summit conference in Alexandria on August 26, 1981. There the two men agreed that the suspended Palestine autonomy talks would be resumed the following month, that trade, cultural, and tourist exchanges would be expanded. It was a poignant, all but forlorn effort. Sadat must have grasped that Begin's reelection had doomed any chance of meaningful self-government in the West Bank. The Egyptian leader was playing for time, loath to admit publicly that his daring initiative for Middle Eastern peace had foundered.

Few in his own country shared expectations of early "peace and prosperity" any longer. But if cynicism fueled growing unrest, Sadat was equally determined to preempt national criticism by throttling it. At his orders, a law was railroaded through the People's Assembly in April 1980 to protect Egypt from writings or statements capable of "generating envy and hatred, or threatening national unity and social peace." The public prosecutor was empowered to bar offenders from public life, to condemn them to "internal exile," to prohibit them from departing the country. The assembly also was stampeded into proclaiming the Sharia—Islamic law—as the "source of all legislation." This latter was a particularly desperate innovation. By the early 1980s, Islamic fundamentalism was gaining renewed prestige and momentum in Egypt, as elsewhere in the Moslem world. In fact, it was Sadat himself who had let this djinn out of the bottle as early as 1973, with his personal decision to mobilize religion as a conservative ally against residual Nasserist socialism. Thus, new mosques began going up everywhere throughout the country, endowed either by the government or by wealthy individuals. Religious traditionalism was encouraged in the universities.

Sadat's effort to drown criticism in a tidal wave of orthodoxy misfired. The qadis despised his materialism and pro-Westernism. They sermonized against his government's corruption and the "betrayal" of Palestine, against its "alliance" with the Americans and the Israelis. Recorded on cassettes, the denunciations sold by the thousands. As it happened, one of the many who listened was an army lieutenant, Khaled Ahmad Shawqi al-Islambouli, age twenty-four. He belonged to a secret group of religious fundamentalists, and his brother was among those who had been arrested in an Islamic demonstration in 1979. For almost a year, Islambouli had been aware that the fundamentalist underground had condemned Sadat to death. Then, in September 1981, the young lieutenant was informed that he would be participating in the scheduled military parade of October 6, in the Cairo suburb of Nasser City, to commemorate the 1973 war. It was a unique opportunity to strike at the president. Islambouli set about studying the route of the intended parade, the location of the bleachers, Sadat's assigned place among the watching dignitaries. At the same time, he recruited three other soldiers for his plot, and arranged to get them into the parade area the day of the ceremony. When finally the passby of artillery columns began that afternoon of October 6, Islambouli's vehicle was nearest the reviewing stand, only 40 yards from the spectators. Hereupon he drew his pistol and ordered his unsuspecting driver to halt.

It was 1:00 p.m. Sadat was enthroned on the ornate bronze dais. At his right sat Vice-President Hosni Mubarak, at his left Defense Minister Abd al-Halim Abu-Ghazala, amid a galaxy of the nation's most important person- alities. Security was elaborate. The president and his entourage were sur- rounded by eight personal bodyguards, by republican guards stationed behind and beside the dais. Minutes earlier, the first of the mounted artillery cais- sons had rolled past, followed by a breathtaking display of aerial maneuvers. With all heads craning upward, no one noticed as one of the artillery vehi- cles braked to a stop near the reviewing stand. It was then that Islambouli leaped out and hurled a grenade at the bleachers. In the confusion, another conspirator, Hussein Abbas Muhammad, stood up in his truck and began firing a Kalashnikov automatic rifle point-blank at Sadat. Muhammad's first shot hit the president in the neck. A third conspirator, Atta Tayel Hemeida Reheil, threw another grenade. Surprise was so complete that for at least thirty seconds there was no effective reaction. Islambouli meanwhile was firing round after round into Sadat. By the time the shooting eased and three of the conspirators were seized, Sadat was dead and many others were slain and wounded. Miraculously, Vice-President Mubarak remained unscathed. For the first time, the people of Egypt had murdered their pharaoh.

When news of the assassination was released later in the day, 43 million citizens continued with the observance of Id al-Adha, the Feast of the Sac- rifice, as if nothing had happened. On the day of the funeral, half a week later, the streets of Cairo remained unusually empty. It was a bleak and furtive ceremony. In fact, it was almost exclusively a Western occasion. From the United States, former Presidents Carter, Ford, and Nixon attended, to- gether with Secretary of State Haig and former Secretary of State Kissinger. Britain sent Prince Charles, former Prime Minister James Callahan, and Foreign Secretary Lord Carrington. From France came President Mitter- and, from Germany Chancellor Helmut Schmidt. And from Israel came Be- gin, intent on paying respect to his "friend" and "partner in peace." It was significant, however, that only one of the members of the Arab League was represented by a head of state, Jaafar al-Numeiri of Sudan. Only two others, Oman and Somalia, sent representatives at all. On the day of the shooting, Radio Damascus exulted: "The traitor is dead. . . . It is a victory. . . . Our comrades in the Egyptian army have avenged us." In Lebanon, people danced in the city squares. In Libya, numerous automobile accidents occurred as drivers careened about the streets, trumpeting their joy.

In Egypt, meanwhile, within minutes of the assassination, continuity was the watchword. Vice-President Hosni Mubarak stepped more swiftly and smoothly into Sadat's shoes than Sadat had into Nasser's. The only candidate in the subsequent presidential referendum, he was awarded a 98.45 percent vote. Afterward, Mubarak pledged not to "budge an inch" from his predecessor's course. Indeed, he reserved his first interview with the correspondent of the Israeli newspaper *Ma'ariv*, and emphasized again his unswerving commitment to peace. Yet, grave uneasiness remained, both in Washington and Jerusalem. Deprived of their most loyal friend in the Arab world, the Americans seemed at pains to prove that Mubarak was their man, too. The United States, declared Reagan, would defend Egypt not only against invasion but also against "internal subversion from external sources." The Sixth Fleet was put on alert, and two AWACS planes were dispatched to patrol the Libyan frontier. Joint Egyptian-American maneuvers, already scheduled for November, would be expanded—to a scale that drew outraged Soviet protests. In Israel, finally, the Begin government welcomed Mubarak's assurances, and expressed its confidence that the new Egyptian president was firmly in command. But it was hardly a secret that, if he were not, the impending terminal withdrawal from Sinai could be postponed indefinitely, and the entire "peace process" then would face collapse. Unspoken in the hearts of the Israeli people, moreover, was the trepidation that this "process" might already have expired with Sadat himself.

CHAPTER VII

"BEGIN II":

THE DJINN OUT OF
THE BOTTLE

A PAYOFF TO THE RELIGIONISTS

Following the tight election of 1981, Begin was obliged to engage in lengthy and intricate negotiations to achieve his precarious one-seat Knesset majority. As always, the Orthodox were positioned to extract the largest concessions. During the years of Labor incumbency, their price had been relatively bearable. The ministry of religions, with its vast bureaucracy, had been consigned as a matter of routine to the chairman of the National Religious party. From the beginning of independence, it had been understood that religious holidays and kosher food would be maintained in the public institutions and affairs of the nation, that matters of "personal status" would be left to the rabbinate. Except for these provisions, however, the Orthodox had never seriously impinged on the delicate pre-1948 equilibrium between "church and state." They had failed repeatedly to ground the national airline on the Sabbath, to win recognition of Torah studies as a central feature of the state school curriculum, to reject as Jews (under the Law of Return) former Gentiles converted abroad by non-Orthodox rabbis. Yet, from Begin, the religionists had expected from the outset to get more. In 1981, they expected to get much more.

Thirty or even twenty years earlier, that expectation would have been unrealistic. The Revisionist leadership had not evinced much compassion for Orthodoxy, a posture Jabotinsky identified with the Jewish fatalism of Eastern Europe, with the "loathsome" ghetto Jew. Begin generally had shared his mentor's bias. Then, after 1977, entering into a coalition deal with the religionists, Begin as prime minister chose to base his accommodation less on political expediency (as had Ben-Gurion and other Labor premiers) than on an inner vision of authentic piety. The approach had been cultivated during his years in the political wilderness. Accordingly, by the time he came to power, his frequent declamations from Scripture, his respectful visits to influential rabbis, his donning of a skullcap at the drop of a biblical quotation, had become second nature. By then, too, the spokesmen of the Orthodox parties declared themselves satisfied with Begin's "positive attitude" to religious tradition. As it happened, that satisfaction was powerfully enhanced in 1977 by the unprecedented award to the NRP of three full ministries: religions, interior, and education and culture. Even the ultra-Orthodox Agudists now declared their willingness to be included in the government coalition—although they could not yet bring themselves to join the cabinet.

Throughout the first Begin government, the religionists were not quite prepared to challenge the "church-state" status quo directly. Likud, they knew, could have sustained its coalition majority without them. Anyway, the NRP historically had not been identified with rabid extremism. Their one perceptible shift of stance in recent years was in foreign policy, specifically on the issue of "Judea and Samaria." Under pressure from the Gush Emunim, we recall, the "integrity of the Land of Israel" became a central preoccupation for the religionists, and on that issue they shared responsibility for the paralysis of the later autonomy negotiations. Yet it was only a matter of time before the NRP would redirect its attention and efforts to the domestic sphere; and in 1981 that time appeared to have come. Conceivably, the party's leadership no longer enjoyed leeway for procrastination. In the recent election they watched their Knesset seats drain away to a dissident Moroccan faction, Tami, headed by the former minister of religions, Aharon Abuhatzeira. Still worse, the fundamentalist Aguda was threatening to preempt the spokesmanship of the Orthodox cause. Under the circumstances, the NRP's leaders sensed then that their only hope of ongoing credibility within the Orthodox bloc was to match the Agudists step by step in the militance of their demands.

It was a realistic evaluation. If the Agudists represented a smaller element among the religious parties, holding a mere 5 Knesset seats, they were also regarded as ideologically "purer" than the NRP. Their hard-core following was concentrated in Jerusalem and B'nai Brak, and their confederative leadership rested in the "Council of Torah Sages," a collegium of seventeen rabbis who were venerated as lineal descendants of ancient and renowned Jewish dynasties in Eastern Europe. Until the birth of Israel, the Agudists had rejected association with "profane, secular" Zionism altogether. Even after 1948, their contact with the Israeli authorities was as minimal as their need for school subsidies and army exemptions could make it. Their pro forma demands for a theocratic state were routinely shot down. With the election of Menachem Begin in 1977, however, the status quo suddenly appeared vulnerable, and the Agudists joined the government coalition. And with the prime minister's deteriorating majority in his first term, his hairbreadth reelection in 1981, they sensed that their moment of vindication had arrived. In ensuing weeks, their widely publicized "exploratory negotiations" with Labor exerted just enough impact on Likud to secure the payoff they wanted. Thus, in his postelection agreement with other coalition partners, Begin was obliged to devote fifty of the document's eighty-three clauses to commitments on religious matters. Many of these promises related to such familiar issues as the incorporation of additional halachah—Orthodox, rabbinical law—into the nation's jurisprudence, allocation of supplemental funds to religious schools, authorization of wider military exemptions for Orthodox girls. But also included for the first time was Begin's undertaking to ground El Al aircraft on Sabbaths and other Jewish holidays, and to "make every effort to assemble a Knesset majority" for amending the Law of Return.

This latter was to the Agudists—and now to the NRP—a litmus test of Begin's reliability. In 1970 the supreme court had ruled that a Jewish husband and his Gentile wife could register their children as Jews by nationality, rather than by religion. The decision touched off a storm of protest among the Orthodox. Ultimately they secured partial relief in the form of a Knesset amendment to the Law of Return. Under the new legislation, halachah thereafter would set the "official" definition of Jewishness. Inasmuch as halachah traditionally had defined a Jew as one either born to a Jewish mother or formally converted to Judaism, the new rubric was not without certain advantages even to secularists. It enabled the government to deny immigration to such questionable elements as the Black Hebrews, a group of American Negroes who claimed the right to settle in Israel under the Law of Return. Yet the official definition eventually opened out more problems than it solved, for it rejected the Jewishness not only of Gentile spouses (many of them, lately, from the Soviet Union), but of their children. Although these intermarried families were warmly greeted in Israel and extended full citizenship, their children were precluded by the rabbinical authorities from marriage with other, "authentic," Jews, or eventually from the right of burial in Jewish cemeteries. The one remaining alternative for the wives and children was to undergo conversion. But what sort of conversion? The original Knesset amendment did not define this process. In Israel itself, plainly, conversion would be performed only by "officially" authorized—that is, Orthodox—rabbis. What of conversions performed abroad, however, under Conservative or Reform auspices? If these now were disallowed by a new Knesset amendment, Israel in effect would be repudiating the legitimacy of Conservative or Reform Judaism in the Diaspora. Such an amendment would represent a gratuitous affront to millions of American and other Jews whose support for Israel was desperately needed.

The Agudists and their NRP partners demanded precisely this amendment. They had always demanded it. Earlier Labor governments had procrastinated, evaded, appointed committees to seek a "compromise." But with Begin's accession in 1977, and his shaky reelection four years later, the Agudists were well seized of their bargaining strength. They were unwilling to accept compromise any longer. The NRP also fell in line. It was under this joint Orthodox leverage, then, that Begin's 1981 coalition document finally included the commitment to "make every effort" to achieve a Knesset majority in favor of the amendment. That majority did not officially materialize, as it happened. Labor rejected the very notion of changing the Law of Return. The Liberal faction of the Likud bloc was not obliged to vote with Cherut on this issue, and it did not. Even so, the imprimatur of legislation hardly was necessary by then. The Orthodox had achieved the "atmospherics" they needed from a dependent and permissive Likud cabinet.

Thus, in practical fact, if not in law, the ministry of religions chose simply not to recognize as Jews those spouses or children who had been converted abroad under non-Orthodox auspices. For years, the ministry had been keeping "black lists" (provided by spies and informers overseas or in Israel) of some 10,000 immigrants whose Judaism was "suspect." While these people could not legally be denied entrance with the other, "authentically" Jewish, members of their families, their lives could be infinitely bedeviled in Israel. Rabbis would not perform marriage ceremonies for them, in some instances would not bury them. Indeed, the impact of this rabbinical inquisition was so chilling for thousands of Soviet Jewish immigrants that it was a major factor in their later defection to the West. Liberal and moderate protests on their behalf usually were unavailing. As Shulamit Aloni, Israel's leading civil rights activist, observed with some anguish, in a published interview of July 1979:

> The fact that a man from Holland who did not convert, married to a Jewish Israeli and the father of a girl, was rejected by the Jewish Agency and was not given the right to purchase a plot in a moshav [cooperative farm] if he does not convert— this no longer makes an impression on anyone. The denials of rights to housing, insurance, advancement in the army, marriage and burial, for reasons of religion and origin are now accepted norms. No one has the energy to get aroused by these subjects any longer.

The religionists' offensive did not spare Israel's security or economy. Under Labor, Orthodox girls had been allowed to claim exemption from military service, provided they underwent strict investigation of their traditionalist bona fides, and contributed an equivalent period of national service in the civilian sector. Now, as early as Begin's first government, the Agudists chivied the NRP into a joint effort to widen the category of exemptions. They succeeded. Under the 1977 coalition agreement, girls invoking the religious loophole were obliged to make only a token appearance before a military review committee. The numbers of exemptees rose from 21 percent in 1976 to 32 percent in 1980. Few of them ever performed alternative service.

In this earlier period, too, the Agudists pressed for an end to El Al flights on the Sabbath. The NRP for its part would have accepted the status quo, under which El Al simply did not publish a schedule of its Saturday flights— which in any case did not go to the United States, with its large and powerful Orthodox constituency. It was known that the company already was operating at a financial loss, due to charter competition and a series of crippling strikes. Indeed, in its first term, the Begin cabinet adopted a hard line against the various unions, even threatening to sell off El Al to private bidders rather than capitulate to intolerable labor demands. The notion of bowing to the religionists would have been equally unthinkable. But here, too, after the nip-and-tuck 1981 election, with the NRP and Aguda operating in close tandem, Begin felt obliged to yield. He agreed to end Sabbath flights. Ironically, it was El Al's labor force this time that protested the reduced schedule. The cancellation of Sabbath flights not only would ground the airline on Saturdays, they pointed out, but would require stopovers that in effect also would halt flights on Fridays. The reduction of the line's competitiveness was certain to produce labor cutbacks. By early 1983, nevertheless, when a temporary court restraining order was lifted, El Al's Saturday flights to Europe were indeed terminated. The company's losses mounted again, at a period of grave economic duress for Israel.

There was hardly an area of the nation's life in which the Orthodox did not move relentlessly on the offensive. Again, as early as Begin's first term, the Agudists launched a campaign to revise Israel's liberal abortion law. The Likud stalled for time, appointing committees and subcommittees to study the matter. But in the aftermath of the 1981 election, the Orthodox no longer could be put off, and the government capitulated. Within the year a new Knesset law ensured that "sociological considerations"—for example, economic hardship—could not be invoked for abortions performed in public hospitals. Although physicians were not barred from performing abortions in private clinics, poorer women had little access to this procedure.

A MOVE TO ORTHODOX VIOLENCE

In the final analysis, the Orthodox exacted their heaviest ransom in the form of patronage, subsidies, and fiscal and military exemptions. The principal conduit for this largess for many years had been the ministry of religions. Except for brief periods in 1958–61 and in the post-1973 Meir and Rabin governments, the NRP leadership had ruled this ministry as its private fief. Its bureaucracy was a vast source of employment for functionaries of the chief and local rabbinates, the religious and kosher-certification councils, the state religious educational system. Now, under Begin, the doors of the treasury were opened even wider for the Orthodox. In 1982 the ministry of religions won a budgetary increase of 390 percent in real terms, at a time when budgets for other ministries were frozen or reduced. "Project Renewal" (Chapter II) allocated a disproportionate share of its social welfare funds to religious neighborhoods. Priority was given to the construction of Aguda kindergartens, schools, and yeshivot. From 1981 on, yeshiva students received government stipends and were placed in the lowest tax bracket, regardless of their actual income. In earlier years, only full-time yeshiva students were exempted from military service. After the 1981 elections, teachers at yeshivot, even teachers of religious subjects in the Aguda school system, shared in this exemption. So now did men over thirty giving Torah lessons in state schools. The agglomeration of these benefits became a national scandal. Editorial criticism was scathing. The government brazened it out.

A more vivid example of Orthodox muscle was the NRP takeover of the ministry of education and culture—a plum that went to the religionists for the first time after Begin's initial victory in 1977. Its new minister was Zevulon Hammer. In his thirties, a graduate of the B'nai Akiva youth movement and of the (Orthodox) Bar-Ilan University, Hammer was a founder both of the Gush Emunim and of his party's Young Guard faction (Chapter I). It was ironic, even astonishing, that under the aegis of this cultured, youthful man, the quality of Israel's education should have deteriorated more alarmingly than at any other period in the nation's history. To begin with, NRP loyalists monopolized every key position in the education ministry, from curriculum planning to teacher training. Thereafter, particular solicitude was devoted not only to state religious schools, encompassing 22 percent of the nation's children, but to Agudist schools, including barely 6 percent on their rolls. Few Israelis were profoundly shocked when the state religious schools' share of the ministry budget climbed by a third between 1977 and 1981. But the Agudist system had always been submarginal, its teachers essentially rabbis or *melamdim* (religious instructors), most of them unfamiliar with European culture or languages, or even with the salient facts of European and Israeli history. Under previous governments, Agudist institutions had offered a bare minimum of nonreligious courses, and thereby managed to claim and receive a trickle of state funds. Now, after 1977, and even more after 1981, the Agudists shared a cornucopia of subsidies undreamed of in all their earlier years. Indeed, they were granted 12 percent of the government's total allotments for education and culture, more than twice their share in the nation's school system.

Yet, even within the state school system, Hammer and his colleagues doubled the teaching time allocated for purely Jewish subjects. The impact of this reorientation more than offset Likud's considerable accomplishment in introducing three additional years of free high school; for attrition in the arts and sciences curriculum not only altered the ideology of Israeli education but warped its quality. So did the reduction of school hours by 15 percent in junior high school and 4 percent in primary schools. The diminution of teaching time in turn was itself a consequence of the government's lavish diversion of educational funds to Agudist and other religious institutions, and of even costlier tax and military exemptions for seminary students and religious teachers. In July 1984, Ora Namir, a Labor member and chairman of the Knesset education committee, reported an alarming drop in the three Rs at the primary school level, in the quality of technical education at the high school level. Israel in 1984 could list only 15 students in school per 1,000 inhabitants. The ratio in Jordan was 17 per 1,000; among the Palestinians, 20 per 1,000. This was a matter of national survival, Mrs. Namir warned. In the previous three years, at a time when military equipment was becoming far more complex and sophisticated, 17 percent of new Israeli army recruits could neither read nor write. They were eleven-year-olds, and younger, when Likud had come to power.

The atrophy of Israeli higher education was hardly less shocking. As early as 1975, budgetary cuts had forced the universities to increase their tuition fees sharply. Those cuts were far more severe under Likud. Teaching salaries in real terms fell by 30 percent between 1977 and 1984. With the freeze on new appointments, heavier teaching burdens deprived younger lecturers of research time. Scientific equipment and library books were in shorter supply. Unquestionably, the crisis was the result, first and foremost, of heavier defense expenditures. But it was gravely exacerbated by Likud's new priorities in education. Indeed, the shift may have reflected not simply a tactical accommodation with the Orthodox, but the Cherut party's uneasy relationship altogether with the nation's intelligentsia—most of whom were known to be Labor or DMC in their political sympathies. One feature of that latent anti-intellectualism was the growth of intolerance among high school students toward unpopular dovish opinions and non-Jewish minorities. The climate was fostered under Hammer's stewardship. "There is an alarming spread of hatred for Arabs and a deepening belief that might makes right," declared Ora Namir, as she ended her report. "Under this regime, our schools have been thrown wide open to chauvinist and anti-democratic influences."

Not the least of those influences was mob intimidation. Here Israel's religious zealots were in the vanguard. Their belligerence surfaced at multiple levels. One was the nation's cultural life, even in its most popular form, archaeological exploration. Nowhere else in the world, conceivably, did there exist as widely diffused a passion for unearthing the relics of the past. Evidence of Jewish provenance in this land was Israel's ultimate legitimization, after all. Yet the excavations occasionally disinterred not only ancient detritus but ancient Jewish gravesites. Here existed still another area of potential confrontation with the religionists. With their memories of Jewish grave desecration by European and Moslem vandals, the nation's archaeologists shared with all Israelis a respect for the dead. They simply reserved to their own trained judgment the actual identification of ancient burial sites.

Then, in August 1981, as archaeological scholars resumed an excavation of the Temple Mount area in East Jerusalem, Israel's Ashkenazic and Sephardic chief rabbis unexpectedly issued a solemn protest. The government must declare the entire southern slope of the Temple Mount a collective Jewish cemetery, they insisted, and order the archaeologists out forthwith. In fact, the demand was far-fetched even for the NRP minister of the interior, Yosef Burg, and for his colleague at the ministry of education and culture, Zevulon Hammer. They rejected it. Whereupon several thousand Agudists hurled themselves into clamorous demonstrations at the site of the dig. On September 18, in a startling display of medievalism, the ecclesiastical "court" of the ultrafundamentalist *charedi*—"God-fearing"—community performed a cabalistic black-candle ceremony, heaping upon the heads of "certain archaeologists" all the curses from Moses to the present. At that point, Hammer caved in. Invoking his ministry's authority over the department of antiquities, he temporarily withdrew the archaeologists' digging license. Legal advice was needed, he explained, on the question of final authority to determine the location of Jewish burial grounds.

Several weeks later, the attorney general provided that advice, asserting that only the education ministry's department of antiquities, not the chief rabbinate, had authority to identify grave sites. Even the possible existence of an ancient cemetery was insufficient reason to halt an excavation, stated the attorney general, if the possibility existed for discoveries of "major scientific and national importance." Immediately, then, the archaeologists declared that they would soon resume the dig. The rabbinate in turn vowed to fight the decision to the end. Several months later the excavation did indeed resume, and so did the demonstrations, almost on a daily basis. Hundreds of religionists lined the approach route to the Temple Mount, shouting and gesticulating. Police cordons were necessary to protect members of the expedition from physical violence—until at last the dig was completed officially in the summer of 1985 (the site is now an important tourist attraction).

Altogether, episodes of intimidation were mounting in Jerusalem. Except for brief episodes in the 1950s, the development in fact was a comparatively recent one. Under Mayor Kollek's lengthy tenure, a certain modus vivendi had been achieved between the ultra-Orthodox and the rest of the city's population. But now, in the Begin era, the climate shifted drastically. Even private citizens were not spared this time. Non-Orthodox residents of Mea Sh'arim periodically had their apartments vandalized, their children harassed. Incidents of religious fanaticism originating elsewhere burgeoned out almost uncontrollably in the Holy City. Thus, in the Orthodox community of B'nai Brak, in July 1977, religionists strung a chain across a public road in an effort to block Sabbath traffic; and a young driver, striking the chain, was killed. When police sought to arrest the guilty parties, 10,000 zealots mounted a protest demonstration. Soon, however, "preservation of the Sabbath" became the rationale for even wider-scale disruptions in Jerusalem.

Their origins actually could be traced to well before the 1967 unification of the city. Mea Sh'arim and others of Jerusalem's religious neighborhoods had long been overflowing with people, with new children. Thus, in 1964, the Klausenberg Rebbe, leader of a right-wing Chasidic sect, arrived in Jerusalem to establish a new, uncrowded, Orthodox neighborhood that would perpetuate the name—Sanz—of his ancestral Hungarian community. The site he finally picked was located on the hilly northern edge of the city, between Romema and the entrance to Jerusalem. Although "Kiryat Sanz" was launched in 1965 with 400 devout families, other religious subcultures later attached themselves to this nucleus, and within a decade there sprang up in and near Kiryat Sanz a kind of Orthodox Bible Belt. Within a 2-mile stretch lived some 2,500 families, a young, dynamic, exceptionally militant population, supporting a dense network of orphanages, yeshivot, and other Orthodox institutions.

In these same post-1967 years, as it happened, the Kiryat Sanz religionists also could see a large non-Orthodox community, Ramot, growing up some 2 miles to the north. At first the development seemed to present no threat to them. But in 1976 tractors appeared at the foot of their hills, and the residents discovered to their dismay that a six-lane highway to Ramot was to be constructed below their neighborhoods. Immediately and reflexively, the Agudist population launched into protests, warning city hall that heavy traffic was likely to move down this highway on the Sabbath. Their anguish registered on Teddy Kollek. An astute politician, the mayor always had shown himself sensitive to the Sabbath peace of Jerusalem's Orthodox inhabitants. Indeed, over the years, he had closed Sabbath traffic on approximately twenty streets, persuading non-Orthodox residents to accept these measures for the sake of living in the Holy City. But Kollek also knew that the prospective Ramot road could not legally be closed, that alternate stretches around Kiryat Sanz were rejected as unsafe by his engineers. Working them-

selves into near hysteria, meanwhile, the Kiryat Sanz residents were drawing Orthodox support from throughout the country. Along the completed
stretches of the Ramot thoroughfare, their partisans hurled stones and brick-
bats, damaging scores of passing vehicles, endangering their drivers. When
the police failed to act decisively, the—non-Orthodox—Ramot dwellers organized their own vigilante groups to open the road, even threatened to
break into Kiryat Sanz with heavy trucks or dogs. The incidents of threat
and counterthreat, intimidation and retaliation, mounted in scope and violence.

As in earlier confrontations in the 1950s between the fundamentalists and
their opponents, this one soon extended beyond the frontiers of Israel. In
March 1981, a crowd of 15,000 Agudists gathered on the Sabbath to protest
the operation of the Ramot road. When they began stoning a passing tourist
bus, the police finally intervened. There were clubbings and arrests. Several
weeks later, then, in New York, 9,000 Satmar Chasidim staged a three-hour
protest before the Israeli consulate. Among their signs and placards were
the familiar slogans of "Zionism—Enemy of the Jewish People," "Free the
Religious Hostages," "Nazi Germany 1939—Zionist Israel 1981." Nearly 1,000
policemen were needed to keep order, but the crowd succeeded in disrupt-
ing rush-hour traffic in mid-town Manhattan even after the barricades were
removed. Ironically, this episode occurred several days after Mayor Kollek
managed to negotiate a "compromise" of sorts in Jerusalem itself. Under the
terms of the agreement, inhabitants of Ramot would be encouraged to drive
on a narrow bypass road during Sabbaths. In turn, a new permanent access
road to Ramot eventually would be built—at great expense and strategic
inconvenience—to circumvent the Bible Belt entirely. Even then, not all
the religionists were appeased. Many continued to protest, threatening and
stoning occasional errant vehicles.

The glowering hatreds, the intermittent violence, far transcended a mere
Kulturkampf. By the early 1980s, the fundamentalist assault on Israel's civic
values approached the threshold of physical strife. But, in fact, those civic
values had never been deeply rooted. The concept of minority rights had
failed more than superficially to penetrate the nation's consciousness. Lacking a system of governmental checks and balances, the Israeli people tended
to assume that democracy was a matter simply of elections conducted on
fixed days and of majority decisions passed by the Knesset. What other free
people, asked Shulamit Aloni, leader of the Citizens Rights party, would
have accepted with such apparent passivity an Orthodox stranglehold on
freedom of religion, freedom to marry, equality before the law? In order to
maintain the "purity of the Jewish people," the "unity of the Jewish people,"
a widespread detection system had evolved in the ministries of interior and
religion, to discover "who is a Jew" and who had alien blood flowing in his
veins.

> Step by step [warned Mrs. Aloni] . . . Israel is being converted from a demo-
> cratic, humanistic state, to an Orthodox, clerical community. . . . Laws with rac-
> ist and discriminatory overtones are beginning to find their way into law books,
> and . . . to become a sort of normative system.

In his lengthy essay, *Between Right and Right*, A. B. Yehoshua, one of Israel's most respected writers, traced the source of this envenomed ideological conflict to the asperities of Jewish history. The Jews had lived in exile for 2,000 years not out of compulsion, he argued, but out of choice. All the explanations they adduced for their Diaspora existence were unconscious rationalization. Economically, surely physically, Jews would have been better off in Palestine. If they were inhibited from return, the cause lay in a deep-rooted folk memory: of the internecine zealotry that had devoured them in antiquity, on the soil of ancient Israel. Terrified of confronting that denouement again in the Holy Land, of reviving the "central and ceaseless struggle between the religious and national systems," Jews by and large had preferred to remain in exile. So long as they were powerless to impose their will on each other, the issue of religious or national domination could be postponed or avoided.

It was only by the twentieth century, Yehoshua continued, that the "neurotic and unnatural" solution to Jewish civil strife, namely, exile and weakness in an incorrigibly hostile Gentile world, proved too great a burden of suffering. The Holocaust was the climax of that agony. The decisive step was taken, then, a return to statehood. But with that return, the old religious-secular confrontation inevitably surfaced again in all its malevolence.

> Today, for example, the Lubavitcher Rebbe living in New York can only ask and try to persuade Jews in New York not to travel on the Sabbath, to send their children to Jewish schools, to eat kosher meat, etc. But should he come to Israel he would be able, and it would be his religious duty, to compel Jews to follow a religious life. The national framework makes coercion obligatory because it makes it possible.

The Orthodox element and its sense of mission were a time bomb for Israel, Yehoshua warned. He sensed a dim, subliminal awareness among his people that, if peace with the Arabs came, an internal "war of the Jews" would erupt again with demonic fury. Indeed, the prefigurations of violence were already evident. "The national component [as distinguished from the religious] in Jewishness must now be strengthened for the sake of survival," he pleaded. Yehoshua's was a cri de coeur that was echoed by tens of thousands of other sentient Israelis, no less aware than he that time was running out.

BEGIN'S "COMPENSATION" FOR WITHDRAWAL

On October 9, 1981, the day of Sadat's funeral, Begin and Hosni Mubarak held a forty-minute private conversation. Each pledged to the other his enduring commitment to peace. Begin promised that all meetings on Palestine autonomy and peace normalization would be held as scheduled, that Israel would evacuate its forces from Sinai by the deadline of April 26, 1982. Several weeks later, Shimon Peres headed a Labor delegation to Egypt to express his party's condolences, and in mid-November Israel's President Yitzchak Navon paid a one-day visit to lay a wreath on Sadat's grave. During the rest of the year, Israeli political and intelligence analysts carefully monitored Egyptian policy statements for intimations of a possible shift in position. There appeared to be none. Notwithstanding the paralysis of the autonomy negotiations, Begin and Mubarak seemed determined to maintain the impression of progress. The former wanted no pressure for a resumption of the Geneva Conference; the latter, no postponement of the scheduled Israeli withdrawal.

Limited segments of the Sinai had of course been evacuated following Kissinger's shuttle diplomacy in 1974–75. Subsequently, an impressive two-thirds of the peninsula were handed back in the immediate aftermath of the Egyptian-Israeli treaty. By late winter of 1981, the United States and a group of West European nations had agreed to provide the minimal 3,000 troops required for the multilateral Sinai force. This task accomplished, Secretary of State Haig emphasized that the Israelis now would have to fulfill their obligation of total withdrawal or risk a major crisis with Washington. In Cairo, too, Mubarak waited and watched with growing concern. There was reason for this anxiety. The vehemence of Israel's 5,000 Yamit settlers, their protests and demonstrations, appeared on Egyptian television. Nor were these *colons* protesting evacuation for reasons of ideology alone. Since 1967, they had transformed their corner of the Sinai Peninsula into a green and prosperous agricultural adjunct of Israel itself. Now they regarded the government's initial compensation offer of $800 million as inadequate. Money surely was not the purpose that had brought the settlers to Sinai. No one appreciated the fact better than did Begin. Nevertheless, a sweetened pot conceivably might be inducement for them to leave. The cabinet's ultimate strategy, then, was to raise its offer by 20 percent—to nearly $1 billion. Even as compensation for land, homes, and chattels, for years of pioneering self-sacrifice, the sum was ludicrously high. Observers noted that an average factory worker would have to save all his wages for seventy years to receive the amount to be paid a single Sinai evacuee. And yet, for members of the Gush Emunim, the payoff was irrelevant. These were the zealots who appeared likeliest to resist evacuation to the end.

In a surprise move on February 26, 1982, therefore, Ariel Sharon ordered the army to seal off the northern Sinai roads. The Gush Emunim had scheduled a massive demonstration of thousands of supporters in Yamit two days later, and the defense minister intended to preempt that effort. Outraged reaction was not long in coming. There were ugly scenes at army checkpoints. Settlers wearing yellow Stars of David baited soldiers with cries of "Nazis!" "Kapos!" and "Yamit will not be judenrein!" Later, nearly 500 grenades were discovered to have been smuggled into Sinai. In ensuing weeks, the army used force to evict hundreds of settlers, most of them engaging in passive resistance. Ultimately, some 2,000 truckloads of dismantled equipment were transported back to Israel, and the largest numbers of Yamit residents departed in a final mass exodus on April 1. Only a small core of diehards remained behind. Most of these took up positions on rooftops. When the troops (unarmed) climbed ladders to the roofs, they were met with burning tires, with fusillades of bricks and other heavy objects. The last of the settlers were not removed until April 23. A series of blasts then rocked the town of Yamit, and only the synagogue was left standing. Two days later, at midnight—fifteen years after the Six-Day War—the last of the Sinai Peninsula reverted to Egyptian hands.

At Sharm es-Sheikh, on April 27, Begin and Mubarak attended a moving ceremony in Israel's former military headquarters at Ofira, which had been left intact for the occasion. Once again both men committed their governments to "peace forever." The following day, the Egyptian president telephoned Begin from Cairo to repeat the pledge. Within the Sinai, only one, seemingly minuscule, issue remained unresolved. This was the fate of Taba, a 600-square-yard wedge of land just south of Eilat. Both sides invoked frontier maps to support their claims to the strip. When direct negotiations failed, the issue was reserved for a conciliation committee, and if necessary even for arbitration. A number of the treaty's other provisions meanwhile went ahead as scheduled. An Egyptian-Israeli checkpost was opened below Taba, as well as a Cairo-Tel Aviv bus route on April 29. Egyptian tourism to Israel remained as meager as in the initial aftermath of the peace treaty. Yet several official delegations paid visits, among them agricultural and industrial missions, a handful of cabinet ministers, members of the ruling National Democratic party. Whatever their frustrations on the Palestine impasse, the Egyptians appreciated that the restored Sinai comprised over 90 percent of the territory occupied by Israel in the Six-Day War.

Tensions between Cairo and Jerusalem were not long in reviving. Mubarak had consented to an early summit meeting with Begin in Israel. The prime minister now declared it a matter of principle that Jerusalem be included in Mubarak's itinerary, as it had in Sadat's. Mubarak refused. In his eyes, the status of East Jerusalem remained a critical adversarial issue. It was that presidential refusal, not the scattering of Egyptian official visitors, that helped set the coldly formal tone of the unfolding relationship. Begin's contributions to the freeze were hardly negligible, of course. His unbudging stance on Palestine was only one of them. Others were the declaration of united Jerusalem as Israel's capital, the saber-rattling over Lebanon, the preemptive strike at the Iraqi nuclear reactor. These and additional displays of ideological and tactical muscularity doubtless were envisaged as compensation for the trauma of evacuating the last—Yamit—foothold in Sinai. Again, in a parallelism with de Gaulle's "grandeur" following the abandonment of Algeria, Begin neglected no speech, no interview, no ceremony, to invoke the mystique of Israeli-Jewish heroism and the historic unity of the Land of Israel. What moderating influences were left to restrain him? His government in 1982 was very different from the one that had taken power in 1977. Dayan, Weizman, Yadin, and Hurevitz were gone. The ruling quadrumvirate now comprised Begin himself, Defense Minister Ariel Sharon, Foreign Minister Yitzchak Shamir, and General Rafael Eytan, the military chief of staff. Committed hard-liners, they in turn were fully supported by the religionists, by the Cherut component of Likud, and by the vast, faceless subculture of poorer, underrecognized Israelis who resonated instinctively and gratefully to displays of national pride and machismo.

The new cabinet's bravura was typically evident on Israel's April 1982 independence day, in the immediate aftermath of the Yamit evacuation. Choosing as its theme "One Hundred Years of Jewish Settlement," the government announced that eleven new military outposts would be inaugurated to mark the anniversary, eight of them on the West Bank. Soon afterward, on May 11, some 200 of Israel's leading personalities, including the president, the prime minister, and other cabinet ministers, were flown to a desert hilltop near the Dead Sea. There they witnessed the reburial of twenty-five skeletons that had been unearthed by Professor Yigael Yadin in a nearby cave fourteen years earlier. Begin had proclaimed the remains to be those of the followers of Simeon Bar-Kochba, leader of the second-century revolt against the Romans. (Yadin himself declared the assertion unproved, and boycotted the ceremony.) The event included a military pageant, replete with drums and flags, and a special prayer by the army chief rabbi, castigating the "evil" Romans. At a time of reduced social welfare benefits, when tens of thousands of Israeli families were living below the poverty line, the entire bizarre spectacle set the treasury back nearly $1 million.

As early as the previous December, anticipating the moment of truth in Sinai, Begin had provided the nation with a rather more substantial offering. This one lay in the far north, on the Golan Heights. There Israel's position was entirely stable. The Syrian-Israeli disengagement agreement of May 1974 remained in force, with Damascus scrupulously fulfilling its part of the agreement. Every six months the UN observers' mandate was renewed with the consent of both governments. Quiet reigned along the disengagement line. Israel was free to settle the Golan as it wished, to reinforce and enlarge existing settlements, to establish new ones without hindrance. Unlike the West Bank or Sinai, the Golan attracted virtually no international attention. The necklace of collective and cooperative farms on this strategic highland was regarded as a legitimate Israeli defense line against an implacably hostile Syrian foe. No Western nation, not even Egypt in the Arab world, was applying pressure on Israel to withdraw from the Golan.

Even so, following Camp David, the 6,000 Israeli settlers in this crucial security zone remained uneasy about their future. During 1979 and 1980, they circulated a petition against a future Golan withdrawal, and more than 750,000 citizens eventually signed the document, including 70 Knesset members. Among the latter were the Labor leaders Peres, Rabin, Allon, and Bar-Lev. For these and other signatories, the security issue was decisive. The Golan's Moslem population in any case had fled to integral Syria, leaving behind only a residual community of Druze, many of whom actually favored annexation by Israel. In March 1981 the right-wing Techiya party exploited the antiwithdrawal sentiment by introducing a Knesset bill to extend Israeli sovereignty to the Golan. It was defeated, but quite narrowly.

Now, nine months later, on December 14, 1981, much to the astonishment of the Labor opposition, Begin appeared in the Knesset and submitted a bill for the application of "Israeli law, jurisdiction, and administration" to the Golan Heights. The prime minister cited a number of rationales for his initiative. One was "historic," namely, that the territory once, in the distant past (although for a period of less than three centuries), had been inhabited by Jews. More fundamental was the long record of Syrian sniping and terrorism along the northern frontier, a display of aggressiveness that had led to the outbreak of the Six-Day War itself, and to Israel's conquest of the Golan. In May 1974, under Kissinger's pressure, the Golda Meir government had agreed to return the entire—additional—area captured in the Yom Kippur War, as well as the district capital of al-Quneitra, which had been reoccupied briefly by the attacking Syrian forces in 1973. But since then, Begin argued, the Damascus government had chosen to "revert to its hostile and belligerent posture." It had forbidden the return of civilians to al-Quneitra, spurned all negotiations with Israel, and refused to accept UN Resolution 242, renouncing force. Worse yet, the Hafez al-Assad regime spearheaded the Arab rejectionist front against Israel, and provided military and political support to PLO terrorism against Jewish and Israeli targets in Europe. The prime minister now quoted before the Knesset a radio report of the previous evening, in which Assad allegedly had declared that he would "not recognize or even negotiate with Israel, even if the PLO does so."

Begin cited this declaration without checking its authenticity. In fact, the statement was refuted within the day by Syria's ambassador to the United Nations, who insisted that his government's view of a comprehensive Middle Eastern peace settlement did indeed leave room for Israel. Assad attached conditions for negotiating, but so had the late Sadat before his visit to Jerusalem. The Syrian explanation was qualified and involuted, but hardly more far-fetched than Begin's. Once again, the prime minister was adopting a compensatory tactic for the impending Sinai withdrawal. The manner, too, in which the Golan Law was passed offered insight into the fragility of Israeli democracy. Begin managed to steamroller the bill through the Knesset in a single day, December 14. The Knesset security and foreign affairs committee had waived the mandatory waiting period between readings, and the bill now passed by a vote of 63 to 21, with 8 Alignment members supporting it. Doubtless, the Knesset's transformation into a rubber stamp would have been impossible had there been an effective political opposition. But the Labor Alignment was caught by total surprise and in total disarray, even as on the Jerusalem issue.

The Golan Law was a violation of UN Resolution 242 and of the Camp David framework. The former had obligated Israel to negotiate a peace with Syria based on withdrawal from "territories" to secure and recognized boundaries, a commitment that was reaffirmed in the 1974 Israeli-Syrian disengagement agreement. At Camp David, Egypt and Israel confirmed that Resolution 242 was envisaged as a basis for peace "not only between Egypt and Israel, but also between Israel and each of its neighbors which is prepared to negotiate peace with Israel on this basis." In the event of negotiations with Syria—one of those neighbors—Israel would have been free to take any position on the Golan it liked, to claim the whole or part of it; but it was not entitled unilaterally to annex the plateau. The damage created by Begin's initiative was very serious, then. Without warning, the all-but-forgotten Golan Heights suddenly became the focus of international attention. From then on, any developments on that vital buffer highland would be carefully scrutinized. The future of the disengagement agreement with Syria no longer was clear, and a valuable propaganda weapon had been given the Arab world and other enemies of Israel. On the Golan itself, numerous once-friendly Druze were incensed. When the military government ordered them to take out Israeli identification cards, many protested by strikes and occasional violence. Their resistance now was added to that of militant Palestinians on the West Bank, of die-hard Jewish settlers in Yamit. In effect, Israel faced trouble on each of its three fronts, and surely the likelihood of strained relations with Egypt, once the Sinai withdrawal was completed. Europe, too, was further alienated. And so, not least of all, was the United States.

From his initial election in 1977, Begin had been intent upon securing Washington's formal and official acknowledgment of Israel as a geopolitical asset to the United States. Previous Labor governments had not ventured to press the issue with the Americans. Indeed, they had been hesitant to take sides in the Cold War. With Begin and his rightists, on the other hand, there had never been a question of obsequiously soliciting the remnants of Soviet goodwill. A formal identity of interests with the mighty United States was far preferable. While Jimmy Carter was not the man to have offered that sort of biased commitment, the new Reagan administration was known to be more receptive. Secretary of State Haig had visited the Middle East in the spring of 1981, seeking a regional alliance against Soviet penetration. Then, before an understanding could be explored with the Begin government, the contretemps broke out on the AWACS sale to Saudi Arabia. The bombing attack on the Iraqi nuclear reactor further chilled relations with Washington. Begin nevertheless persisted in his overtures. Finally, in late November 1981, Haig and Reagan agreed to formalize the concept of American-Israeli strategic cooperation. The concession presumably would assuage Israel at a time when the United States and Egypt were conducting joint military maneuvers. Now, at Sharon's request, and without a formal vote by the Knesset or even by the full cabinet, the prime minister and his closest advisers endorsed the agreement, and Sharon immediately flew off to Washington to sign it.

The understanding was substantially less extensive than it appeared. In its general provisions, the agreement stipulated that Israel would lend its help to the United States should a Soviet-American confrontation occur in the Middle East. Yet, due to Israel's limited resources, that assistance would have to be confined mainly to an exchange of intelligence (which took place anyway), and to the stockpiling of American medical supplies in Israel. Moreover, once the "Memorandum of Understanding" was published in Israel, it was greeted with a blast of Labor criticism. Peres, Eban, and others were quick to point out that the United States was not formally obliged to rush to Israel's aid should the Arabs launch an offensive—beyond earlier, less specific, commitments relating to the 1975 Sinai disengagement agreement and the Egyptian-Israeli peace treaty. Rather, in the event of a Soviet-inspired uprising—for example, in Saudi Arabia—it was Israel that was obligated to provide help to the Americans.

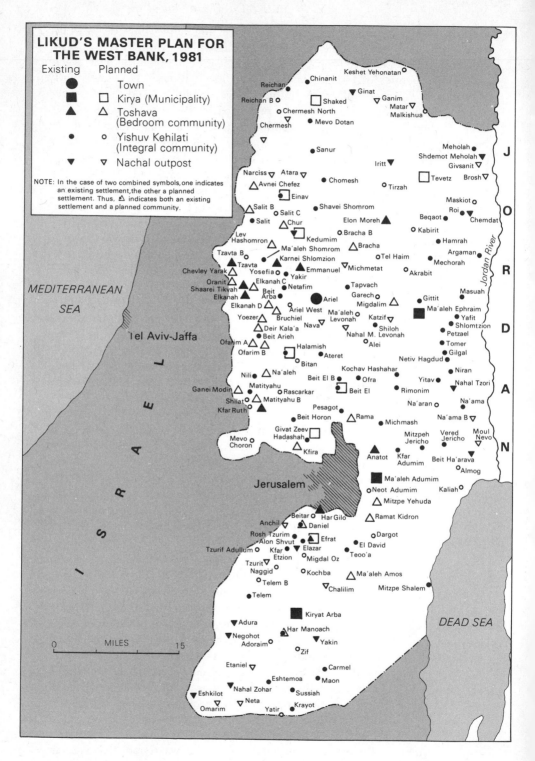

LIKUD'S MASTER PLAN FOR
THE WEST BANK, 1981

Existing   Planned

● ▢   Town

■ □   Kirya (Municipality)

▲ △   Toshava
       (Bedroom community)

● ○   Yishuv Kehilati
       (Integral community)

▼ ▽   Nachal outpost

NOTE: In the case of two combined symbols, one indicates
an existing settlement, the other a planned
settlement. Thus, △ indicates both an existing
settlement and a planned community.

MEDITERRANEAN
SEA

ISRAEL

Tel Aviv-Jaffa

Jordan River

Jerusalem

DEAD SEA

0   MILES   15

Keshet Yehonatan
Chinanit
Reichan
Reichan B
Shaked
Ginat
Ganim
Matar
Malkishua
Chermesh North
Chermesh
Mevo Dotan
Sanur
Meholah
Shdemot Meholah
Givsanit
Iritt
Chomesh
Tirzah
Brosh
Narciss
Atara
Avnei Chefez
Einav
Tevetz
Salit B
Salit C
Shavei Shomrom
Elon Moreh
Maskiot
Roi
Beqaot
Chemdat
Salit
Chur
Bracha B
Kabirit
Lev Hashomron
Kedumim
Ma'aleh Shomrom
Bracha
Hamrah
Tzavta B
Karnei Shlomzion
Tel Haim
Argaman
Mechorah
Tzavta
Michmetat
Akrabit
Chevley Yarak
Yosefia
Emmanuel
Oranit
Yakir
Shaarei Tikvah
Elkanah C
Netafim
Tapvach
Masuah
Elkanah
Beit Arba
Garech
Ariel
Gittit
Ma'aleh Ephraim
Elkanah D
Ariel West
Ma'aleh Levonah
Migdalim
Yoezer
Bruchiel
Katzif
Yafit
Ofarim A
Deir Kala'a
Nava
Shiloh
Shlomtzion
Beit Arieh
Nahal M. Levonah
Petzael
Halamish
Alei
Tomer
Ofarim B
Ateret
Gilgal
Bitan
Netiv Hagdud
Na'aleh
Kochav Hashahar
Nili
Matityahu
Beit El B
Ofra
Yitav
Niran
Ganei Modin
Rascarkar
Beit El
Rimonim
Nahal Tzori
Shilat
Matityahu B
Na'aran
Na'ama
Kfar Ruth
Pesagot
Michmash
Na'ama B
Beit Horon
Rama
Mevo Choron
Givat Zeev Hadashah
Kfira
Mitzpeh Jericho
Vered Jericho
Moul Nevo
Anatot
Kfar Adumim
Beit Ha'arava
Almog
Ma'aleh Adumim
Neot Adumim
Kaliah
Mitzpe Yehuda
Beitar
Har Gilo
Ramat Kidron
Anchil
Daniel
Rosh Tzurim
Efrat
Dargot
Alon Shvut
El David
Tzurif Adullum
Kfar Etzion
Elazar
Teoo'a
Tzurit
Migdal Oz
Naggid
Kochba
Ma'aleh Amos
Telem B
Chalilim
Mitzpe Shalem
Telem
Kiryat Arba
Adura
Har Manoach
Negohot
Yakin
Adoraim
Zif
Etaniel
Carmel
Eshtemoa
Maon
Eshkilot
Nahal Zohar
Sussiah
Omarim
Neta
Krayot
Yatir

J
O
R
D
A
N

Whatever its intrinsic values or dangers to Israel, the "Memorandum of Understanding" in any case was torpedoed by Begin himself on December 14, 1981, when the prime minister rushed through the Golan annexation bill. The effrontery of the move shocked the Reagan administration. At Haig's behest, then, President Reagan announced a "temporary suspension" of the agreement and of some $300 million in projected arms sales to Israel. Here it was Begin's turn to profess outrage. Summoning Ambassador Samuel Lewis to his office on December 20, the prime minister heatedly accused "certain individuals" in the Reagan administration of perfidy and antisemitism. Israel was not a "banana republic," he reminded the ambassador, to be dealt with peremptorily by Washington. Begin then solemnly declared that, although he would honor the commitment to evacuate the Sinai on schedule, he could not be "held responsible" for any failure to reach agreement on Palestinian autonomy. Neither would he offer assurance on Israel's "future plans" should the situation in Lebanon continue to deteriorate. On that tense note, the interview ended. Less than an hour later, Begin's press officer gave the media a verbatim account of the prime minister's outburst. If Begin's diplomacy was a source of embarrassment to his cabinet colleagues, his political judgment conceivably was shrewed—if ruthless. On the very eve of withdrawal from the Sinai, the prime minister once again had projected an image of Israeli self-esteem and sovereign freedom of action.

SHARON'S "PRAGMATIC" SETTLEMENTS FORMULA

As early as 1977, we recall, the Drobles Committee had outlined the new government's territorial strategy. After making appropriate genuflections to security needs, the committee report declared that "state and uncultivated land [in the West Bank] must be seized immediately in order to settle the areas between the concentrations of minority [*sic*] population and around them, with the objective of reducing . . . the opportunity for the development of another Arab state in these regions." The area was hardly impenetrable. The entire West Bank encompassed less than the circumference of greater Los Angeles. Thus, midway into Begin's second term, by 1983, utilizing the various techniques developed by Sharon and the latter's colleagues for redefining "state land," Israel already had acquired direct control of nearly 35 percent of the West Bank's physical terrain and 31 percent of the Gaza Strip area. It was simply the process of Israeli settlement that did not go according to pattern. If there were 5,000 Jews in the West Bank in 1977, as late as January 1981 the number did not exceed 17,500. The shortage was in ideologically motivated settlers of the Gush Emunim type, those prepared to leave the metropolitan areas and live in small, remote, isolated settlements in Arab heartland areas near Hebron or Nablus. The few zealots who made the commitment did little more than stake a symbolic claim to an undivided Land of Israel.

A self-proclaimed "realist," Ariel Sharon understood clearly that, if his vision of "300,000 Jews in Judea and Samaria" were to be fulfilled, the ideological approach would have to be transcended. In fact, he had his new technique ready and well defined. It did not escape the defense minister, or his advisers, that suburbanization already was an established trend in Israel proper. In Sharon's view, then, emphasis more usefully could be placed on "conurbation" areas east of Tel Aviv and Jerusalem, principally extended metropolitan suburbs that bulged over the Green Line into the West Bank. In short, the original, visionary Drobles format could be subordinated to a more "comfortable" pattern of settlement around existing high-demand zones. Here the inducements of urban-quality housing, in close proximity to integral Israel, might be expected to swell the number of settlers by an additional 80,000 in five years, a growth that would generate its own future momentum. So it was, from early 1981 on, that 83 percent of all public investment funds for the territories was directed into conurbation bedroom communities that served mainly Jerusalem and Tel Aviv. Funding now was offered to private builders and purchasers on terms distinctly more attractive than for comparable properties west of the Green Line. Additionally, settlers enjoyed the moral assurance that they no longer were making their homes in "foreign" territory. Israeli administrative services, even Israeli law, followed and protected them.

It was a combination of inducements that succeeded immediately and dramatically. An example was Ariel (once the Arab village of Haris), the largest Jewish settlement on the West Bank and the planned Jewish "capital" of Samaria. By 1982 Ariel was already a sprawling community of 350 families— some 1,250 people—with another 1,650 families committed to move in within two years. Enjoying government guarantees, private contractors were able to provide irresistible mortgage rates, terms so low that they would have been near surrealistic in integral Israel. It was possible for a Jewish family to build a four-bedroom villa in the West Bank, on a spacious plot of land, with well-equipped schools and stunning views, for the price of a small apartment in a crowded neighborhood in Tel Aviv. Like most of Samaria's Jewish settlements, Ariel was essentially a "bedroom" community. The largest numbers of its wage earners commuted each day to Tel Aviv and the coastal plain—a drive of barely forty minutes.

At Nofim, a hilltop villa settlement 4 miles from Ariel, private contractors in 1982 already were touting the delights of dream houses going up in a dream city, of a country club to be built, a road that soon would shorten the drive to Tel Aviv to a half-hour "without passing through any Arab village." In Ma'aleh Adumim, a ten-minute drive east of Jerusalem into the West Bank, development was proceeding at an even faster rate. Two hills, covered with modernistic, brightly colored villas, were nearly ready for occupancy. Although plans existed for a large, high-tech industrial center at Ma'aleh Adumim, this community too was emerging basically as a dormitory town. Many younger Jerusalemites of modest incomes were moving to Ma'aleh Adumim, tantalized by the availability of cheaper homes and good schools. Altogether, it was a long way from that far-off time, only six or seven years earlier, when Gush Emunim idealists had camped out in trailers and army camps, sloshing through mud in winter and jumping over stony tracks in summer, to establish their claim to the land. Israelis no longer were pioneering in the West Bank. They were simply improving their standard of living.

If Jewish settlers found these opportunities economically and aesthetically dazzling, someone had to underwrite the low prices and unlinked mortgages that were virtual give-aways. It was the Israeli government that paid. Estimates of the final bill conflicted widely. In 1980 Minister of Finance Yigal ,Hurevitz argued that total annual expenditure outside the Green Line did not exceed $100 million, or 1.5 percent of the state's budget. The figure was questionable. Indeed, the only thorough study of public outlay in the territories was the West Bank Data Project, a research venture directed by Dr. Meron Benvenisti, the former deputy mayor of Jerusalem, and a sharp critic of Likud's annexationist policies. Under Benvenisti's estimates (published in 1984 and after), direct government capital investment in West Bank civilian programs since 1967 reached $1.5 billion—$750 million under Labor between 1967 and 1977, and $805 million under Likud from 1977 to 1983. Of this figure, perhaps four-fifths currently was being diverted to West Bank housing under the budget of the ministry of housing. Under the rubric of "institutions," the education ministry in 1984 allocated some $14 million to the construction of a school in Ma'aleh Adumim and $22 million for a yeshiva in the Old City of Jerusalem—this at a time when the education ministry's entire construction budget for all of Israel came to barely $100 million and Israeli schoolchildren were facing classes in shifts due to financial shortages. In 1984, too, the ministry of development would be spending $100 million for new water sources. Benvenisti calculated that at least $36 million of this amount would be disbursed in the territories, and possibly much more.

To these civilian expenditures had to be added the numerous tax exemptions for companies and individuals doing business in the West Bank, the extensive monetary and mortgage incentives offered settlers, the salaries paid the hundreds of civil servants who dealt mainly with the settlements. No details were available on the—traditionally cloaked—military capital investment budget. But the wide approach roads built to serve a number of settlements were financed by the defense ministry, although the army maintained its West Bank installations elsewhere. "Special projects," among them the transfer of the Elon Moreh facilities by helicopter to nearby Mount Kebir (Chapter V), also figured in the "silent" expenditures. Not least of all, Benvenisti estimated that private contractors accounted for another $1 billion put into the settlements. If that were the case, total capital invested in the West Bank by 1984 came not to $1.5 billion but to $2.5 billion. Altogether, "authentic" expenditure between 1977 and 1983 was running at a rate almost triple that of earlier years, and probably close to 8 percent of the government's actual budget.

As late as June 1981, Sharon's projected figure of 300,000 Jews living on the West Bank by the end of the decade had seemed grandiose. But afterward, with the moderates out of the cabinet, the Likud government was freer to intensify its settlement efforts, with particular attention devoted to conurbation housing. By December 1982, not less than 103 settlements were operating on the West Bank, 70 of these established under the Begin admin-istration. Even then, they encompassed only about 25,000 Jews. Yet the figure did not include 65,000 Jews who had become residents of the sprawl-ing East Jerusalem neighborhoods since 1967, and particularly in the Begin years. Under this joint category, the number of Israelis living beyond the Green Line climbed to nearly 100,000 by 1983, or 10 percent of the terri-tory's entire population (including the Arabs of East Jerusalem). It was a dense enough nucleus to generate the exponential growth Sharon initially had envisaged. As early as the autumn of 1982, Benvenisti calculated that de facto annexation of the West Bank might already have become irretriev-able. The former deputy mayor was not impressed by the "demographic threat" to the Jews. Between 1967 and 1982, after all, some 350,000 Arabs departed the West Bank, unwilling to live under Israeli rule. Benvenisti estimated, then, that a combination of Arab emigration and Jewish settle-ment would leave almost intact the current demographic ratio of approxi-mately two-thirds Jews and one-third Arabs in Palestine—at least well into the 1990s. Meanwhile, as early as October 1982, addressing the settlers at a new Jewish colony in a Nablus suburb, Minister of Communications Mor-dechai Zippori reassured his listeners: "Don't worry about the demographic density of the Arabs. When I was born in Petach Tikvah, we were entirely surrounded by Arab villages. They have all since disappeared."

Yet it was not demography alone that persuaded Benvenisti and other observers that Israel's rule in the West Bank was becoming irreversible. The sheer quality of life in the new Green Line conurbations was the decisive factor. Few Jewish settlers would have been prepared by then to return to the crowded inner cities of integral Israel. As Sharon and his associates had calculated, decisions about the future of the territories would be foreclosed by hard economic and political facts. Far more than the ideological zealots, the new "conurbanites" would have nurtured a deep economic stake in pre-serving Israel's grip on the West Bank. Each one of the 100,000 settlers beyond the Green Line had at least four relatives in Israel to support his foothold in the territories. The ensuing bloc of voters accordingly might be enough to influence the election of ten or fifteen Knesset members, and thus ensure that proposals for withdrawal and uprooting would face grave legis-lative difficulty.

PALESTINIAN ACTION, ISRAELI COUNTERREACTION

Even as the Begin government accelerated its settlements program, the Pal-estinians for their part groped for techniques of resisting ingestion into a Greater Israel. Over the years a literature of protest developed among them, following the lead of such Israeli Arab writers as Salem Jubran, Tawfiq Zayyad, and Mahmud Darwish. In the Jerusalem-Ramallah area, the literary awak-ening was particularly evident in drama. Numerous little theater groups sprang up, giving performances on stirring nationalistic themes before enthusiastic audiences. Newspapers too came out with avowedly anti-Israel articles and essays. So did three or four new Jerusalem journals. Among the latter, *al-*

Bayaddar, which began publication in 1976, was of high quality, its writers young, innovative, and militantly Palestinian. So long as their articles, poems, stories, or plays did not incite directly to violence, moreover, Israel's authorities permitted them to appear, to lament the harshness of Israeli rule, the absence of a meaningful Palestine entity. Few truly distinguished authors published in these various media—the "brain drain" from the West Bank was reaching serious proportions—but few also lacked for grateful readers.

Behind the printed word, fidelity to the PLO was never disguised. By the latter 1970s, we recall, the guerrilla confederation had tightened its grip on the emotional loyalties of tens of thousands of younger Palestinians. Arafat himself may have favored a moratorium on Fatah violence, at least outside Israel, in a bid to cultivate Western support. Indeed, to project that new, moderate image, he intimated his willingness to countenance a Palestinian state essentially for the West Bank and Gaza. Yet, when he spoke for home consumption, the PLO leader evinced no interest whatever in settling for less than a "democratic, secular state" in the totality of Palestine. Whether abroad or at home, radicals like Saslah Khalef ("Abu Iyad") now spoke for the rank and file of the movement in their plainspoken intention—expressed in a resolution of the Fatah congress in June 1980—to "liquidate the Zionist entity."

If there were differences among local Arab spokesmen, meanwhile, these were principally of emphasis. Elias Freij, mayor of Bethlehem, may have been the most restrained. Together with a number of prominent merchants and lawyers, most of them political veterans or other "notables" under the former Hashemite regime, Freij advocated recognition of Israel and the establishment by its side of an independent Palestinian state. Very likely these men echoed the hopes of the West Bank majority, who sensed that Israel would not disappear, and who were preoccupied with day-to-day issues of family and economic security. Yet, in the atmosphere of PLO intimidation, the public influence of Freij and other moderates was very low by then. In 1979, guerrillas murdered three prominent advocates of coexistence from Ramallah. PLO extremists even imposed a reign of terror over Arab inmates in Israeli prisons, killing and maiming suspected "traitors" among them. In May 1980, six Jewish members of the Gush Emunim community in Hebron were ambushed and slain. When Israel's military government then deported the mayors of Hebron and Ramallah for incitement to violence, the West Bank erupted in a series of riotous demonstrations. Nine Israeli settlers and five suspected Arab collaborators were killed.

As always, intransigence was particularly inflamed among university students. Before 1967, no institutions of higher learning existed on the West Bank. Then, under Israeli rule, four colleges and universities (of distinctly marginal quality) and three teachers' seminaries were allowed to develop. The best known was Bir Zeit, a university for some 2,000 students in the Samarian hills. Indeed, for Palestinians, Bir Zeit became the "Berkeley of the West Bank," a genuine hotbed of radicalism. Most of its students came from the area's refugee camps and smaller villages. Intensely nationalist, they were divided between a neo-Fatah (i.e., pro-Arafat) faction, the Popular Front for the Liberation of Palestine, and a sizeable Communist group. So frequent were their demonstrations by the late 1970s and early 1980s, their stonings of Israeli vehicles and military patrols, that Bir Zeit was closed at least once a year for periods of several days or weeks. Then, on February 15, 1982, a group of Bir Zeit students physically assaulted a visiting Israeli civilian official. The military authorities in turn closed the school for an unprecedented two months. Whereupon new demonstrations of equally unprecedented violence exploded throughout the West Bank. Over a three-week period, fifteen Arab students were killed or severely wounded and several hundred protesters were arrested. By then Bir Zeit and its sister institutions all but dictated the response of the younger Palestinian generation. "Many erstwhile moderates have taken cover behind radical statements," noted the Israeli Druze journalist, Rafik Halabi, "while the few who have stuck to their views have retreated into double-talk or . . . into total silence. No matter who calls the shots these days, the PLO comes out ahead."

The Israeli government was not prepared to deal forbearingly with this unrest. Ariel Sharon now set the tone for Begin's second administration. With his authority over the territories further augmented in his new role as defense minister, Sharon proceeded to orchestrate a shrewd, if complex, carrot-and-stick approach to the Palestinian Arabs. In November 1981, Military Order 947 officially established a civilian administration in the West Bank, under a civilian director. The change, which Sharon described as "liberal" and "humanistic," in fact related logically to Israel's firm stance in the autonomy negotiations. Camp David had called for the "withdrawal" of the Israeli military government as soon as a self-governing entity was freely elected in Palestine. Even before such an election, Order 947 signified now that Israel was imposing its own, preemptive interpretation of this requirement. It accomplished the feat simply by replacing the title, "military commander *of* Judea and Samaria" with that of "commander of Israeli forces *in* Judea and Samaria." For all practical considerations, the officer would retain intact the powers he had enjoyed formerly. The director of the civilian administration in turn was empowered only to introduce subsidiary rules and regulations, subject to the approval of the military commander. It was a tightly limited authority that matched in almost every detail the powers Begin and his colleagues envisaged for the Arab self-governing entity.

Indeed, the "constitutional" change possibly was less even than semantic. It also developed into the facade behind which Sharon would apply a new and far tougher policy to the inhabitants of the West Bank. From 1967 to 1974, Dayan's approach to the territories had been essentially noninterventionist and discretionary. He tolerated an informal dual Israeli-Jordanian rule in the West Bank, the opening of universities, even a certain leeway for political activities. The fact that, even under Dayan, many scores of Palestinians were expelled, houses demolished, and land seized, was generally overlooked as long as the military administration of the time seemed temporary. Then came the Yom Kippur War, the 1974 Rabat Conference, the radicalization of Palestinian youth, and ultimately, with the municipal elections of

1976, the political ascendancy of the PLO. Under Begin's first term, then, the military government, which until 1977 had dealt essentially with routine civilian affairs, was allowed to give more forceful attention to the battle against radical elements. Henceforth the government was very firm in its dealings with the nationalist mayors, turning down their budgetary allocations and licenses, delaying their requests, harassing them increasingly with summonses, interrogations, house arrests. Following Weizman's resignation in 1979, moreover, when Begin acted as his own defense minister, General Eytan was allowed to adopt an even tougher policy, and troops occasionally engaged in quite brutal acts. But it was a hopelessly vicious circle, with both sides becoming radicalized, the gulf in communication between them apparently unbridgeable.

In November 1981, with the introduction of the civilian administration, Sharon sought not only to establish the model for a future Arab autonomous entity, but also to establish links with a possible alternative Arab leadership on the West Bank. Much would depend here upon the ability of the new civilian director. In Menachem Milson, the defense minister found his man. In his forties, a Harvard- and Cambridge-educated professor of Arabic literature at the Hebrew University, Milson was renowned for his broad understanding of the contemporary Arab world. His military credentials were equally impressive, for he was a veteran officer of Sharon's famed Commando Unit 101 and of the Mitla paratroop drop in 1956. Later he had served as chief of Arab affairs in the West Bank military government. Personally, Milson had his own blueprint for evoking Arab cooperation. He had outlined it for the American Jewish magazine *Commentary*. In his 1981 article, "How to Make Peace with the Palestinians," Milson assailed Israel's earlier failure to "play by the rules of the game," the traditional Middle Eastern "game" of rewards and punishments that had functioned throughout the Middle East for centuries, under Turks, British, and Hashemites. As Milson saw it, the military government's lingering hesitation to intervene directly in Arab affairs had enabled the PLO to acquire a political stranglehold on the West Bank population, on those local Arabs—the majority—who normally would have preferred to negotiate with Israel but who were intimidated by PLO bribery, threats, or assassination. The time had come for a radical change of policy, Milson argued, for a total political campaign to eradicate PLO influence in the territories. The technique would be a heavy hand for PLO collaborators, and a parallel encouragement of more moderate interlocutors.

In fact, Milson himself was a moderate. He endorsed the Camp David framework, and anticipated a connection between the West Bank (which he did not hesitate to call the West Bank) and Jordan. But if he rejected Sharon's annexationist vision for the territories, his proposal for eradicating PLO influence seemed useful to the defense minister in the short run. It was. Upon assuming his new appointment, Milson promptly set about administering the "rules of the game." To that end, during 1981–82, his favored clientele among potential moderates were the "Village Leagues." These loose-knit collections of farmers and rural schoolteachers had come into existence three years earlier, ostensibly for the purpose of further modernizing the

West Bank's agricultural villages. Unlike the PLO, which had developed its links almost exclusively among the volatile and nationalist urban elites, the Village Leagues maintained no ties whatever with the cities. Neither were they in communication with PLO headquarters in Beirut. These facts had caught Milson's attention.

The civilian director thereupon opened discussions with Village League spokesmen, most of whom were discreetly pro-Hashemite. Here again, Milson played the "rules of the game." Bolstering the leagues' resources with some $15 million in development funds for roads, electricity, water systems, schools, and clinics, he simultaneously offered their leadership priority in employment, housing, in preferential licenses, in building and commercial permits. Milson did not hesitate to bend municipal services to his purpose. All buses dating from 1956 or earlier were banned from the roads, to be replaced only with the administration's permission; and pro-PLO municipal councils did not receive that authorization. Village League members alone were authorized to issue business licenses and import permits in their districts; PLO supporters were discouraged from applying. Neither did known Arab nationalists receive export permits for their citrus shipments, or visas to Jordan during the summer travel period. To stem the flow of PLO patronage money into the West Bank, travelers were restricted in the sums they could carry over the bridges from Jordan.

The "rules of the game" were not ineffective. The atrophy of PLO funds led several mayors to complain of economic crises in their communities. No doubt the Village Leagues, like the PLO, also were less than fully representative of the West Bank's silent majority. Among city dwellers, they were regarded as bumpkins; among students and other educated townsmen, as quislings. Several Village League supporters, like Yusuf al-Khatib, a Ramallah businessman, and his twenty-three-year-old son, were hunted down and slain by Fatah gunmen. Yet even assassination attempts did not appear to inhibit the organization's growth. By early 1982 Israeli and foreign observers sensed that the PLO's grip on the West Bank, while far from broken, appeared at least to be increasingly contested.

Yet it was then, too, that the leagues also suffered a sharp, unexpected blow of their own. Ironically, it was inflicted not by Palestinian radicals but by the government of Jordan. To succeed, the leagues needed at least the tacit support of Amman. Milson had assumed that Hussein and the latter's advisers would approve Israel's measures to limit PLO influence in the West Bank, to support traditionally moderate, pro-Hashemite elements. Indeed, one of the league spokesmen, Mustafa Dudin, was himself a former minister in the Jordanian cabinet. But in March 1982, under pressure from the joint PLO-Jordanian committee that had guided Amman's policies toward the West Bank since 1978, Prime Minister Mudar Badran signed a decree imposing the death penalty on anyone serving actively in the Village Leagues. The measure seemingly torpedoed the organization's effectiveness. Without Hashemite approval, overt or tacit, no moderate alternative to the PLO appeared to be viable in the territories.

Thereafter, Milson was obliged to use the stick rather more than the carrot. If mayors refused to cooperate, he dismissed them outright and replaced them with Israeli officials. Blacklisted PLO suspects were denied even such necessaries as driving licenses. Other repressive measures included the closure of Bir Zeit University in the wake of campus disturbances, the arrest of several municipal councillors for incitement, the imprisonment of others for real or suspected violence. Three editors were barred from their East Jerusalem offices under the constraints of "administrative detention," a measure that had its legal origins in the British Defense Emergency Regulations of 1945, and which the Hashemite Kingdom, and then the Israelis, maintained intact upon taking over the West Bank. If detentions were far from rare under the Labor government—and particularly after the 1976 municipal elections—they became increasingly frequent and prolonged under Likud.

Yet, for the West Bank Arabs, no feature of the Israeli occupation evoked greater terror than deportation, not even imprisonment or the demolition of homes. It was above all the fear of deportation that prevented far graver outbreaks in the latter 1970s, when the PLO fastened its grip on the territories. "Troublemakers," "inciters" often were peremptorily seized by the army, taken to the border, and obliged to walk across. Until 1969, the principal eviction routes were the Allenby and Damiya bridges over the Jordan. Afterward, however, when the Hashemite government refused to accept deportees, these exits were closed; and Israel subsequently shifted the venue to the Arava Desert between the Dead Sea and Eilat, and in 1974 to the northern border with Lebanon. Heavily criticized by Israeli liberals, the practice of deportation eased somewhat in the last year of the Rabin government. Under Begin, it was resumed, and was linked precisely to the scale of Arab demonstrations.

That scale mounted appreciably following the prime minister's reelection in 1981. Although confrontation was inevitable in a West Bank functioning under PLO intimidation, the accelerating tempo of the government's settlements program and the unrelenting aggressiveness of the Gush Emunim raised tensions to the boiling point. Hebron was a classic example. The growth of the neighboring Kiryat Arba enclave was not proceeding rapidly enough for its founder, Rabbi Moshe Levinger, and his band of fanatics. In 1979 Levinger's American wife led a dozen women in a sit-in at the Hadassah Building, a Jewish property in the center of Hebron that dated back to the pre-1929 period. King David has appeared to her in a dream, explained Mrs. Levinger, and ordered her to restore a Jewish presence in the heart of town. At first even the Begin government regarded the sit-in as provocative. Hebron was an exceptionally devout Moslem community. Not even a movie theater was permitted there. But, once again, the settlers' obstinacy and their network of supporters in Israel prevailed. Backing down under an avalanche of Orthodox and other rightist protests, the government agreed to carry out the the reconstruction of Hebron's old Jewish neighborhood. Immediately Jewish families moved into a number of downtown houses. Beit Romano, adjoining the Arab market, was transformed into a yeshiva, with several Jewish families living on its top floor.

It was supremely in Hebron, therefore, that the irresistible force of Jewish messianism confronted the immovable object of Moslem fundamentalism. The town's Arab population suppurated in anger for nearly a year. Then violence began in the spring of 1980 when a visiting yeshiva student was murdered in Hebron's casbah. The army immediately imposed a curfew and launched into searches and arrests among the resident Arabs. The Jewish settlers of Kiryat Arba, who were not under curfew, ranged through Hebron, indiscriminately destroying Arab property. Mayor Fahd Qawasmeh thereupon vowed to resist the judaization of Hebron at all costs, local Arab violence broke out again, and five more Jews were killed, sixteen wounded. The army's counterresponse was to demolish the buildings that had shielded the Arab ambushers, then to deport Mayor Qawasmeh and two of his associates blindfolded over the Lebanese border. The Jewish zealots meanwhile continued to wreak their own havoc. Over the next days, they attacked Arab schoolgirls, regularly threw rocks at Arab windows, uprooted olive trees in Arab orchards. Other episodes of Jewish vengeance and tacit army acquiescence occurred during 1980 and 1981.

So it was, early in 1982, that the West Bank erupted in a series of Arab demonstrations unprecedented in fifteen years of Israeli occupation. It was on February 15, we recall, that an Israeli civilian official was beaten by Arab students at Bir Zeit University. When the school was ordered closed for two months, rioting broke out at other West Bank colleges. On March 11, then, at Milson's recommendation, the military authorities outlawed the National Guidance Committee, an informal group of mayors and public figures that had coordinated PLO-oriented activities in the West Bank. A week later the Israelis dismissed Ibrahim Tawil, the mayor of al-Bireh, and disbanded his municipal council for refusing to deal with the civil administration. Several days after that, Israeli troops killed an Arab teenager during a confrontation in al-Bireh, and 6 other demonstrators were wounded. Violence in the next month and a half killed 11 more Palestinians and wounded nearly 200. On March 25, Mayor Bassam al-Shak'a of Nablus and Mayor Qarim Khalef of Ramallah were dismissed for "inciting demonstrations," and for serving as members of the late National Guidance Committee. On April 11, Allan Goodman, thirty-eight years old, a recent immigrant from the United States doing his Israeli army service, went on a crazed rampage at the Temple Mount, the site of the venerated al-Aqsa Mosque, killing 12 Arabs and wounding 12 others before Israeli troops subdued him. In the subsequent explosion of strikes and protests throughout the West Bank, 2 more Arabs were killed and some 200 wounded.

However legitimate the security measures, the treatment meted out to Arab and Jewish provocateurs was hardly balanced. On March 16, 1982, an eighteen-year-old Arab youth from the village of Sinjal was shot and killed by Natan Natanson, an Israeli civilian living in the nearby settlement of Shiloh. A week later, Natanson was released on bail. He was never brought to trial. On March 24, Israeli civilians shot and killed another Arab teenager from Bani Naim (near Hebron) after youths from Bani Naim had stoned the Israelis' car. No full-scale investigation of the incident ever took place, no one was brought to trial. On October 26, 1982, another stone-throwing teenager was shot dead by an Israeli from the West Bank settlement of Hinanit. Local police briefly detained the settler, but he too was not brought to trial. Press reports described at least six other cases during 1982 of Palestinians shot or stabbed by Israeli civilian settlers, usually when the Arabs stoned their vehicles or erected road blocks. None of these Jews was arrested or held for questioning; Arab youths involved in the incidents were taken into custody. When Jewish settlers launched grenade attacks on Arab-owned homes and automobiles near Kiryat Arba, on Arab school buildings and homes in the Dheishe refugee complex, there were no arrests. Much of this flailing lawlessness, and its evident immunity to prosecution, could be traced to the emergent political influence of the settlers' regional councils, bodies that had won all but autonomous powers for themselves. As the settlers performed their reserve duty in their own villages on a part-time basis, functioning essentially as an adjunct of the occupying army, they normally received the benefit of the doubt in matters of "self-defense." Their support in Israel was far-reaching. Sympathetic politicians, among them ranking members of the government coalition, could be depended on to intercede for them.

Yet even this mood of right-wing permissiveness was strained, as early as June 1, 1980, by a particularly grisly episode of vigilantism. A month before, six Jews had been killed and sixteen wounded in an Arab ambush at Hebron. It was now the "shloshim"—the Jewish ceremonial thirtieth-day commemoration—of that event. The alleged inciters should be made to pay. Among those marked for retribution were Mayor Bassam al-Shak'a of Nablus, Mayor Qarim Khalef of Ramallah, and Mayor Ibrahim Tawil of al-Bireh—all stalwarts of the recently disbanded National Guidance Committee. As unofficial chairman of the group, Shak'a had made a number of incendiary statements in support of Palestinian nationalism. The Israeli military authorities had sought to deport him on one occasion, but the effort had provoked so vast a protest throughout the West Bank, with all Arab mayors (moderates and extremists alike) resigning, that the order was rescinded. Shak'a was returned to Nablus, a national hero. This time he would not evade retribution.

On June 1, two Jewish settlers in western Samaria drove into Nablus at night and pulled over near Shak'a's home. Under cover of darkness, they made their way to the mayor's parked Opel and attached an explosive device to its engine. Early the next morning, Shak'a left his house, entered his automobile, turned on the ignition. His legs were blown off. Shak'a's colleague in Ramallah, Qarim Khalef, lost a leg from an identical bomb planted in his Cadillac. Simultaneously, acting on a tip, a Druze bomb-disposal expert of the Israeli defense forces examined the garage and automobile of al-Bireh's Mayor Ibrahim Tawil. Brushing against a trip wire, the Druze touched off a plastic explosive buried in a nearby flowerbed. He was blinded. That same day, a booby-trapped grenade exploded in the Hebron market, injuring eleven Arabs, including four schoolchildren.

Accusations and counteraccusations followed. Even Begin felt impelled to condemn the maiming of the Arab mayors. The latter meanwhile refused treatment in Israeli hospitals, and instead were nursed to health in Jordan. Upon being fitted with artificial limbs, they returned to their posts several months later, more embittered and intractable—and more admired—than ever. Yet, as the months passed, little evident progress was made in solving the crimes. It later developed that the army had carried out a preliminary inquiry, then had turned the matter over to the police. At that point, senior political officials attempted to block the investigation. Part of the difficulty was the confusion of jurisdiction in the West Bank, an indeterminate area between military and Israeli civil authority. Even more fundamental, however, was the structural asymmetry of the law. "Two legal systems have been created for the two national groups living in these territories," declared a report issued later by the Tel Aviv-based International Center for Peace in the Middle East, a dovish, humanitarian group of lawyers and academics. "One system is intended for the Palestinian residents of the area and the second for Israelis." A contributor to the report, Claude Klein, professor of jurisprudence at the Hebrew University, compared the West Bank law enforcement system to the old American West under the protection of federal troops. "In case of conflict between settlers and native Indians," he asked, "who would the federal troops protect?" The report, issued in 1984, evinced a growing fear among moderates that vigilantism in the West Bank inevitably would spill over into Israel itself.

The danger was real. Committed to the vision of an "undivided Land of Israel," Begin, Sharon, and their fellow ministers still were unwilling to face the implications of emergent Israeli lawnessness. "There are Arab fanatics and there are also Jewish fanatics," insisted Minister of the Interior Yosef Burg, "but ours are a very small segment of the country." That could have been true. Yet Burg also denied that the law was enforced unevenly in the West Bank, that the police (functioning under his ministry) had been sluggish in investigating the atrocities. It was not until May 1984, after the fall of Sharon as the cabinet's strong man (Chapter X), that arrests were made. Thirty-seven Israelis, most of them Gush Emunim and other West Bank settlers, as well as three army officers high in the military government, were taken into custody and indicted on charges ranging from premeditated murder to membership in a terrorist organization and theft of army property. And when the court hearings finally began in mid-June, the Israeli people witnessed at close hand the Frankenstein monster that had been loosed among them. Many of the defendants were implicated in a large network of atrocities including the maiming of the mayors, an attack on the Islamic College in Hebron and the murder of four students, and a conspiracy, mercifully detected in time, to plant bombs on a number of buses carrying West Bank Arabs to and from Israel during the rush hours.

On the eve of the trial, President Chaim Herzog may have spoken for a majority of Israel's population when he warned that the imputed crimes were a "sickness" and a rebellion against the state, that such deeds "lower us to the subhuman level of the murderous terror organizations that act against us." But if liberals and humanitarians, among them numerous Knesset members, journalists, and academicians, echoed the president's sentiments, there were others of equal eminence who took a divergent view. Speaking in the Knesset, Deputy Speaker Meir Cohen-Avidor (Likud) declared, "My heart goes out to the detainees. These boys are the pride of Israel. They are the best." Several days earlier, Cohen-Avidor had addressed a rally on behalf of the defendants. Earlier yet, he had stated from his post in the Knesset that any Arab who murdered a child should have his eyes gouged out—a statement he repeated as he sought to force his way across a Haifa courtroom to reach five Arabs accused of raping and killing a teenaged Jewish girl. Minister of Science Yuval Ne'eman, chairman of the ultraright Techiya party, had refused to condemn the original maiming of the Arab mayors in 1980, and remained silent now. And so, in effect, did Yitzchak Shamir, who had recently succeeded Begin as prime minister in 1983 (Chapter X). "It is too early to speak about clemency," was Shamir's only response to questioning on the defendants and their alleged crimes. Defense money also was pouring in from right-wing circles both in Israel and the United States.

President Herzog was not wrong, then, in his warning that vigilantist terror threatened Israel with grave internal fissures, and possibly worse. Yet it was a premonitory vision that applied with equal urgency to the full skein of tensions bedeviling Israel, all of them interrelated: the Kulturkampf between Orthodox and non-Orthodox, the festering resentments between the nation's oriental and European sectors, the emergent nucleus of obsessed Jewish *pieds noirs* on the West Bank. The private Israeli citizen may have gone about his affairs, as did the typical Arab West Banker, preoccupied largely with humdrum concerns of family and livelihood. Nevertheless, by the spring of 1982 journalists, academics, belletrists, and other trained observers of the public Israeli scene admitted to a malaise more soul-convulsing than at any time since the immediate aftermath of the Yom Kippur War. It was their mood of growing desperation that Amos Elon captured when he confessed that, were he to rewrite his famed volume *The Israelis: Founders and Sons*, he would be tempted to retitle it *The Israelis: Founders and Losers*. No one claimed that the nation's existence was yet in peril. Its character as a viable democracy, on the other hand, unquestionably was approaching the threshold of its acutest vulnerability.

CHAPTER VIII

THE LEBANON WAR: "OPERATION PEACE FOR THE GALILEE"

A FIXATION WITH THE NORTH

In early June of 1982, Israel's borders with the neighboring Arab countries were reasonably quiet. The southern frontier with Egypt was altogether in a state of peace. If the West Bank trembled in barely suppressed ferment, infiltration from Jordan, at least, had all but ended since 1970. To the northeast, Syria remained irredeemably hostile, but not a single episode of truce violation had occurred along the Golan since the 1974 disengagement agreement, not even since Israel's annexation of this critical buffer plateau in December 1981. Menachem Begin's attention was transfixed, rather, by Lebanon, the smallest and weakest of Israel's neighbors. From Beirut south, this tortured little heterogeneity of sects and factions remained hostage to the PLO guerrilla organizations, and to an overbearing Syrian military presence that extended from the Beka'a Valley directly into the capital. To protect that emplacement, the Syrians only the year before had installed batteries of SAM-6 missiles in the Beka'a. It was an undisguised violation of the tacit understanding between Israel and Syria that until then had allowed Israeli aircraft unhindered surveillance over southern Lebanon.

The crisis ignited by this Syrian action, it is recalled, accompanied by intermittent PLO shelling of Israel's northern border communities and by Begin's dire warnings of air strikes against the missiles, contributed to the overheated atmosphere of the prime minister's reelection campaign. Throughout May and June of 1981, the Israeli air force periodically attacked guerrilla positions in Lebanon, but refrained from action against the SAMs. Unwilling to engage in full-scale war with Syria, Begin permitted the American mediator, Philip Habib, to defuse the confrontation through shuttle diplomacy. But if the prime minister's combination of adamance and restraint proved effective politics, it did nothing to enhance his nation's security. On July 10, 1981, after a six-week moratorium, Israel renewed its air strikes on PLO strongholds in southern Lebanon. When the guerrillas lashed back with a rocket barrage against Naharia, a middle-sized northern Israeli town, air force jets escalated the level of violence in a heavy retaliatory bombardment of PLO headquarters in Beirut, killing some 100 people and wounding at least 600 others. Hereupon the PLO launched into an even more prolonged Katyusha rocketing of Galilee settlements. This time the onslaught all but paralyzed civilian activity in the north. The Israeli army was obliged to take over such elementary services as the distribution of bread. Much sobered by then, and ripe for a truce, Begin gratefullly welcomed the return of Philip Habib. The American mediator in turn successfully used the good offices of the Saudi government to arrange a cease-fire with the PLO.

It was a humiliating experience for Begin. The very notion of engaging in discussions with the guerrilla organization, however oblique, was intolerable enough. But the substance of the cease-fire was even more painful. It made no provision for a restoration of the status quo. Free to replenish their arsenal, to repair and enlarge their military facilities, the PLO leadership subsequently built their front-line strength in south Lebanon to the equivalent of five infantry brigades. Worse yet, Syria was not obliged to remove its missiles from the Beka'a Valley. The threat to Israel's air surveillance was palpable. A rearmed and augmented PLO force conceivably would renew its Katyusha and artillery salvos, its fedayun raids, even its amphibious forays against Israel's sixty-three Galilee settlements and coastal towns—all with indifference to possible air retaliation. For Chief of Staff Eytan, the evidence was compelling that only a larger-scale military operation could eradicate the danger.

A hard-bitten veteran of combat against heavy odds, Eytan was not the sort of man to refrain from pressing his views at the highest civilian level. Begin found him a valuable political ally. Indeed, from the time of Weizman's resignation as defense minister in 1979 until the election in 1981, Eytan himself took over the defense portfolio in all but name. Thereafter, with Begin's approval, he virtually dictated a huge increase in the military budget over the objections of many in the cabinet. Eytan's activist views on Lebanon ultimately prevailed, as well. The heavy air bombardments of 1981 were an example of his approach. And so now was the developing notion of a functional partnership with Lebanon's Christian leadership.

The idea was not new. In his diary entries of 1954, Moshe Sharett, then Israel's prime minister, described his conversations with the recently retired Ben-Gurion and with Chief of Staff Moshe Dayan. A coup d'état had taken place lately in Syria. Ben-Gurion and Dayan now wondered if Israel ought not exploit the Damascus regime's temporary paralysis by invading Lebanon. In the likely event of an Iraqi attack on Syria, noted the former prime minister, "this [would be] the time to arouse Lebanon—that is to say, the Maronites—to proclaim a Christian state." Sharett demurred. Such a move would render Lebanon too weak and divided, he insisted. But from his desert retreat in Sde Boker, Ben-Gurion continued to fire off memoranda, pressing his views, urging Sharett to dispatch emissaries, agents, money—at least to foment a Christian coup in Lebanon, thereby taking that nation out of the Arab League. It was an empty dream, Sharett patiently replied; the Christians no longer were a majority in Lebanon and, in any case, the Greek Orthodox Lebanese wanted no part of a Maronite-dominated state.

Then, in February 1955, the prime minister's diary recorded a visit by a Lebanese emissary, "apparently on behalf of Lebanese President Camille Chamoun. Lebanon would be ready to sign a separate peace [with Israel] if we accepted the following three conditions: guarantee Lebanon's border, come to Lebanon's aid if she were attacked by Syria, and buy Lebanon's agricultural surpluses." The offer struck Sharett and his advisers as less than realistic. It was politely finessed. Still, Ben-Gurion and Dayan persisted, and in May 1955 the latter argued that

> all that is needed is to find a [Christian Lebanese] officer, even a captain. We should win his heart or buy him, to get him to agree to declare himself the savior of the Maronite population. Then the Israeli army would enter Lebanon, occupy the necessary territory and set up a Christian regime allied to Israel.

In despair by then at these hallucinations, Sharett lamented the military's "simply appalling lack of seriousness [in its] whole approach to neighboring countries, especially to the more complex questions of Lebanon's internal and external affairs."

Only three years later, ironically, Israel did provide Lebanon's Christian-dominated government with a token gesture of support, a shipment of 500 Uzi submachine guns, during the brief Lebanese civil war of 1958. General Yitzchak Rabin, then commander of Israel's northern front, also cooperated with President Chamoun in blocking the passage of pro-Moslem infiltrators through northeastern Israel. Two years later, on May 4, 1960, Slade Baker, correspondent of the *Times* of London, conveyed an offer to Shimon Peres, then deputy minister of defense. In London, Baker recently had met with General Amin Shibah, commander of the Lebanese army. Elections to the Lebanese parliament were scheduled for May 12. Anticipating disturbances, the Beirut government had instructed the army to supervise the balloting. But Shibah knew that the officer corps was riddled with Nasserites, men likely to stir up trouble. Could Israel help by creating a "diversion?" Israel needed only carry out a few maneuvers and fire a few shots, Shibah explained (via Baker), and the general staff then would have its pretext for dispatching the Nasserite element away from Beirut to the Israeli frontier. Peres understood. With Ben-Gurion's approval, he cooperated.

All this was by way of overture. In March 1976, Lebanon was convulsed by a grave new civil war. It was on the eighteenth of that month that "Abu Halil," a spokesman for the Phalange—the largest of the Maronite factions—boarded an Israeli missile boat off the Lebanese coast. He was carried to Haifa, and met there with Prime Minister Yitzchak Rabin and Foreign Minister Yigal Allon. Moslem forces were about to overrun the Maronite-inhabited Mount Lebanon range, Abu Halil explained, and Christians were fleeing to East Beirut. The Christian situation altogether was desperate. Again, could Israel help? Allon in fact was willing to provide material support. Rabin also was sympathetic, but preferred that the Lebanese situation first be studied at close hand. Thus, within the week, two senior Israeli intelligence officers were dispatched to Lebanon by sea. There they met secretly with Bashir Gemayel, the youthful commander of the Phalangist militia and son of "Sheikh Pierre" Gemayel, founder-leader of the Phalange party.

The following month Rabin himself sailed up the Lebanese coast. Off the port of Jounieh, his gunboat was boarded by Camille Chamoun, the former president of Lebanon and currently leader of a second important Maronite faction. The meeting was cordial. Chamoun struck Rabin at first as a better potential ally than the Gemayels; for the Phalangists smacked a bit too much of the Nazis, who indeed had inspired Sheikh Pierre when he had attended the 1936 Berlin Olympics as a member of Lebanon's soccer team. On the other hand, the Gemayels had long maintained close relations with Lebanon's tiny Jewish community, and during the Six-Day War their militia had cordoned off and protected the Wadi Abu Jamil, Beirut's Jewish neighborhood. Rabin's interim solution, then, was to help both Maronite groups. During his three-year government, some $150 million in weapons and other supplies were turned over to the two Christian militias. The Israeli officer who managed these shipments, Colonel Benyamin Ben-Eliezer, made numerous secret trips to Beirut in a—largely vain—effort to achieve coordination between the rival Christian factions. Meanwhile, Bashir Gemayel, the Phalangist commander, met periodically with other visiting Israeli agents. Gradually, his eloquence and charm won them over. It was with their help that Bashir quietly reorganized his militia, rebuilt it to a strength of 25,000 men, and at least achieved the balance of power within the Maronite camp. The younger Gemayel also made repeated visits of his own to Israel, exhorting his contacts to intervene more directly against the Syrian garrison in Lebanon.

Then in May 1977 Likud came to power in Israel. Hopeful of a closer association with the new rightist government, the younger Gemayel pressed his case harder with the Israelis. Begin was cautious at first. He preferred to follow Rabin's course of helping the Maronites help themselves. Indeed, the Phalangists' conduct during Operation Litani in 1978 could only have increased the prime minister's misgivings. At the outset of that Israeli incursion, Bashir Gemayel had been asked to provide 800 men to reinforce Major Sa'ad Haddad's pro-Israel militia in the south. He did so only reluctantly, and most of his "fighters" soon deserted. By 1979, nevertheless, Begin was speaking all but openly of Israel's moral obligation to save Lebanon's Christians from a "genocide." Israeli jets were sent on periodic overflights of Beirut as a warning to the Moslems. At General Eytan's request, the prime minister also dispatched additional officers and equipment to Lebanon, and several hundred Phalangist officers were accepted for training in Israel. By 1980, too, Begin was increasingly taken by Bashir Gemayel, and now closed his ears to tales of occasional Phalangist atrocities against Moslems. It was in December of that year that Bashir finally secured the Israeli prime minister's guarantee of an aerial umbrella in the event of a new Syrian attack in Lebanon—and of possible "further" Israeli steps to protect the Christian minority. The blank check was destined to be a fateful one.

ARIEL SHARON'S MIGHTY GAMBLE

Following the 1981 election, as he assumed the portfolio of defense minister, Ariel Sharon soon emerged as the dominant figure in the cabinet, next to Begin himself. With his comprehensive and decisive view of Israel's security needs, the burly ex-general wasted little time asserting his influence over the nation's foreign and defense policies alike. It was Sharon, more than any other Israeli—more even than Begin—who transformed the prime minister's blank check to the Lebanese Christians into a full-blown military scenario.

Born and raised in Kfar Malal, a cooperative farm on the coastal plain, Sharon joined the Haganah while still a teenager. In the war of independence, serving as a platoon commander, he was severely wounded in the battle of Latrun, but returned afterward to combat. With a reputation for initiative and fearlessness, he was a natural choice in the 1950s to lead the elite volunteer Unit 101, the commando outfit that had been formed to combat guerrilla infiltration. Under Sharon, Unit 101 operated like the World War II Desert Rats, mounting reprisal raids across the borders, often deep into Jordan and the Sinai. The commando's daring exploits unquestionably gave Israel a needed boost in morale. But Sharon soon became overly zealous. In 1953, his raid on the Jordanian village of Qibya blew up many houses and a school, leaving sixty-nine Arabs dead. It set a pattern. During the battle for the Mitla Pass in the 1956 Sinai Champaign, Sharon's aggressiveness as brigade commander was responsible for many Israeli casualties. Later, four of his battalion officers accused him of fighting an "unnecessary battle," and demanded his resignation. Instead, he was promoted. As division commander in the Six-Day War, Sharon played a decisive role in capturing the formidable Um Cataf redoubt, blasting open the gateway to the Sinai. Soon afterward, he was charged with pacifying Arab unrest in the occupied Gaza Strip. He fulfilled the assignment, but his methods again shocked many of his troops and upset his superiors. Although an investigation later cleared him of the charge of using excessive force, he was pulled off the job.

By 1973, convinced that his future in the army was blocked, that he would never become chief of staff, Sharon resigned to take up a career in politics. Before the Knesset election could take place, however, the Yom Kippur War broke out. Sharon was called back to action as a division commander. In that capacity, he swiftly organized and executed the brilliant countercrossing of the Suez Canal. But, again, his insubordination, as he pressed the case for the canal operation by going over the heads of his superiors directly to Defense Minister Dayan and to other political leaders, risked a court-martial. Typical of Sharon's indifference to official procedures, or even to public accountability, was an aggressive press interview he gave in the last stages of the fighting. "I am commanding fourteen thousand soldiers and I have to fight with them," he asserted to the newsmen. "But at the end of the war I will screw you all. First I'll cross the Canal and screw the Egyptians, and then I'll come back and screw you all, and you had all better wear helmets." Five days afterward he suggested to a *Ma'ariv* correspondent that "the chief of staff [David Elazar] is guilty of the blunder [lack of preparedness for the Arab offensive] and should be fired at once." Asked later whether he had ever disobeyed a superior's orders, Sharon replied:

> Yes, even in the last war, I tried not to carry out instructions. First of all, I will tell you my own belief in this matter. When I receive an order I treat it according to three values: the first, and most important, is the good of the state. The state

is the supreme thing. The second value is to my subordinates, and the third value is my obligation to my superiors. I wouldn't change the priority of these values in any way.

These words, with their explicit challenge to military regulations, provoked an immediate outcry. Sensing that he had gone too far, Sharon backtracked and claimed that he had been misquoted. The issue was dropped.

By then Sharon was decisively committed to politics as his best hope for advancement. He had joined Likud in 1973, hoping to erase this right-wing bloc's internal factionalism and make it live up to its name as a unified opposition. When the effort initially foundered, he organized a separate party of his own, Shlomzion. It won two Knesset seats in the 1977 election, then rejoined the Cherut branch of Likud soon afterward. For this act of loyalty, Sharon was appointed minister of agriculture in the first Begin government, and subsequently led the vigorous effort that tripled the Jewish population in the West Bank over the next four years. Yet Sharon's obsession with settlements often was manifested at the expense of his responsibilities in Israel proper. During his tenure as agriculture minister, he gave little attention to the collapsing farm economy. Cooperative settlements went bankrupt. Vegetable and flower exports tumbled, and planters were forced to plow under citrus crops that had been productive for decades. If Common Market competition was a major factor in the agricultural crisis, this was small consolation to farmers who believed that their minister had other priorities. They were right. Sharon's preoccupation with the territories was never disguised. In the course of the election campaign of 1981, he organized promotional tours of his handiwork in the West Bank. Nearly 10 percent of Israel's population participated. These tours, and the evidence they provided of the government's success in "reuniting the Land of Israel," played a major role in Begin's reelection.

None of Sharon's efforts was inconsistent with cabinet policy. He was distinguished from other ministers essentially by his style. It was forceful, even brutal, beyond any earlier precedent at his level of government. Once he nearly came to blows with Ezer Weizman. He railed at the "insignificant blabberings" of Liberal party leader Pesach Gruper, and threatened to "strip naked" Deputy Prime Minister Yigael Yadin. Even Begin was not immune to Sharon's taunts of "cowardice" and "charlatanism." It was a confrontational approach in turn that denied Sharon the defense ministry after Weizman's resignation in 1980. Begin assumed the portfolio himself, observing that Sharon "would surround the prime minister's office with tanks if I gave the defense ministry to anyone else." Yet the flamboyant war hero's burgeoning popularity, the evident rapture with which he was greeted—often as "Arik, King of Israel"—in oriental neighborhoods, was a major asset for Begin's 1981 election campaign. Afterward, it was no longer possible to deprive Sharon of his long-cherished plum. He was fifty-three years old upon assuming the defense ministry, a bear of a man with white hair and imposing paunch, a physical bulk that was as intimidating as his temperament. Upon taking up his new post, too, Sharon typically made clear that he intended to run the ministry his way. Rather than place his confidence in the general staff, he organized a "national security unit," turning it into his personal advisory council with its own situation room and special computer. He would develop his grand strategy with his close associates alone.

That strategy was based precisely upon Sharon's forthright assessment of Israel's strength. It was his view that the nation's sphere of influence in the 1980s extended far beyond the Arab Middle East, that Israel was "the world's fourth-largest military power." Perhaps disoriented by the sheer raw might concentrated in the reprovisioned armed forces, the defense minister divulged his geopolitical scenario to his confidants. He intended first to crush the PLO as a military-political factor in Lebanon, then in Palestine, and subsequently to complete the annexation of the West Bank and Gaza. With these goals achieved, Sharon would proceed to unseat King Hussein and give Jordan over in its entirety to the Palestinians, who already comprised two-thirds of the population there. Syria and Iraq were to be destabilized, and the conservative Gulf kingdoms then presumably would be grateful for their deliverance from PLO and other radical Arab blackmail.

First and foremost in the scale of Sharon's not immodest vision, however, was the need to "clean up" the Lebanon situation, to make of Lebanon "an independent state that will live with us in peace." Such a nation required a "responsible" government, that is, a Maronite-dominated regime under the presidency of Bashir Gemayel. "A government of that kind," Sharon informed his confidants in the National Security Unit and in the Cherut party, "cannot come into being [in the scheduled September 1982 Lebanese presidential election] as long as the terrorists control southern Lebanon and two-thirds of the city of Beirut, and as long as the Syrians control whole sections of Lebanon." The implications of this master plan were clear, and they far transcended the destruction of Palestinian guerrilla concentrations along Israel's northern frontier, or even of PLO influence in the West Bank.

For Sharon, then, time was the critical factor. Syria's armed forces already consisted of 3 armored divisions, 2 mechanized divisions, 6 independent armored brigades, 2 missile regiments, and over 400 combat aircraft. If their growth continued at the same pace, they would present Israel with a serious challenge within a few years. By then Israel would long since have evacuated the Sinai, and Egypt's neutrality would be less than certain. The moment to prepare for decisive action was at hand. It was uncertain, conceivably unlikely, that Begin was privy to every feature of Sharon's blueprint, or that he endorsed it in its fullest implications. Nevertheless, the two men unquestionably shared a common appraisal of the Middle East. Throughout the first half of 1982, they functioned virtually as alter egos in their open warnings that Israel would take "immediate action" in response to fedayun attacks on the Galilee communities. Those warnings gained in frequency as occasional infiltrators crossed over. Meanwhile, the air force continued its preemptive strikes at selected guerrilla targets in Lebanon. One such raid, on May 19, 1982, drew a Katyusha response against the Galilee. Although there were no casualties, Sharon immediately authorized a major troop concentration along the Lebanese border and the Golan.

In Washington, reaction to the mobilization was swift. Fearing a major

Israeli counterstrike, the Reagan government promptly dispatched Morris Draper, Philip Habib's associate, to rescue the July 1981 cease-fire. But this time the prognosis for mediation did not appear good. In reply to an urgent appeal from Secretary of State Haig, Begin cabled back: "You advise us to exercise complete restraint and refrain from action. . . . Mr. Secretary, my dear friend, the man has not been born who will ever obtain from me consent to let Jews be killed by a bloodthirsty enemy and allow those who are responsible for the shedding of this blood to enjoy immunity." American intervention alone possibly could not have forestalled Israeli action. Almost simultaneously, however, the president of Zaire, Joseph Mobutu, announced that he was preparing to renew diplomatic ties with Israel (broken during the 1973 war), and that his special representative even then was en route to Jerusalem to inform the Israeli foreign ministry. The news was a possible bellwether of Israeli acceptance throughout Africa. If so, it was hardly the moment to precipitate another international confrontation. Moreover, the prime minister had been warned by Peres that Labor would not support a full-scale war in Lebanon. Begin was given pause. The crisis passed. Troop mobilization was canceled.

Sharon's basic strategy remained unchanged. As early as January 1982 the Israeli defense minister paid a secret visit to Beirut. There he informed Bashir Gemayel that he, Sharon, was prepared to authorize an Israeli advance to the very outskirts of the capital, provided the Phalangists themselves then would exploit their chance to take over the city. Bashir agreed with alacrity. In February, General Eytan and a group of senior officers also visited Bashir at the latter's headquarters in (Christian) East Beirut, where they were greeted lavishly with a full-scale parade and a Phalangist band rendition of the Israeli national anthem. Bashir then entreated his guests to launch their invasion soon, for he needed at least three months to organize his presidential campaign afterward. Eytan was sympathetic. The Israeli commander even discussed his strategy in some detail, including his intention to mount amphibious landings south of Beirut. By the following month, operational plans for the offensive were well advanced. Only General Yehoshua Saguy, chief of Israel's military intelligence, and General Yitzchak Chofi, director of the Mossad—Israel's CIA—were skeptical. The army could not engineer Bashir Gemayal's election through its own "good offices," they protested, and then simply withdraw afterward. But Sharon and Eytan were undeterred. Their senior staff members repeatedly visited Beirut to coordinate arrangements with the Phalange. In the end, the Lebanon expedition would be the most thoroughly planned campaign in Israel's history, with elaborate surveys of terrain, reconnaissance patrols studying all roads and bridges in the south, and combat units practicing maneuvers on a variety of models.

A key imponderable was the reaction of the United States. Here Begin was not insensitive to the suspicions left in the wake of earlier contretemps with the Reagan administration—on the annexation of the Golan, the AWACS sale, the bombing of the Iraqi nuclear reactor. Nevertheless, the prime minister had discerned a certain American acknowledgment of the Syrian threat to Lebanon's survival, even to Middle Eastern peace altogether. Secretary Haig himself had alluded to that danger in his first visit to Israel in March 1981. The United States-Israel "Memorandum of Understanding," which had been quietly reinstated some months after the Golan annexation, appeared to confirm a tentative mutuality of strategic outlook. Then, in February 1982, Foreign Minister Yitzchak Shamir arrived in Washington to sound out likely American reaction to an Israeli move against PLO nests in Lebanon. This time Haig seemed less than discountenanced. If terrorists continued to violate the cease-fire on any front, the secretary conceded, a "limited" Israeli reaction in Lebanon would be understandable.

Haig was well informed on Israeli plans. United States intelligence agencies had pieced together almost all the details of the Sharon-Eytan scenario. Indeed, Sharon himself proved remarkably forthcoming. In late May 1982, following the initial—aborted—Israeli mobilization, the defense minister paid his own "goodwill" visit to Washington. During his meetings, he shocked a roomful of State Department officials by sketching out two possible military campaigns. One was intended to pacify southern Lebanon; the second, to rewrite the Lebanese political map in favor of the Maronites. Clearly, Sharon was putting the United States on notice: one more fedayun attack and Israel would deliver a knockout blow. The threat was too flagrant even for Haig. Unless there were a provocation of "internationally recognized" dimensions, the secretary warned, and an Israeli retaliation "proportionate to . . . such provocation," a move into Lebanon would reverberate seriously in the United States. Sharon was not a man to be intimidated. His reply (in Haig's recollection) was truculent: "No one has the right to tell Israel what decision it should take in defense of its people." Like his prime minister, Sharon evidently had generated an illusory conviction of American forbearance. A stern follow-up warning from Haig to Begin on May 28 also failed to register.

By then, too, Yasser Arafat was equally well apprised of Israel's preparations. Only the timing of the assault eluded him. The guerrilla leader's own strength was not unimpressive. His "military police" and "revolutionary courts" still ruled western Lebanon from Beirut south to the "Red Line" buffer zone. Local PLO commanders ran the municipal governments in Tyre and Sidon, collected the customs in the southern ports, virtually operated Beirut's international airport. Throughout the southern countryside, rival PLO factions vied with each other in billeting their men in villages, expropriating entire buildings in cities, extorting from the wealthy, evoking the terror and hatred of local Christian and Shia Moslem communities alike. By the early 1980s, moreover, Arafat's military infrastructure consisted of five infantry brigades, four artillery and support units, a fledgling tank battalion, even the nucleus of a navy at the Syrian port of Latakia.

Notwithstanding this impressive accumulation of men and weaponry, Arafat was not yet willing to provoke the Israelis. Sobered by the pounding he had taken from Israel's air force in July 1981, he preferred to move cautiously. Within a year he expected to triple his artillery strength, from 80 guns to 250, but until then he would have to reinforce his coastal positions against the threat of amphibious landings. Thus, on May 15, even as his defense preparations were in high gear, Arafat sent Begin a message via Brian Urquhart, assistant undersecretary-general of the United Nations, who had come to the area to discuss a reinforcement of the UNIFIL contingent in the northern sector of south Lebanon's Red Line zone. "I have learned more from you as a resistance leader than from anyone else about how to combine politics and military tactics," the message said. "You of all people must understand that it is not necessary to face me on the battlefield. Do not send a military force against me. Do not try to break me in Lebanon. You will not succeed." Frozen-faced, Begin heard Urquhart out, and said nothing.

The catalyst for invasion was provided in an unexpected site. On June 3, 1982, the De La Rue Group, a British printing and publishing conglomerate, held a banquet at London's Dorchester Hotel. The 400 guests included industrialists and several ambassadors. One of the latter was Shlomo Argov of Israel. When the banquet ended, Argov left the hotel shortly before 11:00 p.m. A young Palestinian, Hassan Said, carrying a shoulder bag, had been loitering on the street outside for nearly an hour. As the ambassador entered his limousine, Said removed a submachine gun from the bag and fired one shot from a distance of 15 feet. Argov was struck in the head and collapsed. The ambassador's bodyguard then pursued Said down an alley, fired his pistol and hit the Arab in the neck. Said was seized, arrested, and confined in a prison hospital. Meanwhile, at 11:20 that same night, another Arab ran toward a parked car near the Dorchester, jumped in and drove off at high speed. A security guard at the nearby Hilton Hotel jotted down the number of the automobile, and gave it to the police. A half-hour later, the police intercepted the car in the Brixton district, where—they soon learned—Said had been living. In addition to the driver and owner, Harwan al-Banna, the automobile contained a passenger, Nuwaf Rosan. With these men were two bags containing guns. In Banna's room in a YMCA hostel, another case was found with an impressive load of automatic weapons, grenades, and detonators. Upon being questioned, the three Arabs confessed to their involvement in the attempted assassination (Argov survived, although he would remain paralyzed).

In court, later, the defendants retracted and accused the police of extorting their confessions. But the circumstantial evidence against them was overwhelming. The trial made clear that the three had planned to hit a wide variety of Jewish organizations and individuals. It was a characteristic pattern of their organization, the Palestine National Liberation Movement. Far from being connected with Arafat, the PNLM's leader, Sabri al-Banna, a.k.a. "Abu Nidal," was bitterly opposed to Arafat's leadership. He had often referred to Arafat as "the Jewess's son" and had made repeated attempts on his life, even as Arafat had pronounced the death sentence on Abu Nidal. The PNLM in fact was trained by Iraqi intelligence—Rosan turned out to be a colonel in the Iraqi army—and its undisguised objective was to provoke the Israelis into an attack on the PLO in Lebanon. When this information was presented to Begin by Israeli intelligence, the prime minister declined to reveal it to the cabinet.

The following day, the Israeli air force began a massive bombardment of PLO targets in Lebanon, battering arms depots and military camps beyond the Red Line the entire distance up to Beirut. The pilots had been well briefed. One major arms cache was located beneath the grandstand of Beirut's soccer stadium. It was destroyed. In the absence of Arafat, who had departed for Baghdad several days earlier to offer his mediation services in the Iraq-Iran War, the reaction of individual PLO factions was to launch a sporadic shelling of Israeli settlements in the Galilee. There were several casualties and heavy property damage. By then, Sharon had rushed back from a scheduled visit in Bucharest. At an emergency cabinet meeting of June 5, he and Begin insisted that the army would have to strike immediately to activate a long-prepared strategy of clearing a 25-mile strip along the Lebanese border. If this were accomplished, the PLO's artillery would be pushed decisively beyond range of Israel's northern settlements. The other ministers were impressed, for the most part. When several expressed tentative reservations that even a limited invasion risked a wider involvement, Sharon blandly assured them that Beirut was "outside the picture," that the limited riposte—"Operation Peace for the Galilee"—would be completed within two or three days, that every effort would be made to avoid a confrontation with the Syrians. The cabinet then gave its approval. So did Peres and Rabin, when Begin informed the Labor leaders of the proposed assault.

A SCHIZOPHRENIC OFFENSIVE

Sharon and Eytan had prepared their strategy carefully. At the outset, on June 6, 1982, three Israeli divisions closed in on the Palestinians from all sides: by a direct advance along the coastal road; by amphibious forces landed between Sidon and Damour; then by a division sent into the central sector, through the Shouf Mountains. The unfolding offensive was envisaged as a classic pincer movement that would encircle south Lebanon, and strike across to the coast. Under no circumstances would the failure of the 1978 expedition be repeated, when the Palestinians had quietly faded away, then subsequently had regrouped. This time they would be cut off entirely, and demolished. At first, too, the operation went according to plan. No resistance was encountered from the UNIFIL force; the hodgepodge of foreign units quickly dispersed as Israel's mechanized columns pounded in from the south. One of the first objectives of the coastal advance was Beaufort Castle, the 1,200-foot-high Crusader fortress that for years had served the PLO as a firing platform. Here the Palestinians had defied all earlier attempts to bomb them out, honeycombing their bastion with passages and cellars. Now, however, an Israeli infantry team climbed directly up the cliff face, with helicopter gunships forcing the defenders undercover. Some 150 Israeli commandos participated in the assault, moving through the interior labyrinth to winkle the Arabs out, and ultimately to kill them all. By the end of the first day, Beaufort was in Israel's hands. The castle then was turned over to Major Haddad's Christian militia.

Other infantrymen moved by coastal road and amphibious landing to the key PLO strongholds at Tyre, Nabatiyeh, and Sidon. With the help of pulverizing air bombardment, the Israelis overran all three towns by the end of the second day. On the third day, June 8, Israeli troops reached Beiteddiq. By then they were already beyond the 25-mile security zone that had been the declared objective of Operation Peace for the Galilee. Within those three days, almost all PLO forces south of Beirut were outflanked, and ensuing hostilities in the coastal area later were essentially a mop-up procedure. From Israel's side, there were few acts of spectacular heroism and little bravura. With the exception of the Beaufort capture, and the imaginative amphibious and helicopter landings, moves were orchestrated carefully and conservatively, exposing Israeli forces to a minimum of casualties. Sharon in the past had been the most audacious of Israel's commanders. But this time, lacking cabinet support for the wider-reaching campaign he envisaged—the link-up with the Phalangists in Beirut, the eviction of Syrian forces from Lebanon— he was obliged to proceed cautiously. By limiting his offensive at first to the narrow coastal and mountain regions, he constricted the army's opportunities for mobile outflanking tactics. It was air superiority that kept the momentum in Israel's favor. Some 80,000 troops were deployed in the offensive. This was a larger concentration of manpower than had engaged the Egyptians in the entire Yom Kippur War. Indeed, it was rather too large for a confined and mountainous area of less than 1,000 square miles. Half the number of troops and heavy vehicles possibly would have been more effective.

As the Israelis drove northward, meanwhile, the Syrians dispatched 16,000 additional troops to Lebanon, bringing their contingent in the little country to nearly 40,000. For the first three days, this force remained comparatively inactive. President Assad was not seeking a confrontation. Neither was Begin. The Israeli prime minister repeatedly made clear his limited purpose in messages to Damascus through Philip Habib, who had been rushed back from Washington to mediate. Assad's restraint was of course characteristic of his policy in the last few years before 1982. The mandate of the UN observer force had been routinely extended since 1974, and throughout Syria's entanglement in the Lebanese civil war its government had remained largely sentient to Israel's warnings of "unacceptable" actions. Nowhere was this forbearance more evident than during Israel's Operation Litani incursion in 1978.

As early as the second day of Operation Peace for the Galilee, however, Eytan's coastal invasion toward Beirut threatened to link up with the Phalange, and thus to isolate the main body of Syrians in the Beka'a Valley from Syrian units in the capital. Another Israeli division was moving through the Shouf Mountains in an effort to sever the Beirut-Damascus highway. The advance through this central axis in fact was Defense Minister Sharon's pivot for his entire Lebanese campaign. If it managed further to separate the two Syrian forces, access would be opened to the Maronites without requiring Israeli entrance into Beirut (a quagmire that all Israeli generals had been trained to avoid). The strategy no doubt evinced a certain logic in foreclosing

Syrian influence in Lebanon—provided it could be fulfilled swiftly. To that end, the Israeli column moving through central Lebanon would itself have to be divided into two parts. One would traverse the Shouf Mountains directly to the Beirut-Damascus highway, the other would advance toward the Beka'a Valley in a parallel move to sever the artery at its eastern entrance. But here lay grave potential danger. The second arm, which was strengthened ultimately to a body of 40,000 troops, and commanded by one of Israel's most redoubtable fighters, Major General Avigdor ("Yanosh") Ben-Gal, almost inevitably would confront the largest body of Syrian troops in Lebanon. At that point, Assad either would have to evacuate this force. from Lebanon completely, as Sharon hoped, or he would have to fight back. It was a serious gamble, one that perhaps even Begin did not anticipate. Sharon did, but kept its implications from the cabinet.

The issue was forced on the morning of June 7, when Ben-Gal's troops attacked Syrian units at Djezzin, a junction controlling access to the southern Beka'a Valley. It was a costly, full-scale battle. Calling up reinforcements, the Syrians launched a heavy counterattack of their own. They inflicted serious casualties on the Israelis, whose armored column was stopped in its tracks. As Sharon and Eytan viewed the situation, the single alternative now for "neutralizing the Syrian presence" was a wider-flanking move against the enemy in the heart of the Beka'a itself, much nearer the Syrian frontier. Yet, for an improved chance of success in this climactic battle, air supremacy was vital, and here no alternative existed but to strike directly against the SAM-6 missiles ensconced in the valley. It was a sobering prospect. As he explained it to the cabinet, however, Sharon chose not to reveal the strategy's full implications. Rather than acknowledge his maximalist purpose of eradicating the Syrian enclave in Lebanon altogether, the defense minister argued that bombardment of the missiles was necessary to "minimize Israeli casualties" in the Lebanese coastal zone. Begin supported this argument. After only brief hesitation, the cabinet members approved the attack, without understanding that they also had endorsed a major ground operation against a powerful Syrian army.

AN ESCALATION OF HOSTILITIES

Although the air operation alone was an exceptionally risky business, Sharon and Begin were determined to have done with it sooner or later. The SAM batteries in the Beka'a Valley, installed in the spring of 1981 (Chapter VI), were intended to protect a line that Damascus regarded both as an indispensable access to its garrison in Beirut and as a natural Israeli invasion route into eastern Lebanon—and thus into western Syria. "To the Syrians," wrote Keith Bulloch, who visited the Beka'a front, "the slim white missiles with their sinister black nose-caps were the symbols of their defiance of Israel. . . . They sited them ostentatiously beside roads and on small hills so that everyone could see them." And they fired them off in earlier months whenever Israel's reconnaissance drones intermittently passed over. That was the Syrian mistake. In the few seconds between the firing and the destruction of these small, pilotless craft, the drones had identified and radioed back to Israel the Syrian radar frequency. Appropriate electronic countermeasures were taken. On June 9, 1982, the Israelis launched their attack. Wave after wave of planes came over, F-15s flying top cover, F-16s carrying bombs, and a Hawkeye command plane, crowded with electronic gear, directing operations and jamming the Syrian radar. The SAMs were launched. They failed to register a single hit. Conversely, seventeen of the nineteen missile batteries were knocked out by Israeli bombs. The following day, another wave of Israeli jets finished the job, destroying the remaining batteries.

Almost as costly for the Syrians was the beating taken by their air force. During the first Israeli attack on the missiles, the Syrians came in with their MiG-25 interceptors. The Israelis shot down 22 of these without a loss. The next day, in the second attack, another 24 Syrian jets were destroyed, again without Israeli losses. In the ensuing days, nearly 50 additional Syrian fighters would be downed without cost to the Israelis. Altogether, 20 percent of the Syrian planes that crossed into Lebanese airspace were destroyed or damaged. Half the pilots of the downed craft also were lost. It was a shattering setback. As early as the night of June 9–10, the Israeli air force was free to ravage the Syrians on the ground. When the latter attempted to move their crack Third Armored Division westward, Israeli planes macerated their vehicles (even as Israel's Merkava tanks exacted a heavy toll of the Soviet-built T-72s). It was after these forty-eight hours of disaster that the Syrians decided to pull out of active fighting entirely, although without official announcement of the fact. Despite its loss of SAM missiles and, ultimately, of more than 100 planes, the Assad regime understood that the war still was being fought in Lebanon, not in Syria, that the Israelis evidently were uninterested in penetrating Syrian territory.

With or without further Syrian participation, however, the war suddenly appeared much more far-reaching than the Israeli cabinet or people had been led to anticipate. So it did to the Great Powers, who entered the scene four days after the onset of the invasion. By then, Assad had sent his defense minister, Mustafa T'las, to Moscow in quest of a Soviet air umbrella. Although the Soviets rejected Syria's appeal for direct intervention, they consented to send equipment and advisers. It was commitment enough to alarm Washington. On the morning of June 10 Reagan dispatched an urgent message to Begin, pointing out that Israel's incursion into Lebanon already far transcended Jerusalem's original announced purpose, and that the Russians were showing signs of edginess. Israel would have to accede to a cease-fire immediately, the president warned. Hereupon Begin convened an emergency cabinet meeting. In their initial, qualified response, the ministers endorsed the formula devised by Begin and Sharon. It was to accept a cease-fire, provided the Syrians withdrew their latest troop reinforcements from the Beka'a Valley, and the PLO withdrew its remaining units beyond Israel's proclaimed 25-mile zone. These were harsh, possibly unacceptable, conditions. Begin's and Sharon's manifest purpose was to win leeway to eradicate

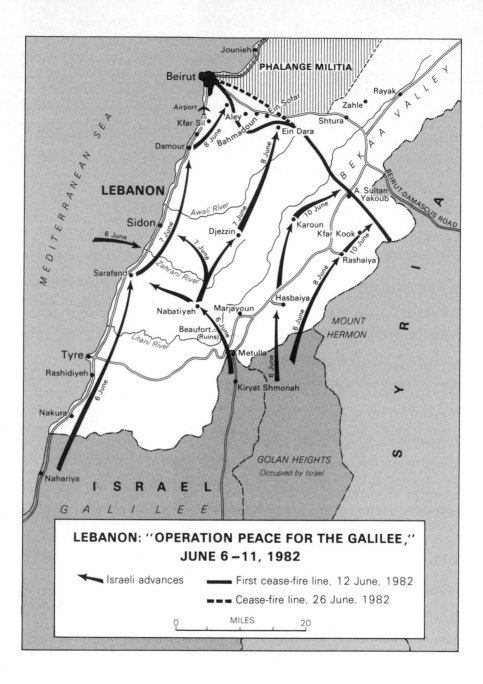

LEBANON: "OPERATION PEACE FOR THE GALILEE,"
JUNE 6 – 11, 1982

Israeli advances   First cease-fire line, 12 June, 1982

Cease-fire line, 26 June, 1982

0   MILES   20

the PLO infrastructure, to gain full control of the Beirut-Damascus highway, and thus neutralize Syrian influence in central-eastern Lebanon. Yet, in the end, after a stern telephone warning from Haig, the prime minister and his cabinet agreed to a cease-fire for noon the next day, June 11. Possibly there would still remain enough time to implant Israeli forces solidly astride the east-west highway.

There was not. Although General Ben-Gal, leading the advance against the Syrians on the central axis, had been given two full divisions and now enjoyed unlimited air support, his troops were obliged to move through winding, undulating terrain and narrow mountain roads. With hundreds of tanks, the Israelis often were reduced to a snail's pace, and the Syrians fought hard defensive battles at key junctions. On one occasion, Ben-Gal's men fell into a serious ambush at Sultan Yakoub, suffering many tanks knocked out and heavy casualties. When a group of Israeli Orthodox reservists panicked, the column had to pull back, leaving behind a number of valuable weapon systems. Syria's forces in Lebanon unquestionably took a thorough drubbing, but they were far from destroyed. Even as they accepted the cease-fire on June 11, their army remained entrenched in fully 35 percent of Lebanon, and in control of two major stretches of the Beirut-Damascus highway. For at least one of Sharon's major goals, it was a less than decisive fulfillment.

There was still hope, nevertheless, of achieving another of the defense minister's objectives. This was elimination of the PLO quasi government in Lebanon. To that end, the Israeli government ignored Yasser Arafat's announced willingness to participate in the June 11 truce, observing that his guerrilla force was not a responsible government but a band of terrorists. Emphasizing the point, Israeli forces on June 12 launched into a sustained air, land, and sea bombardment of Palestinian strongholds south and west of Beirut. The previous afternoon, Sharon had met with Bashir Gemayel in Jounieh, on the outskirts of the capital. It was at this meeting that the underlying fragility of the Israeli-Maronite relationship again was exposed, and again ignored by Sharon. Despite his earlier commitment, Bashir this time gave the Israeli defense minister no assurance that his Phalangists would move on their own against the PLO in Beirut. Sharon was incredulous. "We'll soon be here with tanks," he exploded. "Do something!" In fact, it was the Israelis whom Bashir expected to "do something," namely, the dirty work of cleaning out West Beirut. The young Maronite commander himself remained evasive—although, in the end, he agreed to a link-up of forces outside Beirut.

The bulk of Israel's army on the coastal road meanwhile was engaged in a painful, frontal offensive against Kafr Sil, a wealthy suburb in the hills above the Beirut-Damascus highway. Here the local Syrian garrison waged a tough resistance; for, like the Israelis, they regarded the cease-fire as inapplicable to the Beirut area. The battle of Kafr Sil accordingly developed into the single most brutal of the war, continuing for a day and a half. Israel's crack Golani Brigade required prolonged artillery barrages and aerial bombardment to break the Syrian defense line. Further inland, Israeli paratroops moved at an agonizing pace toward the final ridge before Beirut. Again local Syrian forces did the heaviest fighting. This resistance, too, was overcome only in foot-by-foot combat.

Finally, at 1:00 p.m. of June 13, at a roadblock outside Ba'abda, the eastern entrance to the capital, Chief of Staff Eytan and the first Israeli paratroopers greeted Bashir Gemayel and his father, Sheikh Pierre, together with a contingent of Phalangist militiamen. In Jerusalem, then, Begin jubilantly announced that "the Beirut-Damascus highway has been severed at Ba'abda." Sharon declared only that Israeli troops had "linked up with the Christians" in Ba'abda. No mention was made of Beirut. But whatever the euphemism, the defense minister was not entirely dissatisfied with his accomplishments thus far. Several key objectives of Operation Peace for the Galilee had been reached. The PLO infrastructure south of Beirut had been destroyed. The Syrians had been soundly thrashed, if not altogether evicted from Lebanon. The Israelis had cut a major stretch of the Beirut-Damascus highway. They had linked up with the Phalangists in East Beirut.

There were other features of the campaign that evoked satisfaction. The fighting appeared to vindicate the past decade's heavy investments in Israel's arms industry. In 1973, about half of Israel's weapons were locally made. By 1982, the proportion was two-thirds, and these planes, tanks, guns, communications, and other support equipment had functioned marvelously well. In a single twenty-four-hour period the Israelis had obliterated more SAM batteries than during the entire 1973 Yom Kippur War. They had all but knocked the Syrian air force out of the skies. They had captured vast quantities of PLO equipment. After inventory was completed in October, the list of PLO material included 1,320 armored combat vehicles, among them several hundred—obsolescent—tanks, 82 field artillery pieces, 62 Katyusha rocket-launchers, 215 mortars, 196 antiaircraft guns, 1,342 antitank weapons, 33,000 small arms, and thousands of pieces of communications equipment. Ultimately 4,330 truckloads of this booty were driven back to Israel (much of it to be sold later to other nations). Jerusalem had contended, too, that the destruction of the PLO in south Lebanon was an act of liberation. The claim seemed borne out by local Arab reaction to the Israeli invasion. Foreign journalists confirmed that the inhabitants of south Lebanon were euphoric at the destruction of the Palestinian guerrilla forces. Shia Moslems and Christians alike vied with each other in welcoming Israeli troops, inviting Israeli officers to their homes. When the Israeli army "detained" 9,000 PLO suspects—later reduced to 6,000—Christian and Shia villagers eagerly helped in the task of identification.

Yet the expedition also had proven a more harrowing psychological ordeal than Sharon or Eytan (or Begin) had anticipated. As an example, one of the army's initial objectives was the Rashidiyeh refugee camp, just over the Red Line, on the road to Tyre. First built in 1948 by UNRWA, the camp had grown into a sprawling shantytown, with 30,000 Palestinians impacted there, many of them armed PLO troops. Thus, well after Israel's occupation of the surrounding hinterland, its troops were obliged to spend nine days of concentrated bombing and shelling to subdue Rashidiyeh. Tyre and Sidon also were heavily battered in the Israeli advance, and similar devastation was inflicted on the inhabitants of Damour, a market town 9 miles south of Beirut, where the PLO mounted fierce resistance. The technique of a "smashup" war, of heavy bombardment, had been selected to avoid the casualties Israel had suffered in Operation Litani. To be sure, Israeli troops issued repeated loudspeaker-and-leaflet appeals for the townspeople to flee, and every effort was taken to avoid civilian losses. But the cost to the Palestinians—PLO fighters and noncombatants alike—was still quite terrible, a fact that ultimately registered on world opinion. Estimates later put the number of Palestinian dead in south Lebanon's five main refugee camps at over 6,000. Israeli troops who entered these complexes afterward were aghast at the carnage wreaked by the bombardments. They were shocked, too, by the vindictiveness of the Christian and Shia authorities, who refused the Red Cross permission to set up tent shelters for Palestinian victims.

The price in Israeli casualties was also higher than Begin had imagined. In south Lebanon, much of the Arab resistance was carried out by the refugee camps' "home guards." Tough street fighters, they inflicted severe punishment on Israeli troops moving up the coastal route. In the hostilities raging near Tyre alone, Israel lost more than 100 officers and men. In one particularly ferocious battle at the refugee camp at Ein Hilweh, the home guards defended their enclave as if it were their Masada, blocking the Israeli advance for two days. On June 9 an Israeli brigade fought its way through to the Awali River, linking up with amphibious troops that had arrived earlier; but an additional eight days were required to overcome lingering resistance in the Ein Hilweh casbah. As of June 13, Israel had lost some 130 dead and 600 wounded. Among those killed was Major General Yekutiel Adam, a former deputy chief of staff, whose appointment as director of the Mossad was to have taken effect three days later. Adam was the most senior Israeli officer ever to fall in battle. News of these losses set off widening tremors of concern in Israel. Even then, no one could have anticipated that far worse was to come.

THE LEBANON WAR: "OPERATION BIG PINES"

A SHIFT OF STRATEGIC OBJECTIVES

After a week of war, Israel's army was in effective occupation of southern Lebanon. Its tanks and self-propelled guns surrounded Beirut, trapping some 15,000 Palestinian guerrillas intermingled in the western part of the city among a half-million local Moslems. Now was the time for Sharon to set in motion "Operation Big Pines," his carefully guarded master plan for smashing the PLO infrastructure in Beirut itself, driving its partisans out of the capital and out of Lebanon altogether. As the defense minister saw it, once the Lebanese people were emancipated from PLO intimidation, Bashir Gemayel would be free to conduct and win his presidential campaign, to organize a viable central government, and eventually to sign a treaty of peace with Israel. Not least of all, in Sharon's scenario, the destruction of the PLO quasi government in Lebanon would shatter Yasser Arafat's influence among the inhabitants of the West Bank. The Palestinians' "authentic" spokesmen then might step forward to negotiate with Israel on the basis of the Camp David accords—as interpreted by the Begin government. It was an audacious scheme, one that far transcended the officially proclaimed goal of Operation Peace for the Galilee. Yet it appeared to be fully shared by Lebanon's Christian communities. Their exuberant "shaloms" to Israel's invading forces were echoed by the Gemayel family. On the afternoon of June 13, 1982, Bashir Gemayel and his father, Sheikh Pierre, had awaited Israeli paratroops at the outskirts of Beirut, to embrace them in hearty welcome.

Now was the moment, too, under the protection of Israeli air and artillery power, for the Phalangist leaders to honor their part of the bargain, to send their militia directly into West Beirut and complete the rout of the PLO. But here the agreement hung fire. The Gemayels revealed themselves as curiously indecisive. They pleaded the difficulty of attacking 15,000 combatants ensconced in the city's maze of streets and buildings. For them, in fact, the danger was as much political as military. Only two months remained before the scheduled presidential election. Bashir Gemayel's chances had been mightily improved by the Israeli presence and by the punishment inflicted on the Syrians. There was no need, in his view, to jeopardize a possible future reconciliation with Lebanon's Moslems by engaging in active combat, even against the Palestinians. Despite heavy pressure from Sharon, therefore, the Phalangist leadership refrained from overt military action. They provided the Israelis with a certain logistical and intelligence support, including telephone numbers to tap, sewer maps, the location of key Palestinian installations and observation posts in West Beirut—but nothing more. For its part, the Israeli command soon faced a painful decision. Their troops had pushed too far to leave the PLO enclave intact. Yet the notion of moving directly into the city and exposing troops to bitter street fighting was unacceptable. Accordingly, after a few days of hesitation, Eytan and his staff came up with an alternative, one that gained immediate cabinet approval. It was to uproot the PLO guerrillas by siege, by heavy aerial and artillery bombardment, by interdicting munitions and food supplies.

In retrospect, the decision was unimaginative. Begin's and Sharon's condition for a cease-fire was the departure of all PLO and Syrian troops from Beirut. The demand conceivably might have been met if the Israelis first had sought tacit acquiescence from Damascus, or if they had asked Cairo to mediate. Sensing a chance to shatter the PLO's image as well as its presence, however, the Israeli commanders were determined to seize the opening. Eytan set about protecting his flanks by clearing the pockets of Syrian troops still ensconced in the hills east of Beirut and near the Beirut-Damascus highway. He accomplished the task, but only after two weeks of heavy combat. His effort to break the PLO within the surrounded city itself proved even more difficult. Arafat and his advisers had gauged precisely the Israelis' reluctance to engage in street fighting, or to offend world opinion. Indeed, Begin himself had forfeited his psychological leverage in the first week of the campaign, when he assured the Knesset that the army never would become embroiled in the occupation of a major Arab city. With this insight into the Jews' mentality, Arafat and his advisers now publicly declared their intention of turning West Beirut into a Stalingrad.

But the PLO counterbluff similarly evinced a misreading of Sharon's character, his willingness to break with Israel's historic military tradition of short, sharp reprisals, and to embark upon a war of siege and attrition. Indeed, during the ensuing two months, some 400 Israeli tanks and 1,000 guns poured salvo after salvo into West Beirut. For the first time too in the nation's long series of wars against the Arabs, Israel's air force bombed an enemy capital systematically. Israeli intelligence was able to plot the location of all known PLO strongholds with the help of the elaborate master map prepared by the Phalangists. By the first week of July, some 500 buildings had been destroyed by shells and bombs. In the world's first manhunt by air, the Israelis also nearly succeeded in killing Arafat. On four different occasions, with information provided by their own spies and local informers, they bombed or rocketed targets less than an hour after Arafat's departure.

As Sharon had hoped, the pulverizing siege rapidly became unendurable for the Palestinian guerrillas. In late June, through Philip Habib and other intermediaries, Arafat let it be known that he was prepared to withdraw his men from the city, if appropriate terms and guarantees could be worked out. These were less than clearly defined, however. At first, the PLO leader insisted on retaining an office in Beirut, then on attaching his units to the regular Lebanese army. The Israelis rejected all qualifications. Through Habib, and by their own leaflets and loudspeakers, they warned that evacuation must be total. Throughout July the siege was tightened. Food, water, and electricity were intermittently cut off from West Beirut. Late in the month Arafat produced a "compromise" offer. He would withdraw his men from the capital in a phased, three-step evacuation; but the Israelis for their part first would have to pull back from Beirut and from the Beirut-Damascus highway, then allow the entry of a multinational force as protection against Israeli or Christian reprisals. Again, the reaction from Jerusalem was negative. A one-stage PLO withdrawal was Israel's sine qua non, and there would be no pullback as long as a single Palestinian fighter remained in Beirut. Foreign Minister Yitzchak Shamir informed Habib, too, that Israel's patience was running thin, that the cabinet was prepared to raze Beirut. "In my view, we must not leave a single terrorist neighborhood standing," he warned. To intensify the pressure, Israeli troops on August 4, 1982 took over Beirut's international airport, then began inching forward selectively into the western part of the city. The air force meanwhile carried out its single heaviest bombing of the war—127 sorties in ten hours. Simultaneously, naval vessels offshore joined in the bombardment, launching Gabriel missiles into West Beirut. It was a terrifying crescendo.

Arafat was completely unnerved by then. It was not the Israelis' hard stance alone that eroded his confidence. Where would his troops go? Despite Habib's strenuous exploratory negotiation, not a single Arab country was willing at first to offer the Palestinians sanctuary. The Syrian and Iraqi regimes needed thousands of PLO troops on their soil as they needed the plague. Jordan's King Hussein had waged open war to evict the Palestinians in 1970–71, and was not about to welcome them back now. Even as Habib sought the intercession of the Saudi government, meanwhile, the Israelis tightened the vise, moving additional units into the outlying areas of West Beirut. It was not until August 10, responding to Saudi entreaties, that Syria's President Assad grudgingly consented to accept a maximum of 4,000 PLO combatants "for whom no other refuge has been found." The Hashemite government also relented then, agreeing to accept a token force of 2,000. Others of Arafat's men would be dispersed through North Africa, principally in Tunisia.

That same August 10, Habib submitted the rest of the negotiated blueprint to the Israelis. Under its terms, French, Italian, and American units would enter Beirut, simultaneously with PLO and Syrian evacuation; even as the Israelis would pull back initially to the Alawi River, some 28 miles north of the Lebanese-Israeli frontier. It was a less than ideal arrangement for the Begin government, which still would have preferred the arrival of the multilateral force only in the aftermath of PLO evacuation. But under heavy pressure from Washington, including a personal telephone call from Reagan, Begin acquiesced on August 12, then sold the plan to the cabinet. Simultaneously, however, as a signal to Arafat against a last-minute change of heart, Sharon ordered an unprecedented saturation bombing of West Beirut. It continued for nearly a day, and killed at least 300 people. Reagan was appalled. "Menachem, I think we've been very patient," he warned the Israeli prime minister in yet another telephone call. "Unless the bombing ceases immediately, I'm fearful of grave consequences in relations between our countries." Both government leaders had made their point by then. The bombardment ceased.

THE UGLY ISRAELI

The August 12 saturation attack inflicted a particularly severe blow to Israel's image in the democratic world. Indeed, from the outset, Operation Peace for the Galilee had evoked little international sympathy. On June 6 a UN Security Council resolution demanded an immediate cease-fire and withdrawal of Israeli forces. In Egypt, the Mubarak government harshly condemned Israel's military operation, then announced that it was suspending its participation in the Palestine autonomy talks and postponing "normalized" relations with Israel altogether, so long as Israeli troops remained in Lebanon. Here Mubarak was responding to public opinion. Egyptian press criticism of Israel was spontaneous and virulent. "We perceived the peace with Israel to be the cornerstone for a comprehensive peace in the Middle East," one editor observed bitterly to a group of visiting Israeli journalists, "but for you peace was merely a trick to neutralize us so as to more easily strike at the Palestinian people."

The reaction in other, neutral, lands was hardly friendlier. Western newspaper and television reportage gave a sharply biased account of the war, often accepting PLO figures on Lebanese and Palestinian dead and wounded. Journalists drew portentous analogies between the "genocide" of Beirut and the genocide of Warsaw in World War II. In Norway, traditionally one of Israel's most dependable friends, Prime Minister Kåre Willoch declared his government's support for the Palestinian people's "right to self-determination and national sovereignty." In the Netherlands, a poll revealed a bare 16 percent support for Israel's invasion of Lebanon; 90 percent favored the Palestinians' right to self-determination. The Dutch minister of the interior, E. van Glynn, himself a Jew, proposed the imposition of sanctions on Israel. In Britain and France, there was broad governmental and public condemnation of the Israeli invasion, an attitude Begin did not hesitate to characterize as antisemitism.

As it happened, the Lebanon operation, the siege of Beirut, also evoked concern and confusion within the Diaspora. Although the Jewish communal establishment largely accepted Israel's account of the PLO danger, some of its leaders were torn. As the Joint Israel Appeal in Britain organized an emergency fund-raising campaign for Israel, Lord Marcus Sieff, a veteran Zionist, donated 20,000 pounds to the Red Cross to ease the suffering of Lebanon's civilian population. It was a clear act of political protest. Several other Anglo-Jewish spokesmen now resigned their leadership positions, including Neville Sandelson, M.P., vice-chairman of the British-Israel Friendship Association. In France, Pierre Mendès-France lent his voice to those of two other prominent Diaspora figures, Nahum Goldmann and Philip Klutznick, in an appeal for Israel to end the war and extend recognition to the Palestinians. To a journalist from *Le Nouvel Observateur,* Mendès-France declared that "what [Begin] is doing is a tragedy for the entire world, for his people, and for peace." In the United States, Jews were not less divided, and officials of several communal organizations—among them, the American Jewish Committee, the American Jewish Congress, the (Reform) Union of American Hebrew Congregations—quietly transmitted to Jerusalem the misgivings of their constituencies.

A parallel shift took place in United States government policy. At the outset of the Lebanon invasion, Alexander Haig displayed a certain cautious understanding. On the one hand, he dispatched a firm letter to Begin on June 10, urging a cease-fire. But the secretary of state also regarded Israel's siege of Beirut as a unique opportunity to expel all foreign troops from Lebanon, and thus to establish a stable government in that land and a safer common border with Israel. With the threat of the PLO gone, too, Haig explained at the White House, the Begin government might be persuaded to adopt a more flexible stance in the Palestine autonomy negotiations. "I believed that if we acted quickly and with a sure hand," he wrote later, "the United States could achieve all this." But if Haig initially persuaded Reagan to be forbearing with the Israelis, he soon was undermined by National Security Adviser William Clark and Secretary of Defense Caspar Weinberger. The latter were determined to maintain Washington's credibility among Arab moderates. Making no secret of their wish to exert pressure on Israel, Clark and Weinberger conceivably stiffened PLO resistance. "As talks [on PLO evacuation] continued," Haig recalled with some mortification, "drained of their urgency, it was possible that the Saudis, after interpreting these 'signals' from the United States, had advised the PLO to play for a better outcome. The opportunity for a quick, clean withdrawal of the PLO and an end to bloodshed in Beirut had been destroyed." Additional weeks of Israeli blockade and saturation bombing were required to force a PLO departure—with devastating consequences for Israel's reputation.

Thus, when Begin arrived in Washington to visit Reagan on June 21, 1982, only days before the siege of Beirut, he was received coldly. Neither did the prime minister help his cause during his irascible appearance before the Senate Foreign Relations Committee. Sharply questioned, reminded by the senators of the magnitude of American financial help to Israel, Begin snapped back that the United States could keep its money. Haig monitored these developments with growing chagrin. He sensed that Israel was forfeiting its decisive edge in American goodwill by then. The liberal Catholic journal *America* authentically expressed the national press reaction. Conceding that the Jews and Israel had suffered much, that Israel often was judged by an unfair double standard, the editorial noted too that Israel similarly had benefited by that double standard.

> In our judgment the events of the last two months represent a watershed in the history of Israel, both in its relationship to the United States and in its moral identity before the community of nations. . . . Even [Israel's] enemies had to admire the courage of the Israeli military at Entebbe; even its friends had to be

dismayed by its excesses in Beirut. . . . What has been sown in the rubble of Beirut and the maimed limbs of its children has not been a harvest of peace but a legacy of hatred and violence. The people of Israel should be counted among its victims.

Under the Habib "compromise" formula, meanwhile, the evacuation of PLO troops was scheduled to begin on August 21, 1982, and to end on September 4. Remarkably, all went according to plan. The first Palestinian contingents began their withdrawal precisely as 300 French legionnaires, the first contingent of the multinational force, disembarked at Beirut. Then, after several more days of bravado, of gun-firing parades and emotional goodbyes, and as American and Italian units arrived to join the peacekeeping force, additional Palestinian troops began their departure. They were hooted and jeered by thousands of Christian Lebanese onlookers. The long years of PLO intimidation and terrorism appeared to be ending. Escorted from Beirut harbor by American naval vessels, a Greek ship carried off the first load. Altogether, during the next twelve days, some 14,000 Palestinian and Syrian fighters were evacuated from the capital, either by sea or overland on the Beirut-Damascus highway.

Bashir Gemayel and senior Israeli officials were among those watching the departing conveys. Counting the vehicles, they suspected that at least 2,000 combatants remained behind, as well as weapons caches that were likely to be handed over to Lebanese Moslems. Nevertheless, the Israelis were prepared to honor their part of the bargain, to end the siege, to restore water and electricity in West Beirut, to accelerate their retreat southward. The largest numbers of guerrillas were gone, after all. The prospects of peace even looked bright enough for the multinational force to carry out its own evacuation between September 10 and 12. By then the Israelis fully expected that Lebanon's newly elected government would manage at long last to keep order in its own house, and simultaneously to enter into a treaty relationship with its Jewish neighbor. This, in the end, had been a salient objective of Sharon's grand strategy.

AN OPERATION GONE SOUR

Weeks earlier, in anticipation of the Lebanese presidential election, scheduled for August 23, Bashir Gemayel set about mobilizing the required two-thirds parliamentary majority. His task was a challenging one. Numerous Moslem leaders remained in opposition. So did more than a few Christians, even Maronite Christians, including members of Camille Chamoun's rival bloc. The Israeli government was prepared to cooperate with Bashir. It dispatched a team of Arabists to work closely with him in rounding up his seven or eight missing votes. Thus, in their southern zone of occupation, the Israelis "advised" Shia parliamentary delegates to turn up in Beirut for the election, even provided a helicopter for one elderly delegate in an isolated Beka'a village. In public, Bashir maintained a cool stance on Israel, intimating that a Lebanese-Israeli treaty would require the approval of the "entire" nation. The Israelis understood this restraint. To avoid the impression of overt pressure, they withdraw their troops from the parliament site on August 19.

The mixture of carrot and stick appeared to work. On August 23, two days after the PLO began its evacuation, Bashir Gemayel was indeed elected president of Lebanon by 57 of the 62 parliamentary delegates who attended the session. The rank and file of the nation's Christians were gratified. So, plainly, were the Israelis. It appeared that Lebanon soon would be in a position to follow Egypt's lead to the peace table. "Warmest wishes from the heart on the occasion of your election," read the telegram from Jerusalem to Bashir Gemayel. "May God be with you, dear friend, in the fulfillment of your great historic mission, for the liberty of Lebanon and its independence. Your friend, Menachem Begin."

As it happened, Bashir did not rank a peace treaty at the top of his priorities. His first objective now was to consolidate his strength among all elements of his population; the next, to reestablish ties with the surrounding Arab world. Under no circumstances could the president-elect risk a boycott against Lebanon's shattered economy. Politely, then, he evaded Begin's insistent appeals for an early peace conference. Whereupon the Israeli prime minister pressed even more insistently for an early installment of the Gemayel debt. On September 1, Bashir was flown to Naharia in northern Israel for a postelection meeting with Begin. He was kept waiting in the guest house for two hours. It turned out that the prime minister was detained by a conference with United States Ambassador Samuel Lewis, who was privately conveying the essence of President Reagan's impending new formula for Palestine.

The American proposal left Begin stunned and disbelieving. Once the PLO was beaten and scattered from Lebanon, he had expected Washington quietly to drop its objections to Israel's settlements policy. But Alexander Haig was gone by then—a victim of White House infighting—and George Shultz was the new secretary of state. American leadership had reached full agreement at last on the need for repaired ties with Egypt, Jordan, Saudi Arabia. Ronald Reagan now accepted the view that Israel could not be allowed to profit from the recent carnage in Lebanon beyond its legitimate security needs along its northern frontier, that the time was ripe to unveil a new Middle East peace initiative. This was the message Reagan transmitted to Begin, via Ambassador Lewis, even as he would reveal it publicly the next day in a televised address to the American people. Existing Israeli settlements in the territories might remain, Reagan declared, but additional colonies posed an unacceptable obstacle to peace; Arab inhabitants of East Jerusalem should be allowed to participate in elections to the autonomous entity in the West Bank; there should be "progressive Palestinian responsibility" for internal security, for the land and its resources. While Reagan conceded that an independent Arab state in the West Bank was not an appropriate solution to the Palestine question, a joint Hashemite-Palestinian confederation struck him as a reasonable alternative. Unequivocally, if belatedly, then, Washington was putting its own firm interpretation on the original Camp David accords, and making plain its rejection of Begin's scheme to exploit the Lebanon victory by ingesting the West Bank.

As Ambassador Lewis delivered Reagan's message, he suggested gently that the president's approach in fact served Israel's advantage, for it challenged the Rabat Resolution of 1974 (in which the Arab states recognized the PLO as "the sole legitimate representative of the Palestinian people"), by acknowledging Jordan's preeminence among the West Bank Arabs. Begin was not placated. He viewed Reagan's letter as a mortifying blow, for it grasped the ultimate territorialist purpose of Operation Big Pines, and precisely rejected it. Later, publicly, Begin would direct his criticism at three features of the American scenario. It was discussed in advance not with Israel, but rather with Jordan and Saudi Arabia, neither of them a party to the Camp David accords. By reiterating the view that East Jerusalem Arabs should participate in elections for the West Bank administrative council, Washington rejected the nonnegotiable concept of a united Jerusalem as Israel's capital. Reagan's insistence, finally, that the West Bank be closely associated with Jordan (in fact, the president's one important departure from the Camp David format) wiped out the five-year transition period stipulated by Camp David, and conflicted with the United States' role as "honest broker" between Israel and the Arabs.

But of course it was the substance, not the form, of the Reagan plan that infuriated the prime minister. Begin had repeatedly emphasized to the Americans that destruction of the PLO in Lebanon would eradicate that organization's ability to intimidate West Bank Arabs, and that "more moderate" local leaders, those prepared to consider an "acceptable" form of autonomy, would emerge at last. The prime minister was horrified that the United States should now be torpedoing that "acceptable"—in effect, Israel-dominated—Palestinian entity. "We handed Lebanon to Washington on a silver platter," observed one of Begin's cabinet colleagues, "and it wants to take the West Bank away from us."

In a high dudgeon, then, after his session with Ambassador Lewis, Begin proceeded to his conference with the awaiting president-elect of Lebanon. Accompanied by Ariel Sharon, he entered the guest house, still fuming, and greeted Bashir Gemayel. After a few introductory amenities and an exchange of champagne toasts, Begin got down to business. "Where do we stand on the peace treaty?" he asked abruptly. In some embarrassment, Bashir proposed a nonaggression pact as an alternative. Begin stiffened. Alluding to Major Sa'ad Haddad, the pro-Israel Christian militia leader in southern Lebanon, the prime minister observed that Haddad at least knew where his bread was buttered, that his was an example to be followed. Bashir immediately countered that he intended to put Haddad on trial for treasonable dealings with Israel—although aides suggested that the major would be let off with a pro forma condemnation. When the outraged Begin cut in with the "suggestion" that Haddad be appointed Lebanon's defense minister, the two men began shouting at each other. Sharon then interjected that Israel had Lebanon in its grasp, and Bashir would be wise to do what expected of him. "Put the handcuffs on!" cried the latter, holding out both arms. "I am not your vassal!" The meeting ended abruptly. Returning to Beirut, the young president-elect was furious that Begin had treated him "like a bellboy." Soon afterward, when Casper Weinberger visited Lebanon, Bashir informed the American secretary of defense that Begin's churlish behavior had "set me free" from all earlier commitments to Israel. Bashir hardly needed the pretext of Begin's peremptoriness, of course. From the outset of Phalangist-Israeli contacts, he had intended to use the Jews for his own purposes. His militia's passive role in the fighting had made that much clear. Now he would maneuver to augment his government's bargaining power.

For their part, Begin and Sharon were not likely to abandon their investment in the Lebanon campaign. In overt violation of the Habib withdrawal formula, Israeli troops moved 600 yards north in Beirut, ostensibly to "demine" roads. The advance put them near the impoverished Moslem neighborhood of Bir Hassna, overlooking the Shatila refugee camp. There they dug in. The gesture was intended equally as a warning to Bashir Gemayel and as a response to the Reagan initiative. Publicly, meanwhile, Begin continued to express gratification at the "unqualified success" of Israel's "campaign for peace." As far back as his Washington visit, on June 21, the prime minister had assured a gathering of American Jewish leaders that "the day is near" when an independent Lebanon would sign a peace treaty with Israel. A week later, following a particularly heavy bombardment of Beirut, Sharon had informed a group of reporters that Israel was "forging a triangle of peace" in the Middle East, linking Egypt, Israel, and Lebanon with open borders. In a television interview that same month, Begin asserted that Operation Peace for the Galilee had "healed the nation of the trauma of the Yom Kippur War."

The illusion of healing was punctured even sooner than Begin's expectation of a "friendly, neighboring" Lebanon. From the moment Israeli columns veered toward Syrian positions on the Beirut-Damascus highway, Shimon Peres and most of the Labor Alignment leadership began expressing their reservations. Here they were somewhat in advance of public opinion. During those early weeks the nation remained confused, not entirely certain of the objective it was being called upon to support. The mobilization, the music, the bulletins—it had happened just a bit too often. Yet, as late as the third week of June, polls revealed that 93 percent of the Israeli public considered the operation justified. Begin and Sharon even enjoyed a substantial increase in popularity, reversing their decline of previous weeks.

It was among the troops that disenchantment set in most rapidly. Their casualties—more than 300 within the first ten days of fighting—were much heavier than had been anticipated. Soldiers at the front discerned no logical goal in the offensive as it pushed far beyond the 25-mile zone. Hirsh Goodman of the *Jerusalem Post* was surrounded by men from elite combat units and berated for "mindlessly repeating official explanations [for the war] that we all knew to be false." By the time Israeli forces reached the suburbs of Beirut, a number of senior officers were appalled by their orders to make ready for a possible offensive against the city's western—Moslem—sector. One of these was Colonel Eli Geva. The precocious son of a distinguished retired general, Geva at thirty-two was the youngest brigade commander in the Israeli army. It was his tank column, then blockading PLO forces from the south, that would play the decisive role in an assault on West Beirut. In mid-July, after six weeks at the front, Geva suddenly asked General Eytan to relieve him of his brigade command and allow him to serve as a mere tank officer. Should orders be issued to lead his brigade into the city, Geva explained, he could not as "a matter of conscience" expose his troops and Beirut's civilian population to the heavy casualties that were sure to ensue. Shocked by the request, Eytan, then Sharon, and finally Begin himself sought to dissuade Geva, but unsuccessfully. In the end, the prime minister accepted Geva's "resignation," thereby ending a career that almost certainly would have led to a ranking staff position. The episode was given widespread publicity in Israel, and evoked both respect for Geva and deepening reservations about the war.

Among other officers and troops, those misgivings reflected not merely Israeli losses, but the thoroughness of Eytan's "smashup" offensive against Lebanese towns and Palestinian refugee camps alike. "Slowly the reality strikes home," recalled a non-com of the capture of Tyre. "It's not possible that we did such devastation! No shop not blasted, buildings crumpled under their roofs, boats sunk in the bay . . . and above all the overwhelming stench . . . of rotting bodies in the village." Reserve officers were stunned by the attitude of the high command, even of government ministers. When Ya'akov Meridor, the minister appointed to help Lebanese war victims, arrived at Sidon, he was asked what should be done with the Palestinian refugees. Gesturing to the north, he responded: "Push them out and don't let them come back." To many reserve officers, Meridor's approach reflected the military command's, an apathy, even brutality, that infected everything and everyone. They discerned it in their own troops, not a few of whom engaged in widespread looting. A number of these officers accordingly decided not to remain silent. Once on furlough, they submitted detailed reports to Begin, to other members of the cabinet, to the media. Subsequently, they appeared in public gatherings to express their opposition to the war, then formed a movement called "Soldiers Against Silence" to demand the resignation of Sharon and an end to the war altogether. As early as July 3, 1982, moreover, the Peace Now movement (originally founded in 1978 to protest Begin's apparent foot dragging on peace negotiations with Egypt) organized a demonstration in Tel Aviv. An estimated 100,000 citizens participated. In Israel's history, no such protest had ever been mounted in wartime. That same month, Abba Eban captured the anguished national mood in an article published by the newspaper *Ma'ariv.*

> Nobody who listened to [Begin's initial projected war aims] could have predicted that within seven weeks the war would be in full swing; hundreds more Israelis would be killed or maimed; thousands of civilians in Lebanon . . . have lost their lives; a moral torment would have seized many Israelis in and out of the army; a series of bombardments of civilians would have transformed Israel's reflection in the consciousness of this generation; the Egyptian peace treaty would be in question; world opinion would have been lost; and the United States, without whom

the military victory cannot be translated into political gain, would be so sharply alienated. . . . The shattered bodies in the Beirut hospitals, the buildings fallen on scores of mangled civilian corpses, the piled up garbage breeding rats, the children with amputated limbs, above all, the Israeli soldiers on their endless stretchers and funeral beds. . . . [T]hese six weeks have been a dark age in the moral history of the Jewish people.

The auguries were not good for the prime minister. After the siege of Beirut, a public opinion poll revealed a sharp increase of Israelis willing to relinquish part of the West Bank in exchange for peace, from 37 percent in May to 51 percent in August; those unwilling to cede an inch dropped from 58 percent to 47 percent. Even many Likud members now shifted on the issue, de facto annexation, that itself had been a major inducement for the Lebanon campaign. Begin hurriedly set about taking remedial steps. To add to his ruling coalition, he persuaded the three deputies of the right-wing Techiya party, those former Likud members who had broken with him on Camp David and the peace treaty with Egypt, to rejoin his government. Their quid pro quo was his formal assurance that the West Bank settlements would remain forever under Israeli rule. It was a small price for the harried prime minister to pay. In other respects, the government's defensive tactics rang increasingly hollow. Shortly after the July 3 Peace Now rally, Likud organized a huge counterdemonstration of its own in Tel Aviv. Some 200,000 citizens attended, many carted in from other parts of the country. Addressing the rally, Begin noted that in all previous wars there had been a consensus among parties; and if there were not this time, Labor's opposition status was responsible. Sharon meanwhile excoriated the Alignment for dissent within the army. At his request, the attorney general threatened to prosecute the publisher of a Mapam article that appeared at the front, criticizing the war. The threat later was withdrawn, but the government banned front-line distribution of newspaper editorials opposing the war.

It was the outcry of Jewish compassion for Arabs that Begin and his colleagues were at a particular loss to understand. Committees to provide food, clothing, and money for the Lebanese victims were springing up in Haifa and other cities, a number of them under Histadrut—labor federation— sponsorship. At the beginning of the war, the prime minister referred to Palestinian terrorists as "animals on two legs." Later, in an appearance before the Knesset's foreign affairs and defense committee, Eytan termed the Arabs who demonstrated in the West Bank "drugged roaches." These were not accidental utterances, nor were they meant to describe a specific policy against Palestinians in Lebanon or on the West Bank. Rather, they had become integrated into a terminology that was used by politicians and generals alike, and certainly by the Likud's most vociferous oriental supporters. In his numerous public appearances, Begin strove to place Israel's attacks and raids within the vast context of military horror. If the Israelis had bombed Tyre, what about Coventry? If they had razed Sidon, what about Dresden? If the opposition insisted on seeking a political settlement for the Middle East, the prime minister invoked the ghost of Chamberlain and Munich. If the Israeli television network screened some footage of a dead Lebanese child, Begin did not lack the 1,500,000 Jewish children sent to the ovens by the Nazis, or—as a last resort—the pathetic memory of his own family. The strategy was less than successful. "Are we really to view the miserable refugee camps as Munich and Nuremberg?" wrote Professor Ze'ev Manowitz, an authority on the Holocaust, in an article published in the newspaper *HaAretz* early in August.

> Are we to understand that the flattened hovels outside Sidon represent the Palestinian Dresden? Are we to see thousands of old people, women and children, bereft of all and exposed to the elements, as the paragons of a master race? Are we really to see Beirut as Berlin? There may be political method here, but it in no way diminishes the madness. . . . In transforming a justified punitive action and preventive measure into total war, without regard for the price to be exacted, Begin has lost touch with reality and is pursuing phantoms born in the greatest tragedy that ever befell our people. Whatever its final outcome, the epitaph to be placed upon the war in Lebanon will read: Here lies the international stature and moral integrity of a wonderful people. Died of a false analogy.

THE SHOCK OF SABRA AND SHATILA

Had conditions in Lebanon stabilized following the PLO departure, the Israeli people might yet have absorbed the carnage and turmoil of the war. But Sharon was not prepared to forego a major objective of his strategy. On September 12, 1982, in a five-hour conversation with Bashir Gemayel, he received the president-elect's assurance that the Phalangist militia would "clean out" the remaining 2,000 PLO guerrillas still hidden in West Beirut's refugee camps. Such a move unquestionably would have violated the recently negotiated Israeli-American agreement, which envisaged law and order as the prerogative henceforth of the Lebanese army. But the American marines and other—French and Italian—contingents of the multinational force had departed Lebanon some days before, convinced that their job was done. The terrorist nests appeared to be fair game now for quiet disposal by the president-elect's own militia. Then, only two days later, the scheme collapsed. At 4:00 p.m. of September 14, Bashir turned up at a party branch office for his weekly lecture before a Maronite women's group. The previous afternoon, a member of a Phalangist pro-Syrian dissident faction had planted a bomb there. Now, at 4:10, as the president-elect was addressing the women, a huge explosion demolished the building. Dozens of occupants were killed or mutilated. One of the dead was Bashir himself.

The shock experienced by Bashir's followers registered almost as forcefully among Israel's government leadership. Even before this catastrophe, Phalangist officers were conducting a low-key "liquidation campaign" against PLO and Syrian agents in areas under their control. But now, within hours of Bashir's assassination, a mortified Sharon instructed Eytan to seize the key junctions commanding West Beirut, "to restore order," then to allow the Phalangists entrance directly into the refugee camps. The directive was not committed to writing, nor was Begin informed of the conversation between Sharon and Eytan. Whereupon, at 5:00 p.m. on September 16, 1982, Israeli troops began moving into West Beirut. They encountered little guerrilla opposition. At this point, Sharon informed the Phalangist commanders that they were free to destroy the remaining PLO infrastructure. "I don't want a single one of the terrorists left," he emphasized. The defense minister insisted later that he had cautioned the Phalangists against undisciplined attacks on innocent civilians. The warning could hardly have been realistic. Christian hatred of the Palestinians was a fact of life in Lebanon. Five years earlier, during the Lebanese civil war, PLO troops had stormed the Christian town of Damour, slaughtering hundreds of civilians. Now, in September 1982, male survivors from Damour formed the core of the Phalangist militia that prepared to enter the Palestinian shantytowns of Sabra and Shatila. As Israeli troops invested the area, they surrounded the two camps, then opened an access way for the Phalangists. Several hundred of the Christian militiamen moved directly into the refugee warrens, until they dropped out of sight of Israeli observers.

That same evening of the sixteenth, shortly before the onset of the Jewish New Year, Israel's cabinet met in Jerusalem. A number of the ministers were vexed that they had not been consulted on the army's entrance into West Beirut. Sharon, who attended the meeting, said nothing about Sabra and Shatila. Yet, by then, Israeli intelligence officers watching the camps from nearby roofs picked up snatches of radio conversation between the Christian militiamen. Aware that something was wrong, that indiscriminate violence might indeed be occurring, they communicated their suspicions to army headquarters. No one alerted General Amir Drori, the northern front commander. It was only several hours later that a group of Israeli military correspondents also learned of firing in Sabra and Shatila, and managed to reach Drori on their own. The commander immediately ordered the Phalangists out of the camps. At the same time, Ze'ev Schiff, military correspondent of *HaAretz,* got through to his friend Mordechai Zippori, the minister of communications (and former deputy minister of defense). Zippori in turn put in a call to Yitzchak Shamir, warning the foreign minister of a possible atrocity. Shamir did not bother to investigate.

By afternoon of the next day, American and other foreign correspondents similarly were alerted to rumors of a massacre. So was Morris Draper, the American mediator, who urgently dunned the Israeli government. Sharon, too, finally was apprised of an evident slaughter in Sabra and Shatila. The defense minister did not bestir himself. Remarkably, while General Drori's orders were being interpreted and passed down through echelons, additional Phalangists entered the camps to participate in the "actions." Thus it was, on September 17, that a second night of bloodshed and terror settled over the camps' streets and alleys. It was not until early the next morning that Sharon finally visited Sabra and Shatila personally, saw the evidence of mass killings with his own eyes, and ordered the Phalangists out. Yet four additional hours passed before the last of the militiamen departed. By then their work was well and truly completed. They had bulldozed nearly all the bodies and dumped earth over them.

On the evening of September 18, the Israeli cabinet met to inquire into the killings. They heard Sharon, Eytan, and Drori insist that the Phalangist officers simply had "lost control of their men," that the moment information of misconduct had reached them, the Israeli commanders, they had intervened to drive the Maronite militia out. Whether or not the ministers were satisfied by the explanation, they decided to close ranks, and issued a statement that an atrocity had been perpetrated by a "Lebanese unit" that had entered the camps at a point far distant from Israeli positions. Later Begin reportedly characterized the episode as "Gentiles killing Gentiles." The dissemblement was a grave miscalculation. By late afternoon of the eighteenth, news of the massacre was being reported on foreign radio and television programs. The remains of some 230 men, women, and children already had been found in Sabra and Shatila, and additional scores of bodies were being disinterred almost every hour and photographed by newscameramen from many different countries. Estimates of the total number of dead soon ranged between 1,000 and 2,000. Even earlier, the Egyptian government had conveyed to Jerusalem its deep embarrassment at Israel's offensive into a neighboring Arab land. Now, with the revelations of Sabra and Shatila, Cairo announced that it was instructing its ambassador to Israel, Sa'ad Mortada, currently on home leave, to remain in Egypt for "consultations." On September 20, 1982, the United States, Italy, and France stated that the original multilateral peacekeeping force would be returned to Beirut within the week.

In Israel itself, a thread of credulity and toleration appeared to snap over that Rosh HaShanah weekend as photographs of the bloodstained corpses appeared on local television. The war thus far had not produced a clear political or military victory. Now apparently it had reduced Israel to the level of its Arab neighbors. The confrontation was unbearable. Almost immediately, public demands were launched for an independent commission of inquiry. A conference of senior military commanders openly and vigorously expressed its resentment of Sharon. In turn, stunned by the mounting criticism and unrest, the defense minister was forced to annul the call-up of an entire reserve corps. Abuse poured in from every sector of the country. Press outrage was almost unanimous. The "expulsion of the chief of staff and the defense minister from the ranks of the decision-makers is a primary condition for us to be able to look ourselves in the eye and the world in the eye," wrote *HaAretz*. Labor's organ, *Davar*, warned that "every minister who . . . does not resign immediately acts as if he gives his post-factum approval to the slayings." Mapam's *Al HaMishmar* called on the entire government to resign to "cleanse the people, the nation, and Zionism."

It was of interest that significant numbers of religionists now joined in the condemnation. Minister of Education Zevulon Hammer felt obliged to appear on television, exhorting Israelis to preserve "fundamental Jewish values." Other NRP ministers joined Hammer in pressing for an inquiry into the massacre. To Begin's consternation, a Likud minister, Yitzchak Berman, resigned from the cabinet to protest the government's foot dragging on the inquiry. Berman was accompanied by his Likud colleague in the Knesset,

Dror Zeigerman, by the director of the government press office, and by the commander of the senior military staff college. Still another remarkable intervention at this point was that of Yitzchak Navon, president of the state. Bravely transcending the ceremonial limits of his office, Navon appealed for a "thorough and impartial judicial inquiry" into the massacre, and subsequently disclosed that he would resign if his demand were not met. Then, on September 24, 1982, a protest demonstration was mounted by Peace Now, Soldiers Against Silence and members of the Labor Alignment. From every corner of Israel, 400,000 people—nearly a seventh of the nation's Jewish population—poured into Tel Aviv's Municipality Square and overflowed blocks around. No such explosion of public outrage had ever occurred in Israel's history.

Hereupon, deeply shaken, Begin relented. On September 28 he announced that he was appointing a commission of inquiry under the chairmanship of Justice Yitzchak Kahan, president of the supreme court. The three-man commission began its proceedings within days, and over the next weeks took detailed testimony from Israelis and Lebanese, from local and foreign journalists, and from every other possible source. In all, sixty sessions were conducted, fifty-eight witnesses heard, hundreds of documents studied. Begin himself, as well as Sharon and high military officials, appeared before the body. (The prime minister's answers to the inquiry were vague: "I don't know. . . . I'm not sure. . . . I don't recollect.") Finally, in February 1983, the commission issued its report. Its conclusions were paraphrased in historical terms:

> When we are dealing with the issue of indirect responsibility, it should also not be forgotten that the Jews in various lands of exile, and also in the Land of Israel when it was under foreign rule, suffered greatly from pogroms perpetrated by various hooligans. . . . The Jewish public's stand has always been that the responsibility for such deeds falls not only on those who rioted and committed the atrocities but also on those who were responsible for safety and public order. . . .

The report made clear that Begin had not participated in the decision to bring the Phalangists into the camps, that he had received no news about those orders until the evening of September 16. Nevertheless, for "two days after the prime minister heard about the Phalangist entry, he showed absolutely no interest in their actions in the camps." The defense minister's role was exposed as even less ambiguous. In contrast to the Agranat Commission of 1974, which had refused to condemn Dayan, the Kahan Commission was unsparing in its evaluation of Sharon's behavior. "As a politician responsible for Israel's security affairs," declared the report, "and as a minister who took an active part in the war in Lebanon, it was the duty of the defense minister to take into account all the reasonable considerations for and against having the Phalangists enter the camps, and not to disregard entirely the [possibility] . . . that the Phalangists were liable to commit atrocities." An equivalent responsibility was imputed to Eytan. The chief of staff had concurred with Sharon in allowing the Christian militiamen entrance to Sabra and Shatila, and later was informed that excesses were occurring. Yet he neither investigated this information nor stopped the atrocities. "We determined that the chief of staff's inaction . . . constitutes a breach of duty . . . incumbent upon the chief of staff."

The chief of military intelligence, General Yehoshua Saguy, together with the officers directly on the spot—the area commander General Amir Drori and his intelligence chief General Amos Yaron—were condemned no less bluntly. The report observed, too, that Foreign Minister Shamir "erred" in not having bothered to investigate the information conveyed to him by Mordechai Zippori. Summarizing its findings, the Kahan Commission made no recommendation about Begin and Shamir, beyond determining their responsibility. Yet it suggested that Begin "consider" the advisability of removing Sharon from office. Eytan was spared a similar recommendation only because his term of service in any case was due to end in several weeks. But the commissioners urged the dismissal of Generals Saguy, Drori, and Yaron.

The report, its prelude in antigovernment demonstrations, its aftermath in dismissals, came too late to preempt outraged world reaction to the massacres. If press criticism of the Lebanon war altogether had been harsh, this time condemnation of the Sabra and Shatila episode was unprecedented. Later a certain belated editorial admiration was expressed for the innate decency of the Israeli people in having forced the commission hearings. Now, however, the European office of the Anti-Defamation League counted at least forty-four incidents of antisemitism during 1982, a three-fold increase over the previous year. Upon police advice, the chief rabbi of Great Britain wore a bulletproof vest when venturing out. In traditionally pro-Israel Amsterdam, police protection was required for a conference on Zionism and antisemitism. The World Conference on Soviet Jewry, scheduled to convene in Paris in October 1982, was moved to Jerusalem because of a "hostile climate." Even in Australia, in January 1983, two bombs exploded at the Melbourne Jewish Center, and shots were fired at a Zionist youth camp. Begin's Likud supporters in turn seized upon these events as self-fulfilling proof of the justice of their cause, evidence that the world was inherently and remorsely against Jews, that objections against Israel's policies were simply another manifestation of an old disease. Within Israel, Begin's hard-core constituents still remained loyal. "What about the Syrians? What about the Iraqis?" their protests went, as a cross-section of oriental Jews was queried in Jerusalem's Machene Yehuda market. Although these supporters plainly did not favor killing Arab women and children, they argued that Jews had not actually perpetrated the massacre and therefore should not be held accountable for it. The subtlety of indirect responsibility was beyond them.

Not so for much of the rest of the nation. If the prime minister was prepared to brazen out world opinion, even the forceful criticism of world Jewry, he was unable to ignore the avalanche of hostility among hundreds of thousands of his fellow citizens. In public opinion polls, 50 percent of the population supported the Kahan Report. Only 30 percent found it unfair. Even as the cabinet debated its findings, and Begin held fast against dismissing Sharon—who himself refused to budge—crowds began gathering outside the prime minister's office. Many were supporters, shouting slogans in defense of Sharon (the defense minister actively encouraged their efforts). But far many more were critics, hurling invective against the government. Nor were they placated by the dismissal of Saguy, Drori, and Yaron, even by the announcement that Eytan would not be reappointed chief of staff. They wanted Sharon out, Shamir out—Begin out. On February 10, 1983, the Peace Now committee mounted yet another large demonstration, this one to protest Begin's evident contempt for the Kahan Report. Several thousand marchers paraded through the streets of Jerusalem with placards condemning the government. Suddenly, from a group of heckling Likud supporters, a disturbed young man of Iraqi descent, Yona Avrushmi, threw a grenade into the crowd. Ten marchers were wounded in the blast, one of them the son of Minister of the Interior Yosef Burg. One marcher was killed. He was Emil Grunzweig, a thirty-three-year-old kibbutz member and paratroop officer who had fought in Lebanon, and who recently had been conducting graduate research on a project for Arab youth. Grunzweig's funeral the next day in Haifa, attended by thousands of sympathizers, including many Israeli Arabs, became a scene of bitter antigovernment recrimination. Finally, after almost a week of widening public indictment, the prime minister's office announced a "compromise." Sharon would step down as defense minister but remain in the cabinet as minister without portfolio. But several weeks later, possibly as a secret addendum to the understanding, Sharon was appointed to the key ministerial defense committee.

The "compromise" hardly placated a major sector of the Israeli population. By then the nation was polarized more acutely than at any time since the German reparations crisis of 1952. In the wake of the Sabra and Shatila massacre, a young writer, Zvi Atzmon, concluded a bitter poem with the words:

> In the Land of Israel arose the Jewish people,
> laden with all its history, two thousand years old,
> and as long as there is inside my heart a soul,
> I am guilty.

Never before had a war been so debated in purely Jewish—in contrast to Israeli—terms. Begin defended it, and the casualties inflicted on a civilian Arab population, by invoking repeatedly the images and memories of World War II and the Holocaust, by equating the PLO with the Nazis, the assault on Beirut with the Allied assault on Berlin. Yet, when news of Sabra and Shatila provoked sweeping condemnation, the prime minister once again invoked the dormant memories of Jewish history, terming the outcry a "blood libel" against the Jewish people. Even earlier, it is recalled, these analogies were wearing increasingly thin for many Israelis. Thus, Yad vaShem, the Holocuast Memorial Shrine, became a focal point of national controversy. Next to its gate, a survivor of the Warsaw Ghetto and Buchenwald staged a hunger strike against the government. Protests became serious enough for General Eytan to order the cessation of all army visits to Yad vaShem. A bereaved father, who had lost his only son in Lebanon, published an open letter to Begin:

> I, remnant of a rabbinical family, only son of my father, a Zionist and Socialist who died a hero's death in the Warsaw Ghetto revolt, survived the Holocaust and settled in our country and served in the army and married and had a son. Now my beloved son is dead because of your war. Thus you have discontinued a Jewish chain of age-old suffering generations which no persecutor had succeeded in severing. The history of our ancient, wise and racked people will judge you and punish you with whips and scorpions, and let my sorrow haunt you when you sleep and when you awaken, and let my grief be the Mark of Cain upon your forehead forever!

TRAPPED IN THE QUAGMIRE

Bashir Gemayel had not yet been buried when the Gemayel patriarch, Sheikh Pierre, announced that Bashir's older brother Amin would seek to assume the murdered president-elect's place. The reaction in Jerusalem was guarded. It was understood that Amin had been unenthusiastic about the Israeli connection from the beginning, and had remained in contact with Damascus. With considerable reservations, then, the Begin government watched as Lebanon's parliament elected Amin Gemayel president on September 21, 1983. His majority of 77 votes was unexpectedly large. The thirty-two-year-old Amin may have come to office untrained for the tasks before him, but evidently the Greek Catholic, Druze, Sunni, and Shia communities regarded him as an acceptable alternative to his iron-fisted brother. In fact, Amin was a weak alternative. Notwithstanding his reputation for flexibility, he possessed no independent power base among the Phalangist party and militia, and thus never fully controlled any of Lebanon's welter of religious factions, not even his own. Sensing his vulnerability, moreover, Amin was all the more conscious of the need for returning to a more traditional pattern of national politics, the alliance between a Maronite president and the Sunni Moslems of the Greater Beirut area. It was in these circumstances that he found himself hard-pressed to accommodate Israel's demands.

In Jerusalem, the cabinet struggled between two schools of thought on Lebanon. While Sharon remained defense minister (this was the period when the Kahan Commission was still conducting its hearings), he continued to think in terms of Operation Big Pines, that is, of delegating full responsibility for the security zone in south Lebanon to the Lebanese government itself. That government in turn, presumably an ally now of Israel, even under Amin Gemayel, would sign a peace treaty with the Jewish state. But in an opposing viewpoint, Minister of Communications Zippori argued that Israel should place no faith in the Gemayel regime and should simply hold out in southern Lebanon on its own, even if its presence conceded the parallel right of Syrian forces to remain in the Beka'a Valley. The Israeli cabinet, obliged to choose between these rival positions in the autumn of 1982, initially supported Sharon's maximalist approach. Thus, on October 11, Foreign Minister Shamir took the Sharon plan to Washington. Secretary of State George Shultz promptly endorsed it, somewhat naively assuming that Amin Gemayel soon would impose his authority on the whole of splintered Lebanon.

On his own, afterward, Sharon conveyed the plan to Beirut. The formula envisaged a staged withdrawal of all forces—Israeli, Syrian, and PLO—with the Palestinians first departing the Beka'a Valley, the Syrians following, and the Israelis pulling back to a line 25 miles from Israel's border. Simultaneously, a multilateral force would occupy the zones of Syrian and Israeli withdrawal, at least until agreement could be reached on demilitarization and until Israeli and Syrian forces carried out their final departure. Each step of evacuation would be protected by negotiated security arrangements. On the political level, meanwhile, the formula proposed the establishment of an Israeli mission in Lebanon. The office would enjoy diplomatic status and forge a "normalization of relations on the basis of open borders."

Amin studied the document carefully. Then, early in December, 1982, he sent word to Sharon that while he was prepared to appoint a "high-level" committee to negotiate with the Israelis, the plan itself was unacceptable, even as a basis for discussions; Lebanon's Moslems, and assuredly the other Arab nations, would never tolerate it. Incensed, Sharon departed immediately for Beirut. Meeting first with Sheikh Pierre, he warned the Gemayel patriarch that Amin would have serious difficulties governing Lebanon if he rejected the Israeli blueprint. The old man was not intimidated. "If we open one gate to Israel, we will lose twenty gates to the Arab world because of it," he countered. Like the late Bashir, Amin was unable to accept any private deal with the Israelis that risked isolation from the Arab hinterland, particularly from Syria.

Notwithstanding these inauspicious preliminary encounters, "high-level" Lebanese-Israeli committee meetings did in fact begin on January 3, 1983. The Lebanese delegation included representatives of that nation's various religious communities, and was led by Dr. Antoine Fattal, a seasoned Maronite diplomat. David Kimche, director-general of Israel's foreign ministry, headed his nation's team. The American emissaries Philip Habib and Morris Draper spelled each other as chairmen. The discussions themselves were held alternatively in Khalde, Lebanon, and in Kiryat Shmonah (later in Naharia), Israel. At the outset, the Lebanese took a hard line. They rejected the very notion of a continued Israeli military presence on Lebanese soil, even on a "temporary" basis, or of accepting Major Sa'ad Haddad's militia as an Israeli surrogate. Nor were they prepared under any circumstances to sign a formal peace treaty. Plainly, the shadow of Syria was hovering over the Lebanese delegation. Syrian troops maintained their hold on more than a third of Lebanon's territory, after all. Damascus still was capable of arousing enough Moslem support in Lebanon itself for a war of attrition against the Gemayel government.

The Americans shared Israel's concern at this unexpectedly firm Lebanese response. They too had anticipated nailing down a treaty that would vitiate both Syrian and PLO influence in Lebanon. With the Syrians and the PLO removed as a disruptive force in Lebanon, it was Secretary of State Shultz's hope that the United States at least could serve as tutor to the young Christian president, and gradually transform his little country into a stable, pro-American nation, at peace with Israel. If that could be achieved, Lebanon would join Egypt as a shining example for the rest of the Arab world of the benefits to be derived from peaceful relations with the Israelis.

Accordingly, in February 1983, Shultz dispatched an envoy to Tunis, Arafat's temporary haven, in a secret effort to negotiate the evacuation of PLO forces from the entirety of Lebanon under protection of the existing American-French-Italian multinational force. The discussions unquestionably violated the spirit, if not the letter, of the long-standing American pledge to shun contact with the PLO. In any case, they failed. Arafat was not interested. The one result of Shultz's venture was to provoke Hafez al-Assad. The Syrian president struck back. Inciting a dissident Fatah rebellion against Arafat, he punished the PLO leader for his "impudence" in flirting with the Americans. Then Assad successfully maintained pressure on Amin Gemayel to resist concessions to Israel. This was accomplished by arming the Druze militia in Lebanon's Shouf Mountain range. A fiercely independent religious community, the Druze in turn posed a serious threat to Amin's tenuous hold over the Lebanese interior. Not least of all, Assad protected his nation's foothold in the Beka'a by reinforcing his garrison there to 1,200 tanks and by accepting (on Syrian territory) new batteries of Soviet long-range SAM-4 missiles, together with Soviet crews to man them. Shultz's plan for "neutralizing" Assad manifestly had backfired.

For weeks, then, Israeli-Lebanese negotiations continued to be deadlocked. In some degree, Lebanon's position actually was stiffened by the U.S. ambassador in Beirut, Robert Dillon, whose overt distrust of Israel and "American Jewish political pressures" occasionally resembled a classical phobia. Shultz himself then visited the region in late April of 1983, determined to tie down an Israeli-Lebanese agreement through shuttle diplomacy. In fact, he succeeded. After several trips between Jerusalem and Beirut, he persuaded the Begin government to abandon its insistence on a formal treaty and to accept the concept of "unofficial" relations, the kind Israel had maintained for years with Iran. A joint supervisory committee would take the place of diplomatic missions. For his part, Amin Gemayel quietly assured the Israelis that he had President Assad's promise "in writing" to pull all Syrian units out of Lebanon, on condition that the Israelis similarly withdrew. After some hesitation, Begin and his advisers finally consented. This was the format that the Israeli and Lebanese delegations signed on May 17, 1983, in separate ceremonies in Khalde and Kiryat Shmonah.

Although it was far from a peace treaty, the document contained some notable language. Both sides acknowledged their desire "to ensure lasting security for both their states," asserted that they "consider[ed] the existing international boundary between [them] inviolable," and confirmed that "the state of war between [them] has been terminated." Article 4 declared:

1. The territory of each party will not be used as a base for hostile or terrorist activity against the other party, its territory, or its people.
2. Each party will prevent the existence or organization of irregular forces, armed bands, or organizations the . . . aims and purposes of which include incursions or any act of terrorism into the territory of the other party, or any other activity aimed at threatening or endangering the security of the other party. . . .

An annex on security arrangements clarified the dimensions of a buffer region north of the Israeli border and (without mentioning his units by name) affirmed that Major Haddad's militia would be recognized as a Lebanese "auxiliary force" and "accorded a proper status under Lebanese law." The agreement unquestionably was less than Begin or Sharon had hoped to achieve. Yet it appeared to have salvaged the essence, if not the formal lineaments, of peace. So long as the new Gemayel government maintained order and kept armed bands away from southern Lebanon, Operation Big Pines could be reckoned at least a partial success.

Even in its qualified and limited form, however, the accord was doomed. It was Assad who helped ensure its demise. Proclaiming the agreement to be a "deal of capitulation," the Syrian president warned all elements in Lebanon not to honor it. He rejected with contempt the very notion that Syria would pull its forces out of Lebanon in return for an identical Israeli commitment. He, the president of Syria, had not been consulted on the document's provisions, after all—Amin Gemayel's assurance to the Israelis notwithstanding—and accordingly he was not bound by it. Long before the Israeli invasion, for that matter, Assad had regarded his army's presence in Lebanon as vital: specifically to inhibit the Beirut government from following Egypt's example in making peace with Israel; and, even more fundamentally, to exert Syria's continued influence in Lebanese affairs, and thereby Syrian prestige in the Arab world at large. Far from discerning advantage in a parallel Israeli evacuation, Assad regarded a trip-wire confrontation with the Israelis on the Golan and in Lebanon as even more useful. It helped mobilize Syrian nationalism, and thus diverted internal opposition to his regime.

Yet, if the Lebanese-Israeli accord remained dead on the paper, it was by no means Syrian opposition alone that killed it. There was virtually no likelihood that Amin Gemayel's government could have forged enough unity in Lebanon to preempt outside influences in that country. From their bailiwick in the Shouf range southeast of Beirut, the Druze mountain villagers invariably had resisted incursion—Christian or Moslem—on their traditional enclave. The Sunni Moslems of West Beirut, the Shia of south Lebanon, historically had maintained a similar resistance to infringement of their communal autonomy by a Maronite-dominated government. Nor, in the end, would even a Phalangist prime minister have tolerated indefinitely the quasi independence of Sa'ad Haddad's Christian militia in the south. Without an ongoing Israeli presence in Beirut itself, finally, there appeared no guarantees against a revisionist PLO infiltration from the north to the Lebanese capital. Sharon's maximalist vision of Operation Big Pines had overreached itself from the outset.

ISRAEL'S BALANCE SHEET IN LEBANON

Frozen in their tracks now by a concomitant Syrian military emplacement, Israeli troops each day were exposed to the depredations of PLO and Lebanese Moslem snipers. As early as October 3, 1982, when discussions first began with the Amin Gemayel government, 6 Israeli soldiers were killed and 22 wounded in an ambush outside Beirut. Even the Shia of the south, who had not initially challenged the Israeli presence, began joining in the attacks. Elsewhere, Druze snipers in the Shouf range also began taking their toll of Israeli personnel. Then, on November 11, an explosion rocked Israel's military headquarters building in Tyre. In the single worst military disaster in Israel's history, 74 soldiers and civilians were killed outright, with another 27 wounded. It was uncertain whether the blast was the result of a bombing or of a gas leak, but its aftermath was intensified public demand to clear out of Lebanon. Yet, even then, Sharon (still in office at the time) held firm, insisting that the army would not depart until the newly elected Beirut regime signed a treaty with Israel. During these first six months of military occupation, 463 Israeli soldiers and other security personnel were killed and some 2,500 wounded. The losses mounted almost daily.

So, too, did the anguish in Israel. In December 1982 an opinion poll revealed that 41 percent of the public considered the war a mistake. On June 4, 1983, a mass Peace Now rally in Tel Aviv was attended by 150,000 persons. Meanwhile, a group of military reservists founded still another organization, "Yesh G'vul"—There's a Limit. In their opposition to the war, the members held discussion workshops, circulated press releases, helped organize demonstrations. Other protesters evaded assignment to Lebanon simply by arranging to perform their duty elsewhere, or by finding ostensible excuses—a pregnant wife, a failing business—to avoid service altogether. The omens for the country were not good.

In late June 1983, finally, the question of withdrawal was addressed by Sharon's successor as defense minister, Moshe Arens. Reared and educated in the United States, a former Technion professor of aeronautical engineering and more recently Israel's ambassador to Washington, Arens was a Cherut stalwart and a hard-liner. Yet even he understood that an immobilist posture could not withstand the daily attrition in casualties, the political disorder in Lebanon, the growing disaffection at home. Arens favored a pullback of Israeli troops from the Shouf Mountains to the Awali River line. The move would be unilateral and no longer attendant upon a reciprocal withdrawal of Syrian forces. Ironically, the one government to question the proposal was the United States, which feared that the Syrians would exploit the chaos likely to follow an Israeli departure. On July 4 George Shultz flew to Israel to seek postponement of the withdrawal. He was unsuccessful. The Israelis executed their move exactly two weeks later, pulling back about 20 miles from their advance positions to the Awali River. The maneuver did not seriously affect their security. Although the army relinquished its grip on the Beirut-Damascus highway, it maintained control of the strategic Jebel Barukh heights, overlooking the Beka'a Valley and large parts of Syria. The cease-fire line with the Syrians thus far remained unchanged.

For Lebanon itself, however, the consequences of the withdrawal were indeed very grave. In the wake of the departing Israeli army, Druze irregulars in the Shouf promptly descended upon sixty isolated Maronite villages, slaughtered at least 1,000 civilians and left perhaps 50,000 homeless. The Syrians meanwhile began probing westward from their enclave in the Beka'a. Soon only the Lebanese army (religiously fractured, and all but worthless even as a home guard) and the Phalangist militia (effective principally against unarmed Palestinian refugees) remained as a "shield" for Beirut. As Shultz had feared, protection of the capital now devolved mainly on the few hundred American marines and French legionnaires of the multinational force, the contingents that had been rushed back to Lebanon in the aftermath of the Sabra and Shatila massacres. At the time, President Reagan had assured Congress that "there is no intention or expectation that the United States armed forces will become involved in hostilities." The marine contingent would remain only long enough to create "an environment which will permit the Lebanese armed forces to carry out their responsibilities in the Beirut area."

But as the months passed, the United States, like Israel, discovered that the task of fortifying the Gemayel government was all but hopeless. Damascus was systematically inciting Lebanon's Druze and Shia communities, and the marines soon found themselves targets for these fractious elements. By late August 1983, the American unit guarding Beirut airport came under such intensive fire that Reagan was obliged to send a naval task force to the Lebanese coast. Before long the battleship *Missouri* was firing its huge guns at Druze positions. Far from inhibiting antigoverment attacks, the naval bombardment drew heavier fire on the isolated American contingent. Then, in late October 1983 a suicide bomb attack killed 241 marines. At this point, Congress decided that the American presence in Lebanon had become counterproductive. It surely had done nothing to bring unity or security to the little nation. In February 1984 the last of the marines were withdrawn (the French had departed weeks earlier), and finis was written to a major American diplomatic and military fiasco.

Israel watched these events in dismay. The multinational force had been Jerusalem's last hope of preserving the functional integrity of the Gemayel government, of allowing the Israeli army an organized, staged evacuation. Now, with the disintegration of security in central-southern Lebanon, there was no way to avoid a swift, even precipitous, departure. The daily hemorrhage in Israeli lives was becoming a Chinese water torture. On November 4, 1983, an Israeli checkpoint was blown up by a booby-trapped truck, killing an additional twenty-nine soldiers and wounding many others. By then Jerusalem understood that the question no longer was one of withdrawal or of shoring up the Gemayel government, but of salvaging at least a degree of security for Israel's northern frontier—the initial rationale for Operation Peace for the Galilee. Conceivably the Shia of southern Lebanon might yet be persuaded to cooperate with Major Sa'ad Haddad—and after Haddad's death (from cancer) in 1984—with his successor, Brigadier Antoine Lahad. Together, these elements might form a cordon to protect the southernmost region of the Awali River "security" zone against Palestinian guerrilla infiltration. It was wishful thinking. The cordon remained exclusively a Christian one. Indeed, by the winter of 1983–84, Defense Minister Arens suspected that the modest 1,800-man Lahad militia faced discreditation if it operated exclusively as an Israeli auxiliary force; that the residue of ill feeling against Israel would become irremediable without a full evacuation of the security zone. Arens was right. Early in March 1984, Begin's and Sharon's most coveted desideratum for Operation Big Pines went up in smoke. Under Syrian pressure, the Amin Gemayel government formally canceled its May 17 accord with Israel.

Was the war then an unmitigated disaster for Israel? In fact, it was not without certain short-term advantages. It appeared for the while to have removed PLO artillery from range of the Galilee. It destroyed the PLO state-within-a-state in Lebanon. Stripped of its impressive administrative, political, military and financial infrastructure in Beirut, the guerrilla confederation no longer was able to offer day-to-day guidance to its substantial following on the West Bank, as it had for more than a decade. The resumption of Voice of Palestine broadcasts from the distant capitals of Algeria, Iraq, and North and South Yemen, and publication of the official PLO weekly in Cyprus, were poor substitutes for the Beirut operation.

In the West Bank, too, there was evidence that the Lebanon invasion may have benefited Professor Milson's campaign to undermine Arafat's influence (Chapter VII). Stunned by the maceration of PLO forces in Lebanon, the West Bankers all but closed themselves in their homes, leaving their towns and villages in eerie silence. Then, as Israel's siege of Beirut tightened, several known PLO partisans in the West Bank began stating openly that negotiations with Israel were possible after all. In late July 1982 a majority of the West Bank mayors, including several who had been dismissed for their PLO activities, publicly welcomed Arafat's recent implied acceptance of UN Resolution 242. The Village Leagues, not long before almost fatally undermined by Hashemite repudiation, seemed now to be fighting back with growing confidence against PLO intimidation. Indeed, the number of their branches revived impressively following eviction of the PLO from Beirut, and by the summer of 1983 plans already were under way for organizing a Federation of Village Leagues. Against this backdrop of mounting success, it was all the more ironic that Menachem Milson himself should have resigned as director of the civil administration following the Sabra and Shatila massacres. He had also been outraged by the government's peremptory rejection of the Reagan plan for Palestine, which he, Milson, favored. Even so, the threat of a PLO reign of terror on the West Bank seemed by then to have been mitigated.

There were other gains. One was apparent as early as the spring of 1984 when a mission entitled the "Lebanese Christian Agency" rented quarters in a modern Jerusalem apartment building, opposite the pine groves surrounding the Knesset. The members of this mysterious entity professed to speak for "Petit Liban," the traditional heartland of Maronite communities extending through East Beirut, up the coast to the outskirts of Tripoli, and along the inland region known as "The Mountain." Since Bashir Gemayel's assassination, it happened that there had been a growing divergence among the Phalangists. The supporters of Amin, led by Sheikh Pierre and members of the party's central committee, argued that the single effective way to preserve the remnants of Maronite power was to accept a diminished role in an essentially Syrian-dominated Lebanon. But another group led by Fady Frem, an engineer who succeeded the late Bashir as commander of the Phalangist militia, rejected this approach as defeatist. Frem and his partisans insisted that Syrian influence in the land could effectively be resisted through a close association with Israel. Once it became evident, then, that Amin soon would cave in to Syrian pressure to abrogate the May 17, 1983 accord with Israel, Frem sent Pierre Yazbeck, his press representative, to Jerusalem to rent office space for a "Petit Liban" mission.

Although gratified by this development, the Israelis urged Yazbeck to widen the base of his mission to include other Christian denominations. The emissary agreed to try. He was at least partially successful. In mid-March 1984, at Yazbeck's initiative, delegates of fifteen Christian denominations and associations convened to form a National Christian Council. The assembly issued a statement condemning the "Syrian direction taken by Parliament," and denouncing abrogation of the May 17 agreement with Israel. Yazbeck meanwhile confidently assured Israeli journalists that the formal "cantonization" of Lebanon was all but certain, that the emerging Christian and Druze cantons would maintain their own "diplomatic" relations with Israel. If some Israeli officials were privately skeptical, Yazbeck, unperturbed, proceeded to affix the Petit Liban "flag" to his Jerusalem office, a cedar superimposed on a red crucifix. Thereafter, he organized a symposium of Lebanese and Israeli academics in the Christian enclave of south Lebanon. "Israel wants normalization with Lebanon," Yazbeck boasted then and later. "We shall give her normalization. Even including a travel program to enable Israelis to visit our sector and vice versa. I'm planning to establish ferry service between Haifa and Jounieh. If we show the Druze and the Shia that it's possible to defy Syria, perhaps they will follow our example." But in the interval, with Lebanon crumbling away, the Christian "cantons" became hostage increasingly to neighboring pro-Syrian militias. Few Israelis were optimistic that the link with "Petit Liban" would flower into an important relationship, or even survive at all. For the while, however, it was a fragile blossom salvaged from the rubble.

Otherwise, the long-term results of the Lebanon war were painfully coun-

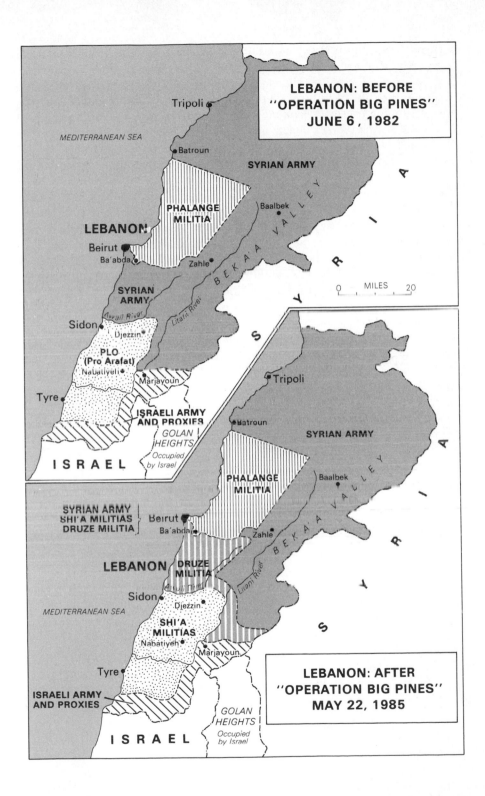

LEBANON: BEFORE "OPERATION BIG PINES" JUNE 6, 1982

Tripoli

MEDITERRANEAN SEA

Batroun

SYRIAN ARMY

S Y R I A

PHALANGE MILITIA

Baalbek

LEBANON

Beirut

Ba'abda

Zahle

B E K A A   V A L L E Y

0   MILES   20

SYRIAN ARMY

Awali River

Litani River

Sidon

Djezzin

PLO (Pro Arafat)

Nabatiyeh

Marjayoun

Tyre

ISRAELI ARMY AND PROXIES

GOLAN HEIGHTS
Occupied by Israel

I S R A E L

LEBANON: AFTER "OPERATION BIG PINES" MAY 22, 1985

Tripoli

Batroun

SYRIAN ARMY

S Y R I A

PHALANGE MILITIA

Baalbek

B E K A A   V A L L E Y

SYRIAN ARMY
SHI'A MILITIAS
DRUZE MILITIA

Beirut

Ba'abda

Zahle

Litani River

LEBANON

DRUZE MILITIA

Awali River

Sidon

Djezzin

SHI'A MILITIAS

Nabatiyeh

Marjayoun

Tyre

ISRAELI ARMY AND PROXIES

MEDITERRANEAN SEA

GOLAN HEIGHTS
Occupied by Israel

I S R A E L

terproductive. More than 600 Israeli lives had been lost by the spring of 1984, with three times that many wounded, and the toll continued to rise. The nation's military and civilian morale was seriously undermined. If the PLO infrastructure had been shattered, a new, and possibly graver, military threat—from Syria—was looming over the northern frontier. The war had left Assad's armed forces intact, and significantly more dominant than before the Israeli invasion. Their air losses notwithstanding, the Syrians had halted Israeli armor midway down the Beirut-Damascus highway, and afterward remained ensconced in fully 35 percent of Lebanon. With Soviet help, too, they had invested an immense effort in improving their antiaircraft system. Any future Israeli assault in central-eastern Lebanon was likely to be much more difficult and costly. Neither had Begin achieved another cherished goal. He had anticipated that the campaign would divert international attention from Israel's creeping annexation of the West Bank. It had not. In the immediate aftermath of the PLO withdrawal from Lebanon, the Reagan administration was not deterred from unveiling a scenario for the West Bank that was gall and wormwood to the Israeli prime minister. And, perhaps most insupportable of all, Israel's image throughout the world, a painfully achieved amalgam of heroism and humanism, had been grievously, if not irretrievably, disfigured among well-meaning Christians and Jews alike.

CHAPTER X

THE TRANSITION

FROM THE BEGIN ERA

A LEADER DEPARTS

Even as the mournful costs of the Lebanon expedition were becoming apparent, and Israeli forces were retreating to their temporary buffer zone along the Awali River, Menachem Begin was drawing his own conclusions from the recent war. He had been scheduled to visit Washington in late July 1983. Suddenly, on the nineteenth of that month, he canceled the trip, citing "personal reasons." It was possible that the cancellation was influenced by the death several months earlier of his wife, to whom he had been intensely devoted. But the accumulation of public pressures and criticism almost surely exerted their influence. On August 28, physically wasted and visibly depressed, Begin announced his intention to step down as prime minister, and on September 15 he formally submitted his resignation to the president of the state.

Begin's was a compromised legacy. The peace treaty with Egypt was a magisterial diplomatic and political accomplishment, one perhaps none of his Labor predecessors could have achieved—with or without a Sadat as bargaining partner. It was conceivable, nevertheless, that the Lebanon war, the costliest and most far reaching of his initiatives, was intended as compensation for abandonment of the Sinai. Those close to the prime minister believed that he had launched the invasion to be remembered in history as the author of his own military triumph, to be ranked with the Six-Day War. It had all gone wrong afterward, of course. But so did the wider pattern of his incumbency. The charge of squandered material and moral resources applied not only to the $1 million-a-day cost of the Lebanon war, but also to the artificially sustained value of the Israeli shekel, the dissipation of hard currency reserves on cheap consumer goods and foreign holidays, and not least of all to the heavily subsidized Israeli presence on the West Bank. Until Lebanon, however, the erosion of Israel's substance at least had not been measured in the lives of young soldiers, in polarized, even shattered, public morale.

As in an earlier career fixated by territorialist dogma, Begin's vision was tunneled from the outset of his premiership by need for a free hand in an undivided Land of Israel. It was a mystic romanticism that found its closest regional precedent in Eleutherios Venizelos's World War I dream of a revived "Kingdom of Ionia" in western Turkey. If the Greek leader's fantasy proved catastrophic in 1922, Begin at least might have exploited the advantage of hindsight. By the 1970s and 1980s, a parallel obsession of ruling over an alien, West Bank, majority (a proportionately far larger demographic hinterland than the Turkish population in Venizelos's hallucinatory Ionia) was as much a political anachronism as an economic albatross. Could Begin eventually have "worked through" that territorialist fixation—as had Ezer Weizman and Moshe Dayan before him? Would he have dared face the electoral consequences of such a volte-face? Neither in his first nor his second cabinets was there evidence of that kind of incipient moderation. Nor of that sort of political courage.

Indeed, the prime minister's image as a strong, if stubborn, leader may altogether have been a myth. In fulfilling his irredentist idée fixe, Begin at Camp David had shrunk from a forthright commitment to vacate the Sinai settlements. Rather, he had left the decision to the free vote of the Knesset. His effort to negotiate separate votes on the Egyptian-Israeli treaty and Sinai settlements issue, to evade the personal stigma of abandoning the Yamit cluster, similarly represented less than moral fortitude. Jimmy Carter related that, upon reaching an understanding with Begin, it was necessary to wait as the latter submitted the agreement to a painstaking clause-by-clause discussion by the cabinet and Knesset. No one was going to pin an unpopular concession on Menachem Begin alone. And in the final denouement of the Sabra and Shatila massacre, in the aftermath of the Kahan Report, the prime minister was unable himself to accept the responsibility of dropping Sharon. Again he left the matter to a cabinet vote.

Yet, ultimately, it was Begin's most egregious failure of perspective and judgment that his inflammatory rhetoric was applied not alone to the nation's latent chauvinism, to its more than latent ethnic tensions, but to an interpretation of the Jewish experience that at times approached paranoia. From beginning to end, in his diplomacy with foreign leaders no less than in his military impetuosity, he adverted to the trauma of the Holocaust. The brooding memory of that horror unquestionably was locked in Begin's mind and heart. But its tactical application to current issues was shrill, and ultimately fatiguing. Whether it was the bombardment of an Iraqi nuclear reactor or of a Lebanese village, the denunciation of Yasser Arafat or of Helmut Schmidt, the martyrological parallels delivered repeatedly to Anwar al-Sadat or to Jimmy Carter—Begin's indiscriminate invocation of the Holocaust may have been his most flagrant transgression. By word and deed, he politicized, and thereby trivialized, the single most tragic chapter in Jewish history.

NEW ELECTIONS ON THE ECONOMIC PRECIPICE

No natural successor was at hand when the prime minister stunned his party and nation by announcing his intention to resign. Four days later, on August 31, 1983, still in shock, the Cherut central committee met to choose its new chairman. The contest devolved essentially into a two-way race between David Levy, the minister of housing, and Yitzchak Shamir, the foreign minister.

The latter was chosen by a vote of 435 to 302, and assumed the prime ministry on September 15. It was not unlikely that Shamir's career parallel with Begin was a factor in his election. Born Yitzchak Yezrenicki in tsarist Poland, he also was trained in law at the University of Warsaw. Upon immigrating to Palestine in 1935, he similarly joined the Revisionist Irgun Z'vai Le'umi, and his underground activities against the British won him three prison terms. In 1948 Shamir became a partisan of the ultraright Lech'i organization, which was suspected of responsibility for the assassination of Count Folke-Bernadotte, the United Nations mediator. Although he did not make his peace with the "moderate" Cherut faction until 1970, he rose swiftly in Begin's confidence afterward. Thus, when Likud came to power in 1977, Shamir was made speaker of the Knesset, and three years later succeeded Dayan in the foreign ministry. There at first he belied the dire predictions of ministry professionals and foreign diplomats alike. His quiet, affable manner, his willingness to listen, offered a welcome contrast to Dayan's lonely gruffness.

After his initial year and a half in the foreign ministry, however, Shamir's hard edged Revisionism surfaced again. He remained as unshakable on the issue of the West Bank as did Begin. He supported the Lebanon war, and his culpability—by indifference to evidence—for the Sabra and Shatila massacres was duly revealed and condemned by the Kahan Commission (Shamir's first reaction to the report was to criticize it harshly, to urge his fellow ministers "not to be hasty" in acting on its recommendations). Now, acceding to the premiership at the age of sixty-nine, a diminutive, brush-moustached man with a ready smile, Shamir maintained his even-tempered, low-keyed manner. By the same token, he evinced not the slightest willingness to budge on issues of Israel's "legitimate security." At a time when the new defense minister, Moshe Arens, took a firm line against Jewish vigilantism, Shamir offered little moral support.

Conceivably, the Knesset's moderates and the public at large would have forgiven Shamir his militance had he come up with a firm plan to extricate the nation from the Lebanon quagmire. But he did not. Neither did the new prime minister appear capable of grappling with the most recent economic crisis. As it happened, 1983 witnessed a number of collapses on the Tel Aviv stock exchange. In late January, rumors of pending government steps against mutual funds caused a sharp drop in the market. Then, in mid-May, the announcement of an unprecedented single-month inflation rate of 13 percent, combined with a warning from the Bank of Israel that stern monetary action was needed, touched off a renewed wave of selling on the exchange. In a trend that would continue for the rest of the year, billions of shekels were withdrawn from savings schemes to purchase dollars or durable goods. Israel's currency became increasingly worthless.

By then the massive printing of money was a consequence also of the Lebanon war and the government's social welfare policy, which shifted to undisguised populism in 1981, to cheaper housing for young couples, to lower import fees on foreign goods. By July 1983, as a result, with the annual inflation rate approaching 400 percent, Finance Minister Aridor's fiscal adventurism came under sharp attack even from his own Likud colleagues, who demanded tighter budgetary restrictions. Aridor did in fact promise to move in this direction. Yet before major steps could be taken, Begin suddenly resigned, and the ensuing political turmoil interrupted further economic deliberations. The stage accordingly was set for a denouement that later was termed Israel's "economic Yom Kippur," a blow that struck exactly ten years after the 1973 war.

In late September 1983, reports began to circulate that the government was unable to secure further international loans, that drastic curtailments soon would be imposed on foreign currency transactions, that a major devaluation of the shekel was imminent. As a hedge against these anticipated government measures, the public early in October began selling off large quantities of its dollar-linked blue-chip bank shares (their value artificially and unconscionably sustained by the banks themselves over the years as a lure for inflation-proof investment) to raise money for an apparently safer investment in dollars. The commercial banks in turn, encountering this wave of selling, and frantic to maintain their—inflated—share values on the stock exchange, were forced to import dollars from their subsidiaries abroad. Yet by then, with the public's appetite for dollars insatiable, the banks soon exhausted their hard currency reserves. Panicking, their management turned to the government for help. They got it, after a fashion. On October 9, trading in bank shares was suspended on the Tel Aviv exchange. The emergency measure was followed by an all-night cabinet meeting (the first session under Prime Minister Shamir). Early the next morning the treasury announced a 23 percent devaluation of the shekel and a 50 percent increase in the price of subsidized goods. The shock of this sequence of events was eased somewhat by an announced plan for transforming limited amounts of bank shares into interest-bearing government debentures, dollar-linked and redeemable in five years. Yet the formula was publicized too late to rescue the tens of thousands of Israelis who already had disposed of their bank shares at a fraction of their original value. Hundreds of millions of dollars in savings were lost.

The week's trauma was not yet over. On October 13, Y'diot Acharonot headlined a story that the treasury was planning a "dollarization" of the Israeli economy, a linkage across the board of the Israeli shekel to the United States dollar. Under questioning, Finance Minister Aridor admitted that the report was substantially correct, that he intended to submit the proposal at the next cabinet meeting. When the public intensified its already wild rush to the dollar, however, opposition to the plan mounted on all sides. Another urgent cabinet session was called. Then, fifteen minutes before the session was due to begin, Aridor resigned, and Prime Minister Shamir went on television that evening to calm the nation by announcing the end of the "dollarization" scheme. Aridor's harshest critic, Yigal Cohen-Orgad, was chosen as the new minister of finance.

With the run on the dollar finally easing then, the stock exchange was reopened for limited activity on October 20, 1983. Trading in bank shares resumed three days after that. Although the shares swiftly lost an additional 35 percent of their nominal value (and nearly double that in dollar value), this time at least there was no panic; for the treasury in effect took over from the banks, pumping some $150 million into the stock exchange to block an even more precipitous decline. The infusions continued until the market recovered. In an additional effort to restore stability, the government barred Israelis from purchasing more than $3,000 in foreign currency for travel abroad. A week later the cabinet passed a series of even more draconian revenue-raising measures, including newer taxes almost everywhere. It is possible that these steps eventually would have salvaged the nation's tottering economy, but on November 15 the public was stunned by the news that Israel's currency reserves had dropped below the $3 billion "red line."

Reeling before this latest threat, and lacking Begin's charisma, Shamir did not appear capable of reversing the dramatic erosion of public support for the government. Opinion polls suggested that Likud had reached the height of its popularity in the first week of the 1982 Lebanon campaign, when it might have obtained a clear Knesset majority. Following the refugee camp massacres, however, the government's reputation plummeted. By January 1984, in the depths of the economic doldrums, and seeing no clear prospects for disengagement in Lebanon, Israel's voters (polls revealed) would have awarded 24 more seats to the Alignment than to Likud had an election been held then. The groundswell of public unrest could not be contained indefinitely. At last, in March, after several harsh disagreements over economic and social policies, a tiny coalition faction, the religionist Tami party, gave the Labor opposition the majority vote it needed for an early election. The date was fixed for July 23.

A REVERSION OF THE POLITICAL GUARD

As the politicians geared up for the hustings, a major departure from earlier campaigns was the absence of Menachem Begin. Suffering from an acute melancholia that later was revealed to be medical rather than contritional, the former prime minister by then had withdrawn completely from political life, even from a place on the Likud Knesset list. Begin's disappearance in turn may have contributed to "the cleanest election in [Israel's] thirty-six-year history," as it was characterized by Justice Gavriel Bach, chairman of the central election committee. Both major blocs appeared to take pains to avoid confrontation. In fact, both were afflicted by internal, structural weaknesses. In April 1984, on the eve of the campaign, Cherut underwent a bitter fight for the party leadership between Shamir and Sharon. Eventually Shamir received 56 percent of the central committee vote; but the narrow victory hardly bespoke a vast and unswerving following. Within the same period, too, the Cherut leadership forced the smaller La'am and Liberal parties to accept a diminished quota of "safe" positions on the Likud list. Again, as in the struggle between Shamir and Sharon, the bargaining was tough and left wounded feelings. It was at least a consolation for Likud that the Labor Alignment was plagued by an identical divisiveness. Thus, the Mapam component similarly was forced to accept a reduced number of "safe" positions on the Alignment's Knesset list; even as Labor was further polarized by a wider spectrum of clients than Likud's, including hawks and doves, Marxists and capitalists, non-Europeans and Ashkenazim, religionists and secularists, Arabs and Druze, kibbutz and moshav members, and a select number of women. Assured places had to be found for them all.

Pluralism was the name of the game when twenty-six lists registered for the election in June 1984. Thirteen of these parties were entirely new. They represented a number of ethnic factions, lists of Indian and Georgian Jews, and others that were formed by single individuals who earlier had been associated with major parties. The latter included Arieh Eliav, a distinguished former secretary-general of the Labor party; Ezer Weizman, the 1977 Cherut campaign chairman and ex-defense minister; Yigal Hurevitz, earlier of La'am and an ex-finance minister in the Likud government. Nor did political Orthodoxy escape this splinterization. It is recalled that, for some two decades, the National Religious party had been the Knesset's third largest bloc, after Labor and Likud, and traditionally it had disposed of not less than 12 Knesset seats. But the defection of oriental religious constituents from the NRP in 1981 halved that number and led to the establishment of Tami, the first avowedly Jewish ethnic party elected to the Knesset in thirty years. By 1984, lingering discontent among other Orthodox non-Europeans generated an even more vociferous religious faction known as Shas (Sephardi Torah Guardians), this one an offshoot of Agudat Israel. Thus, many orientals—particularly Moroccans—who had voted for Tami in 1981 resentfully switched to Shas following the conviction (for embezzlement) of their "kinsman," Aharon Abuhatzeira, Tami's leader, and a former minister of religions. Other schisms in the Orthodox camp were the result of disagreement on the West Bank issue, or simply of personal rivalries. Eventually the old Orthodox bloc of NRP and Agudat Israel would be reduced to a mere 6 Knesset seats between them, while Shas and Tami won 4 seats and 1 seat, respectively.

Labor's strategy meanwhile was aimed at winning over the large number of disappointed Likud supporters and moderate "floaters." Profiting from their mistakes in the 1981 campaign, Peres and his colleagues scaled down their attacks on personalities this time, and concentrated directly on Likud's record of mismanagement and economic failures. To be sure, with its own diverse constituency of hawks and doves, Labor could not afford to challenge Likud directly on such "national'" issues as West Bank settlements and Arab affairs. As a consequence of growing public anguish over Lebanon, rather, both blocs promised to withdraw: Likud, when security conditions permitted; Labor, within six months. Meanwhile, Peres did not disguise his preference for the Allon Plan frontiers, but made no commitment to uproot existing settlements on the West Bank; and under no circumstances would he countenance a return to the 1967 borders, the establishment of a Palestinian state west of the Jordan, or negotiations with the PLO. Shamir and his colleagues in turn exploited themes that had succeeded in earlier campaigns, portraying Labor as the camp of privilege (a code word for Ashkenazic elitism), of corruption and dovish appeasement. And, indeed, although ethnic issues were invoked less virulently this time than in 1981, the continuing polarization between oriental and European Jews was apparent on election day. More than 70 percent of the Likud vote came from orientals, more than 70 percent of the Alignment's from Ashkenazim. Once again, Labor had failed to expunge its negative image among the non-Europeans.

These lessons were revealed after the fact, however. Until a week before the election, polls showed the Labor Alignment leading by 10 Knesset seats. The margin narrowed only at the last moment. Then, following the election, as votes were tallied and retallied, it became clear that the Alignment had won 46 seats, Likud 41. On the one hand, Likud had dropped 7 seats from its earlier Knesset strength. On the other, the shift represented less than a distinct victory for Labor; for most of the 100,000 voters who eventually broke with Likud ended up with other parties of the right—with Techiya (5) and Kach (1). To be sure, the 3 seats lost by the Alignment from its 1981 showing went to its dovish satellite party, Shinui, and thus were "mobilizable" for coalition bargaining purposes. So, conceivably, were the 4 seats of the Citizens Rights Movement, the 3 of Ezer Weizman's Yachad party, the 1 of Yigal Hurevitz's Ometz faction. But the 6 Arab seats, going to the two anti-Zionist Rakach and Palestine Land parties, were regarded as too compromised to be exploited for a potential Labor government. In truth, both major blocs had fought themselves to a standstill.

As the days passed, and as the contending leaders wheeled and dealed in their efforts to put together a viable coalition, the best chance initially appeared to lie with the Alignment. Weizman and Hurevitz reached agreement with Peres, and this collective move to the center allowed President Chaim Herzog (himself a Laborite) to nominate Peres as prime minister-designate. The move was premature. It soon became evident that Peres was incapable of forming a government without the religionists, and the latter apparently had learned to be comfortable with their current and generous—Likud partners. Shamir and his colleagues then presented a compromise proposal. It was for a 1967-style "wall to wall" coalition that at least would ensure a common hard stance on Jewish settlements in the territories. The offer did not appeal to Peres. And so the interparty negotiations continued at Jerusalem's King David Hotel, with no apparent solution. By the final week of September, however, even Peres had to admit the need for accommodation. As he saw it, the priority now was to acquaint the nation with him in the role of prime minister. With his image as a statesman more widely established, many things could happen later.

Accordingly, over the bitter opposition of the Alignment's 6-member Mapam faction, and of the 3-member Citizens Rights Movement, both of which declined to enter into a marriage of convenience with Likud, and of the 5-member right-wing Techiya group, which in turn disdained contact with the Alignment, an understanding was reached at last on September 14, 1983, for a "national unity" government. Even without the parties of the far left and right, the new constellation would encompass the largest number of established parties—97 Knesset members in all—and most of their leaders would sit in the cabinet. Peres himself would serve initially as premier, while Shamir would revert to his earlier position in the Begin government as foreign minister. After twenty-five months, the two men would reverse roles, with Peres becoming foreign minister and Shamir prime minister. Rabin would serve as defense minister under both premiers, other Laborites as health and education ministers, while still other ministries would be reserved for such new Labor allies as Weizman and Hurevitz. Elsewhere, beyond the foreign ministry, a number of Likud stalwarts, including Ariel Sharon, similarly would be granted portfolios, among them, the finance and commerce ministries. So would Yosef Burg of the NRP (who switched from the ministry of interior to the ministry of religions) and leaders of the Shas (Sephardi) religionists. The new regime was a strange, bicephalous apparition, with Labor's minuscule numerical advantage insufficient to permit fundamental policy changes, either in the territories or in church-state relations. However tightly limited, the new prime minister's freedom of action would be directed essentially to the crisis of the raging inflation, to the quest for a way out of Lebanon. But for most Israelis, this was priority enough.

PERES: THE MAN AND THE IDEOLOGY

After years of waiting in the wings, Shimon Peres at long last assumed office at the age of sixty-one. Throughout his lengthy government career, sobriquets like "implementer," "technocrat number one," and "preacher of scientification," had been attached to him both as compliment and criticism. In truth, no Israeli, not even Dayan or Sharon, had ever achieved as wide-ranging an influence on the nation's public life without actually holding the title of prime minister. Born in Poland in 1923 (né Perski), the son of middle-class parents, he moved with his family to Palestine eleven years later, where he joined a kibbutz. His political career began at the age of eighteen, when he was appointed northern regional coordinator for the "working youth" movement. Afterward, as a disciple of the eminent Labor theorist Berl Katznelson, the young man climbed swiftly in the ranks of the Mapai party. In the early 1940s Peres served in the Haganah, working closely with his friend, Moshe Dayan. By the end of World War II, he had risen to the position of Haganah director of manpower, a post that also obliged him to deal with arms acquisition and manufacture. It was the threshold of a historic expertise.

In 1952 Peres was appointed deputy director-general of the ministry of defense; and in 1954, director-general. Thirty-two years old, he was a clear favorite by then of Prime Minister Ben-Gurion, who saw in him and Dayan the kind of young pragmatic "doers" who offered the best alternative to Labor's windy idealogues. In some respects, this early period in the defense ministry was Peres's most brilliant. It was then, almost single-handedly, in 1955–56, that he engineered the arms deal with France that laid the basis for the Suez-Sinai collaboration, and afterward for the joint weapons research-and-production that transformed Israel into a major military power. It was Peres too in these same years who played a central role in the development of Israel's nuclear research capacity. Later, as the relationship with France cooled, the young director-general orchestrated even more extensive arms deals with Germany and the United States. At the same time, Peres's close association with Ben-Gurion created a bond that transcended the political and approached the filial. When the Old Man departed the prime ministry in 1963, and afterward broke with the ruling Labor gerontocracy to found the dissident Rafi faction, Peres (and Dayan) risked their futures by accompanying Ben-Gurion into the political wings.

The opportunity for revival did not come until 1967, following the Six-Day War. It was Dayan's participation in the initial "wall to wall" government that facilitated Rafi's return to the Labor party. In ensuing years, Peres filled a series of cabinet positions, as minister of transport, of immigrant absorption, of communication, of information. Although these were essentially peripheral appointments, they allowed him to repair his ties with his former Labor colleagues, to demonstrate his administrative skills in keeping the party viable during the shaky aftermath of the Yom Kippur War and the Agranat Report. So it was, upon Golda Meir's resignation in April 1974, that Peres emerged as a major contender for the party chairmanship. His bid then was premature. The party veterans ensured that Yitzchak Rabin succeeded Mrs. Meir as chairman and prime minister. Afterward, too, we recall, the open rivalry between Rabin and Peres (the latter serving as defense minister) became one of the juiciest feuds of Israeli political life. Neither man was above petty rancor during the tough infighting. Yet the declining fortunes of the Rabin government did not appear to tarnish Peres's image seriously. Opinion polls showed him a clear favorite over Rabin. In the Labor party convention before the 1977 election, he received 1,404 votes to Rabin's 1,445. Had Rabin not been prime minister, Peres unquestionably would have been offered his party's chairmanship. He got it anyway, of course, once the premier's illegal bank account in the United States was disclosed. Following Rabin's "leave of absence," it was Peres who campaigned for the prime ministry—and lost in the 1977 upheaval.

Thereafter, fending off his predecessor's challenges to regain the party leadership, Peres fared much better in the 1981 election. Indeed, he nearly unseated the Begin government. During his subsequent years as opposition leader, he worked tirelessly to rehabilitate the Labor party, to wipe out its debts and rebuild its branches. In his extensive travels, also, he cultivated an aura of statesmanship, meeting with Sadat and Ceausescu in Europe, with Brzezinski and Mondale of the Carter administration in the United States. Despite grave Labor objections to the abandonment of the Sinai, Peres held his party in line to support the Camp David accords and the peace treaty with Egypt. His articulateness (which some cynics described as glibness) impressed many voters. So did his evident breadth of culture, his taste for literature and music. Above all, Peres's intellectual ambit embraced a passion for science. It was his conviction, fully shared by Ben-Gurion in the days of their earlier association, that the nation's future depended upon skilled manpower and the imaginative exploitation of technology, upon the "scientification of Israel." It was as the supreme technocrat, then, the pragmatic "doer," that Peres now assumed the prime ministry in September 1984. A trim, well-tailored man of ruddy complexion and a full head of glossy hair, he stood at the maturity of his powers and experience. Notwithstanding the hybrid nature of his mandate under the coalition agreement, he was confident of selling both himself and his program to the Israeli people—once the levers of power were in his hands.

THE THREAT OF A DETERIORATING SOCIETY

I. *Political Dysfunctionalism*

The difficulties Peres inherited clearly were older and more complex than those of an indeterminate election. A nation already rent by ethnic, religious, and territorialist cleavages was seeking to function under a corroded electoral process. Any party that polled at least 1 percent of the vote (19,000 in 1984) won representation in the 120-member Knesset. In a classic study of the electoral systems of twenty western democracies, political scientist Douglas Rae found Israel's to be the most fractionalized. In no Knesset since the establishment of the state were less than ten parties represented, and in none did a single party win a majority. For its defenders, Israel's method of elections closely approximated the ideal democracy. Indeed its roots lay within the pre-1948 Zionist experience, when proportional representation allowed the World Zionist Organization to encompass the rich diversity of Jewish political opinion, to keep every Zionist party involved in the struggle for the Jewish homeland. If their votes then were devoid of coercive power, at least they were not entirely wasted, as in nations with single-member districts and majority victories.

Yet, as matters developed, the Israeli citizen after 1949 was not selecting a government, but authorizing an intricate process of coalition bargaining. No voter knew until after the election, and usually after weeks of intensive horse trading, for whom and what he had cast his ballot. With the various political factions themselves often coalitions, Israeli government soon became a politician's nightmare of double- and triple-helixes. All major cabinet decisions were influenced by the need simply to assure stability, and the result was a built-in disincentive to decision making and risk taking. In his earliest cabinets, Ben-Gurion had sought to revise the system, but none of the smaller parties appeared interested in committing suicide. Later attempts were equally unsuccessful, including Yigael Yadin's Democratic Movement for Change. Among their other objections to reform, Begin and the Cherut leadership in 1977 feared that constituency elections would broaden representation for Arab voters, who were heavily concentrated in the Galilee and other specific geographic areas. As before, the temptation to delay and equivocate, to seek minimalist solutions to difficult problems, remained unavoidable.

Still another emergent structural weakness in Israel's political system was a diffusion of the lines between military and civilian authority. The state's founding fathers had been very clear in their intention to depoliticize the armed services. Ben-Gurion accordingly purged most of the rightist (former Etzel) officers after 1948, together with leftist Palmach-Mapam officers. But the very diligence with which the prime minister selected his top-commanders from "nonpolitical"—that is, Mapai—factions was itself a political act. Indeed, from Ben-Gurion through Begin, political "reliability" was a decisive factor between two candidates for military office. Thus, General Ezer Weizman, a Cherut partisan, was passed over as chief of staff in the Eshkol government for General Chaim Bar-Lev, a Laborite. In 1969, a candidate for the chief of military operations branch was General Yeshayahu Gavish, associated with Dayan, who still was identified with the Rafi party; while the other leading contender, General David Elazar, was affiliated with Achdut HaAvodah, a less "threatening" faction under Labor, and accordingly one that enjoyed the support of Chief of Staff Bar-Lev and Prime Minister Golda Meir. Elazar got the job, becoming chief of staff a few years later.

The interpenetration of military and politics worked both ways. From the nation's earliest years, political activity served as an important second career for retired Israeli officers. The phenomenon was particularly widespread after the 1967 war, when military heroes were "parachuted" into the Knesset and cabinet almost from the moment they took off their uniforms. Men such as Weizman moved directly from the armed forces to the cabinet in 1969, as did Bar-Lev in 1972. Even earlier, Dayan served as minister of agriculture and then minister of defense. In 1974 Rabin was premier and Allon was deputy premier and foreign minister. In 1977 Yadin was deputy premier, Dayan foreign minister, Weizman defense minister, and Sharon minister of agriculture in charge of the settlement of the territories. Altogether, between 1949 and 1981, of some 190 generals of various grades, over a third took up political careers or assumed important civilian administrative directorships that were traditional political plums, including senior appointments in the foreign ministry, government corporations, the Jewish Agency, and Histadrut.

After 1967, it was specifically this opportunity for postservice career advancement that fostered the revived politicization of senior military personnel. A common pattern was for an officer to be a member of Labor and active in the party's "servicemen's department." But there were cases in which officers used more forthright methods. General Elazar, commander of the northern front, swiftly formed a relationship with the kibbutz movement in order to ensure both his promotion to chief of staff and a political basis for himself in Labor upon his eventual retirement from the army. The most flagrant case in point assuredly was Ariel Sharon. It was in 1969 that that the burly former general demanded the post of southern front commander. He was rebuffed by Yitzchak Rabin, the chief of staff. Sharon then approached the Liberal party, in an effort to negotiate a "safe" place on the latter's Knesset list. Apprised of this development, Pinchas Sapir, Labor's strong man, persuaded Rabin to relent and give Sharon the southern command.

Four years later, Sharon again threatened to resign from the army and join the political opposition—currently, Cherut—if he were not appointed chief of the operations branch, the traditional stepping stone to chief of staff. This time his bluff was called. Prime Minister Golda Meir allowed him to resign, and Sharon in turn formed his own Shlomzion party as a tactical move to an eventual leadership role in Cherut. Yet, even as a ranking Likud figure, serving as minister of agriculture in the Begin government, Sharon continued to press for greater visibility. He wanted the portfolio of defense, upon Weizman's resignation in 1980. A year later he got it, a payoff for his crucial role in Likud's 1981 reelection campaign. Following the Lebanon war, however, Begin had occasion to reevaluate the grim consequences of that reward. Thus, upon Sharon's resignation in the aftermath of the Kahan Report in 1983, his successor, Moshe Arens, made a serious effort to "de-Sharonize" the defense ministry and armed forces. But in a nation as highly factionalized as Israel, it was a question whether politicization in the military could be checked more than briefly.

A development not less threatening was political polarization on the—civilian—right, a consequence specifically of the West Bank issue. For rightist critics of the Camp David accords and the peace treaty with Egypt, no euphemisms, no "legalisms," were acceptable as substitutes for the plainspoken reality of an undivided Land of Israel. Concentrated increasingly in the Techiya party, this intractable element was not to be confused with the chauvinists of the Orthodox parties, for whom territorial maximalism was recently superimposed baggage. Dr. Yuval Ne'eman and his colleagues based their opposition to Begin's "concessions" at first on security concerns, on the danger of a PLO state emerging in Palestine, of Israel reduced again to its vulnerable pre-1967 borders. Had Israel been caught within those frontiers during the Yom Kippur War, Ne'eman argued, its situation would have been fatal. Mere alteration of the old Green Line, therefore, even demilitarization of the West Bank, would never adequately guarantee the nation's safety.

Techiya's position, if extreme, at least appeared grounded initially in "respectable" strategic considerations. But afterward the party's obsession with security led it into an undisguised annexationist stance.

> Which is morally and ethically better [Ne'eman asked], once it becomes inevitable that our sons should hold the passes in Samaria: should they be there as foreign occupiers, as soldiers residing in a base in a foreign country . . . or should they feel it is theirs by right, and it is their homes that they are defending? . . . Meanwhile, world opinion would begin to understand that the only "legitimate rights" of the Palestinian Arabs are the right to live in peace [as resident aliens] with the Jewish majority in Israel.

The facile rationale under which Palestinian Arabs were legislatively to be transformed into *dhimmis* suggested the ease with which even a Western-educated Jewish intellectual could espouse crypto-Fascist ideas. In the 1984 election, then, the Techiya party won an unexpected five Knesset seats, and for the first time emerged as a major force on the Israeli political scene.

Curiously, the burgeoning impact of the Techiya group, with its membership of scientists, lawyers, and writers (among them, the eminent novelist and former left-winger Moshe Shamir), tended to be overlooked by Israeli and foreign observers. Their attention was transfixed, rather, by a much noisier and more lurid phenomenon. This was the Kach party, whose founder was an American-born rabbi, Meir Kahane. Ordained in a Brooklyn yeshiva, swiftly dismissed afterward from his first and only pulpit, Kahane eventually secured a position as a columnist for the *Brooklyn Jewish Press.* It was during his journalistic period in the 1960s that Kahane, son of an ardently Revisionist father, organized the Jewish Defense League as a "counterterrorist organization." The JDL's tiny membership in fact scrupulously avoided physical confrontation with antisemites, but its belligerent posturings evoked colorful newspaper reportage. In 1972, with an aggressively manipulated image as a "fighting Jew," Kahane moved his base of operation to Israel. There, two years later, he founded his truculent Kach party.

The name, meaning "By Force," was taken from the motto of Begin's pre-state Etzel underground group, and its approach initially was suggested by its full page newspaper advertisement shortly before the Knesset election of 1981, proposing a five-year prison sentence for any non-Jew engaging in sexual relations with a Jewess. Yet Kahane's "program," beyond its manifest purpose of calling attention to himself and electing him to the Knesset, soon was even more baldly stated. It was to force the Arabs of the "Land of Israel" (that is, of Israel and the West Bank) to emigrate to their "own lands." "We'll show the world we Jews can be bastards, too," Kahane shouted at a party rally, as he proposed Jewish terror to counter Arab terror. "The Arabs must leave!" Lest there were doubt on this point, Kahane organized threatening marches upon Arab towns in Israel. The police invariably were waiting to block a confrontation, as police had waited in the United States before similarly theatrical JDL gestures. Kahane himself was periodically arrested, held for administrative detention, once even sentenced to nine months for incitement. But the man long since had achieved his goal of national publicity.

Thus, if Kahane failed in his bid for the Knesset in 1981, his demagoguery soon evoked resonance well beyond his party's hard core of Moroccan lumpenproletariat and Orthodox American Jewish immigrants. In the West Bank town of Kiryat Arba, a community guarded by barbed wire, searchlight beams, and gun-bearing zealots, the Kach yellow-and-black flag hung from many windows, even as Kahane's tense, bearded face appeared on numerous posters. One of the latter's adherents in Kiryat Arba was a certain James E. Mahon, Jr., a Christian-born neo-Nazi from the United States. Discharged from the American army after Vietnam, with numerous physical and mental scars, Mahon hung around the Nazi fringes, acquiring the nickname "Crazy Jimmy" from the Washington police, before suddenly converting to Judaism and going off to "fight for the Holy Land." He took the name Eli HaZe'ev— Eli the Wolf. Thereafter, in Israel, he received a prison sentence for acts of violence against Arabs, and shocked even the militant Kiryat Arba settlers by his lust for blood and weaponry. Late in 1981 Mahon was one of six Kiryat Arba residents killed in an Arab ambush. It was no longer an eyebrow-raiser afterward when General Eytan, the military chief of staff, spoke at his funeral, as did Israel's Chief Rabbi Goren and Foreign Minister Shamir.

An Israeli army detachment fired three volleys over his grave. Kahane, then serving his prison sentence, was released on parole to attend the funeral. The gesture of respectability, both to "Crazy Jimmy" and to Kahane, was not lost on the Israeli public. "The Golem has overpowered its creator," lamented an editorial in *HaAretz*.

Nowhere was the threat more evident than in the 1984 Knesset campaign. As the electioneering reached its apogee in June and July, several episodes rivaled the hooliganism of 1981. Amnon Rubinstein of the moderate Shinui party, seeking to address an audience in Petach Tikvah, was obliged to run a verbal gauntlet of Kahane's yellow-shirted Kach followers and finally to take refuge behind the barricaded doors of his party office, unable to give his speech. Peres underwent several comparable experiences; in Tiberias he was physically attacked and had to be spirited away by the police. And when the election finally took place, Kach received 26,000 votes, more than enough this time to assure Kahane himself a seat. It was an appalling development for hundreds of thousands of Israeli moderates. President Herzog refused to include Kahane in the traditional presidential reception for designated members of the Knesset. Elsewhere, the world press devoted an inordinate amount of space to Kahane's election, regarding it (possibly with a faint subliminal satisfaction) as the ultimate "legitimization of bigotry" in Israel.

If it was that, it also represented another, more invidious, threat to the Jewish state. Fixating public attention with his raucous demagoguery, blatant appeals to racism, hooligan band of followers, and his penultimate electoral success, Kahane fulfilled an entirely unintentional role: that of a diversion from the Techiya party's much larger delegation of five Knesset members. Had Professor Yuval Ne'eman and his educated, comparatively well-mannered bloc of colleagues personally charted this bizarre electoral juxtaposition, they could not have been more effective. The image they projected surely was the opposite of moderation. By contrast to Kahane, however, it appeared at least to be one of respectability.

II. *A Patchwork Economy*

Peres stepped into an economic chamber of horrors that would have intimidated the leader of a far wealthier nation. During the seven years of Likud rule, Israel's foreign currency debt had nearly doubled, from about $12.5 billion in 1977 to over $23 billion in the summer of 1984. Indeed, by then it was the highest external debt per capita—over $5,000—of any state in the world, and the cost of servicing it amounted to not less than 37 percent of the national budget. Israel's annual trade deficit had reached $5.5 billion by 1984, and its annual inflation rate exceeded 400 percent. The source of this unfolding disaster remained largely apolitical. By the end of the 1970s, the price of oil was more than twenty times higher than at the beginning of the decade. Without a significant increase of energy consumption, and even before abandoning the Sinai wells, Israel was paying a fuel bill that had climbed from $774 million in 1978 to $2.1 billion in 1980—and then shot up even more dramatically in 1983 and 1984. Agricultural depression further exacerbated the financial plight. As late as 1983, farms and groves still provided 9 percent of Israel's export earnings, but the figure was much less than a decade earlier. Plainly, Sharon's neglect of agriculture, his diversion of funds to the West Bank settlements, took their toll. But so did potentially more lucrative, industrial alternatives for kibbutz settlements; the world recession, which lowered demand for several of Israel's most profitable winter fruits and vegetables; and the European Common Market, which recently had admitted Spain and Portugal, with their highly competitive citrus groves.

Yet, more than any other "objective" factor, it was still the cost of defense that was breaking the back of Israel's economy. During the 1950s, at a time when the nation devoted the largest part of its resources to the absorption of hundreds of thousands of immigrants, defense spending consumed between 6 and 9 percent of the GNP. The ratio climbed steadily. The watershed change was the Six-Day War. In ensuing years, the enlargement of the standing army and reserve elements, the massive expansion of military industries and the vast investment required for more sophisticated weapons systems, the cost of the War of Attrition, then of the Yom Kippur War—all produced an even more spectacular leap in defense expenditures to nearly half the GNP by 1980. The same upward movement was evident in the military's percentage of the government budget. At its nadir, in 1961, defense expenditures comprised 19 percent of the budget. Its peak, in the Yom Kippur War, was almost 50 percent, a figure that was nearly reached again during the Lebanon invasion. No exact cost for the Lebanon war and occupation was produced by the government, but the direct outlay could not have been less than $2.5 billion. Nor did even this sum include the economic impact of long and repeated mobilization of reservists, many of whom were obliged to spend three or more months a year away from their civilian jobs. If these indirect costs were tabulated, the final reckoning may have approached $5 billion.

There were less "objective" influences, however, that added their burden to the nation's overheated economy. One was the Begin government's intensified settlements program. In all its dimensions, we recall, that program had consumed some $1.5 billion a year in public funds, and conceivably an additional $1 billion in private investment. This was the equivalent of over $150,000 for each person who moved into the West Bank—at a time when Israeli development communities such as Kiryat Shmonah were decaying in industry and population for lack of government funds. Additionally, David Levy's program for subsidized public housing, Zevulon Hammer's unprecedented diversion of funds to religious institutions, Yoram Aridor's efforts to maintaining the (import) buying power of the shekel—all further eroded the nation's hard currency reserves. The wound inflicted by this combination of imperialism and populism was further exacerbated by Israel's widening vocational imbalance. By the early 1970s, agriculture encompassed a bare 6 percent of the work force, and only slightly more of the GNP; while industry accounted for 24 percent of the work force and 23 percent of the GNP. By contrast, the government and the Jewish Agency alone employed 38 percent of the work force. Business and finance, transportation, storage, and communication together employed another 20 percent, while tourism accounted for the final 12 percent. Israel's economy, then, was becoming increasingly service-oriented, with its "productive" branches overshadowed by the commercial, financial, and communal sectors.

By the 1970s and 1980s, the dream of Israeli families for their children tended to mirror that of Diaspora families, for white-collar business and professional careers. A consequence in turn of the developing Jewish labor shortgage in the factories and on the farms was a heavy infusion of West Bank and Gaza Arabs, who took up the slack not only in menial "black" work but in construction, even in industry. So it was, in updated form, that the vocational pattern of a Jewish squirearchy was being replicated nearly a century after the original Zionist pioneering settlement. Those Israeli workers who performed their jobs, meanwhile, whether in industry, the professions, or the services, failed to match the productivity levels of other modern societies. In a period of acute manpower shortage, they did not doubt that their unions would protect them, that the time-honored Histadrut work norm of seniority would shelter them from the Western challenge of merit hiring and firing. It was of course a vocational ethic less than likely to salvage the nation's economic competitiveness. By the early 1980s, Israel was given sharp reminder that make-work inefficiency no longer was supportable. As part of its reciprocal trade agreement with the European Economic Community, the government was obliged gradually to end its restrictions on Common Market imports. In 1985, too, a free trade agreement was negotiated with the United States, opening the market of each country to additional imports from the other. Henceforth, local products would survive or perish in an open battlefield—equally abroad and in Israel itself.

Thus far, capital transfers from overseas had buttressed Israel's standard of living. The single most important source of that largess was U.S. government aid. By 1983, Israel had received nearly $23 billion in assistance from Washington. The sum comprised an unusually high percentage of American security assistance to all nations throughout the previous decade. Some $12.3 billion of this money took the form of outright grants; the remaining $10.7 billion, of long-term, interest-bearing loans. Following the 1979 Egyptian-Israeli treaty, moreover, Israel became altogether the single largest recipient in the world of U.S. economic and military assistance. The relationship stood in dramatic contrast to Israel's poverty-stricken early years, even the period from 1964 to 1973, when the nation's share of U.S. overseas assistance did not exceed $1:6 billion, or just 2 percent. Yet it was precisely this recent effulgence of American aid that had begun to function as a kind of narcotic for Israel, postponing the structural changes that alone would have allowed the country to face serious new economic challenges from abroad. By the same token, the threat of reduced infusions from Washington became a diplomatic weapon against the nation. In the event the largess stopped or declined, private Jewish philanthropy or investments from abroad would not begin to make up the difference. Indeed, since the late 1970s, contributions from exclusively Jewish sources, from the United Jewish Appeal or from equivalent funds elsewhere, and loans through Bonds for Israel, represented less than 20 percent of official American government assistance; and these Diaspora transfers covered barely 3 percent of Israel's import bill, 5 percent of its total budget.

Meanwhile, Israel still appeared to be awash in affluence. The monetary populism of Yoram Aridor had left shops bulging with imported consumer items. If the nation's foreign currency reserves were dropping alarmingly, the average Israeli, and surely Begin's grateful oriental constituency, had seemed hardly responsive to the long-range implications of the fact. But with the Lebanon war, and its prodigious outpouring of funds, even the least sophisticated Israeli began to suspect that the piper soon would have to be paid. Indeed, heavier taxes already were being imposed, including a value-added (sales) tax of 15 percent, and a wide spectrum of new imports on property, automobiles, fuel. Food and transportation subsidies were reduced sharply even before the Peres government took office. The gap between rich and poor widened still further during these last Begin-Shamir years, however disguised by cheap imports and subsidized apartment construction.

Would the nation accept the disciplined austerity needed to return to a sounder financial basis? The prospects at best were mixed. Until the 1984 "government of national unity," Israelis had become accustomed to automatic wage hikes linked to an endlessly rising cost of living, builders' payments linked to inflation, bank interest rates and other costs linked to the dollar. Now, faced at last with a certain governmental unity of economic purpose, the Histadrut displayed a tentative willingness to forgo automatic wage increases. That forbearance in turn, linked with a rigorous new program of price controls, helped sharply to reduce the rate of inflation, from 30 percent to 2 percent monthly by the end of 1986. The question was how long such restraint would endure. Over the years, under Labor and Likud alike, increasing numbers of Israelis of all backgrounds had learned to survive not only by automatic wage hikes and strikes, but by cheating on their expense accounts, their contractual obligations, their taxes. An entire underground economy had developed. Professionals and technicians moonlighted exclusively for cash, often at the expense of their salaried jobs. "Black money"—cash or hidden currency accounts overseas—was rapidly becoming the norm for business and professional dealings in Israel. These economic pressures, together with mounting contempt for the law and for the viability of Israeli democracy altogether, were fostering a potentially lethal illness in the nation's midst. There was no elegant euphemism for the disease. It was crime.

III. *Crime*

In 1979 a survey carried out by the Public Opinion Research Institute revealed that Israelis were more concerned about crime than about any other of the country's manifold social problems. There was reason for their anxiety. Violation of the law was becoming endemic to a society that had learned to adjust to the burdens and obligations of citizenship often by avoiding them altogether. The malady affected everyone, public officials as well as private

ISRAEL'S ECONOMY IN THE LIKUD ERA, 1977-1984

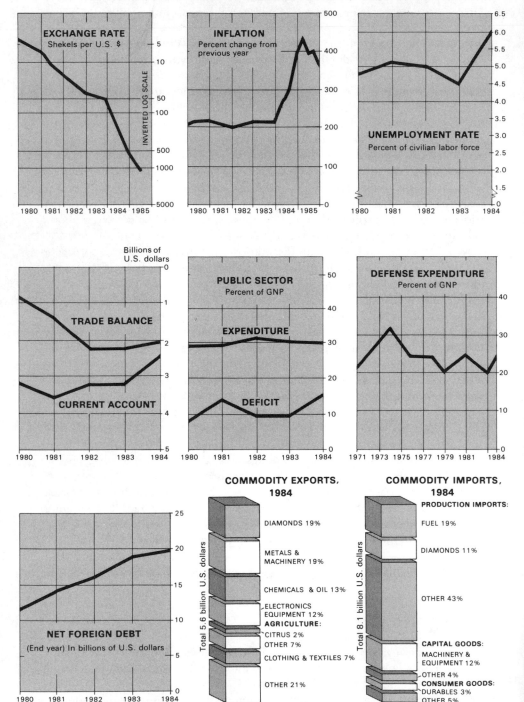

EXCHANGE RATE
Shekels per U.S. $

INFLATION
Percent change from previous year

UNEMPLOYMENT RATE
Percent of civilian labor force

TRADE BALANCE

CURRENT ACCOUNT

Billions of U.S. dollars

PUBLIC SECTOR
Percent of GNP

EXPENDITURE

DEFICIT

DEFENSE EXPENDITURE
Percent of GNP

NET FOREIGN DEBT
(End year) In billions of U.S. dollars

COMMODITY EXPORTS, 1984
Total 5.6 billion U.S. dollars

DIAMONDS 19%
METALS & MACHINERY 19%
CHEMICALS & OIL 13%
ELECTRONICS EQUIPMENT 12%
AGRICULTURE:
CITRUS 2%
OTHER 7%
CLOTHING & TEXTILES 7%
OTHER 21%

COMMODITY IMPORTS, 1984
Total 8.1 billion U.S. dollars

PRODUCTION IMPORTS:
FUEL 19%
DIAMONDS 11%
OTHER 43%
CAPITAL GOODS:
MACHINERY & EQUIPMENT 12%
OTHER 4%
CONSUMER GOODS:
DURABLES 3%
OTHER 5%

citizens. Indeed, the former set the tone for the latter. It is recalled that in 1974 the managing director of the Israel Corporation was indicted, and eventually convicted, for having transferred millions from the corporation abroad for personal investment. In later months and years, the chief of the Egged transportation company's insurance department in Haifa, a former treasurer of the government housing corporation, the director of customs, a high official of the Bank of Israel, the former manager of Israel's oil company operations in Sinai, the director of the Histadrut medical insurance program, the director of Israel's coins and medals corporation, a former senior national insurance (social security) official—all were arrested and sentenced to prison for embezzlement or fraud of one variety or another.

Horrified by these transgressions, the cabinet in August 1977 appointed a committee headed by Police Chief Michael Bochner to draw up a "crime map" of Israel. Five weeks later the committee issued a 184-page report. It revealed that growing numbers of organized groups were specializing in white-collar economic crimes that depended upon active assistance in official circles. If these elements had not yet penetrated the upper levels of government, they were making important connections at the intermediate echelons. Thus, they circumvented customs laws, entered legal businesses with "black money," exploited state transactions for illegal gain, illicitly obtained various licenses from local authorities. The alliance of criminals with municipal officials was a particularly unsavory one. In September 1977 the deputy mayor of Ashkelon, Baruch Abuhatzeira, was indicted for bribe taking. In 1979 the former mayor of Netania was arrested for fraud while serving in office and simultaneously acting as chairman of a bloc of land development companies. In 1981 the director of welfare projects in the municipality of Rosh HaAyen was found guilty of fraud.

No sector of the public domain remained untouched by peculation. In 1974 several highly placed officials in the defense ministry were arrested, later prosecuted, for bribery and fraud. These included the director of the ministry's electronics department, who had funneled ministry contracts to his private company in the United States; the director of the ministry's transportation department, who had exploited his position to buy and sell military automobiles for private profit; the chief of the maintenance division, who had solicited and accepted gifts. Ultimately fourteen defense ministry personnel were convicted in the squalid affair; but it was evident that officials of the ministries of industry and transport also were involved. Within the armed services themselves, at all levels, cases were periodically uncovered of graft, embezzlement, of rampant, wholesale theft of property, including weapons and ammunition.

Nor was there qualitative distinction to be made between the venality of secular and religious officials. The vast apparatus of religious courts, kashrut commissions, religious councils, and burial societies proved an especially lucrative source of income—legal and illegal. In November 1979 a police investigation of Jerusalem's religious courts determined that numerous rabbis were taking bribes for "expediting" divorce and conversion processes. Seven suspects were arrested, prosecuted, and eventually convicted, together with Rabbi Moshe Mizrachi, deputy chairman of Tel Aviv's religious council. The corruptibility of rabbinical courts by then was so widely acknowledged a fact of Israeli life that it generated a folklore of pungent anecdotes. Had it not been concern for accusations of "antireligious bias," the legal authorities would long since have investigated and prosecuted even more vigorously. But in 1980 one malefactor was simply too prominent to be overlooked. He was Aharon Abuhatzeira, minister of religions, subsequently leader of the Tami (Sephardi Religious) party, and brother of convicted embezzler Baruch Abuhatzeira, the former deputy mayor of Ashkelon. In 1981 the minister was tried and found guilty of stealing public money from a trust fund he had managed before joining the government, thereby becoming the first cabinet minister ever convicted of a felony.

If this was the example set by public officials, it was less than surprising, in an age of exorbitant taxes and high inflation, that private citizens should also have cheated on their taxes, embezzled, and defrauded in their business affairs. The courts were backlogged for years with thousands of these cases, the defendants including business and professional people of every description. By 1976 the overworked police were solving a meager 17 percent of business crimes. This was white-collar crime only. More sinister yet was the widespeard growth of hard-core felonies. The multimillion-dollar watermelon cash market, as a single example, was a source of extortion and large-scale laundering of "black money." Many of the distributors had criminal backgrounds. Dominating the wholesale market, they managed to squeeze farmers and retail sellers unmercifully. Their main source of income, however, was the protection racket. In Tel Aviv's Carmel Market, thugs "hired" themselves out as "watchmen" of the fruit and vegetable stalls. Few vendors refused their services, and those who did experienced mysterious fires, even bombings. The police were helpless to break the racket. Stall owners rarely brought charges, fearing a closer examination of their business volume by the tax authorities, or even harsher retribution by the extortionists.

The newest and most sinister category of hard-core felonies was drug trafficking. Its growth as a major income producer dated essentially from the occupation of Arab territories after the Six-Day War. Israelis, many of them non-Europeans with an intimate knowledge of Arabic and Arabs, swiftly developed a lucrative connection with the Middle Eastern drug trade. Much of the purchased hashish eventually was smuggled out to Europe, but not a little remained in Israel itself. Juvenile probation officials testified before the Knesset education committee that drugs accounted for the shocking growth of robberies and mayhem among young people, particularly among oriental youths in slum neighborhoods and development communities. The ramifying impact of this delinquency was tragic. Military officers witnessed it in the erosion of discipline in the ranks, the mounting incidence of aggressive behavior against officers. Eventually drug-tainted misfits either were rejected for the military or were dismissed from the armed forces outright. Afterward, they were obliged to carry identity cards stating that they had not done their national service for reasons of "social delinquency," a stigma that would deprive them of access to government jobs, even to drivers licenses.

Ironically, many young men bore their cards with bravado. They were members of "Commando 24," they boasted (Paragraph 24 of the National Service Law, which applied to their cases). In several outlying towns not less than half the youths proudly carried these cards. It was a development conceivably more dangerous than the Arab enemy.

So was the growth of violent crime. The Bochner Report of September 1977 gave a harrowing account of extortion and intimidation in the produce market, in restaurants, bars, and a growing variety of smaller retail establishments. Terrorization occasionally included the murder of witnesses and informers. Police officials were appalled by the large quantities of illicit weapons and ammunition circulating in Israel, a consequence in turn of mounting theft in the armed services. Altogether, as early as 1974, one out of twenty Israelis was the victim of a crime, and the annual rate of increase in the next three years averaged 11 percent. But the number of violent crimes was rising even faster, up 100 percent from 1973 to 1975. It was little consolation that virtually all murders were among criminals themselves. By 1980, only the United States, the Netherlands, and West Germany among developed countries listed higher per capita murder rates. Hardly less ominous was the declining solution rate for these, and other, crimes. In 1960, the figure stood at 56 percent. In 1984 it had dropped to 29 percent. Even of "solved" crimes, less than a third resulted in convictions. At that rate, a person's chance of success in legitimate business was much smaller than in crime.

"We are at five minutes before midnight," warned Professor Shlomo Shoham, a noted criminologist at Tel Aviv University, in 1984. "The symbiotic relationship between the criminal class and the authorities that exists in the United States had not yet developed here, but we are on the verge of transition to this relationship." The roots of the problem unquestionably were entangled in the nation's economic difficulties, the historic Jewish distrust of—unfriendly—government, the frequently scarred Holocaust survivor and North African immigrant mentality, the callousness to violence resulting from an endless sequence of wars and guerrilla assaults. Yet the low priority given to crime fighting in Israel's history could not be attributed simply to the country's understaffed police resources, to the diversion even of this limited personnel to the struggle against Arab terrorism. A more fundamental explanation may have been the inability of Israel's founding fathers to admit that a phenomenon as un-Jewish as crime could ever rear its ugly head in the miraculously restored Third Jewish Commonwealth. No provision had been made for this contingency.

Was the problem unsolvable? Had the cancer spread too far? Not when the will existed to fight back. An example was provided by the reaction of Israel's farmers to large-scale thievery in agriculture. It happened that the agricultural bloc and its lobby in the political parties and Knesset was one of the most powerful, arguably *the* most powerful, in Israel. When this group put its foot down and demanded action to stem the millions of dollars lost annually to farm-produce thieves, it won a major commitment of funds for police reserves. Soon the number of thefts was reduced dramatically. Nothing less was required for other sectors of the economy and government. Professor Shoham had stated the facts with unsparing candor. For the little Zionist nation, riddled increasingly with corruption and violence, it was indeed five minutes to midnight.

IV. *Emigration*

Well before the 1973 Yom Kippur War, evidence was growing that the most fundamental of Zionist tenets would have to be reevaluated. This was the assumption that Jewish statehood would resolve the anomaly of Jewish dispersion. The vast majority of Jews in the free world plainly had no intention of uprooting themselves to settle in the Jewish homeland. Conceivably, the painful fact might have been assimilated by Israel's leaders. But it was far more difficult to confront an even more sobering reality. Emigration from Israel was reaching calamitous proportions. As early as 1969 at least 250,000 citizens had departed. Subsequently, both the momentum and quality of that exodus increased. Earlier departures in the 1950s and 1960s had usually represented an efflux from a previous wave of immigrants. In later years, however, emigration included a far higher proportion of veteran Israeli citizens, even of native-born *sabras*. Many of the latter were young, often single, and about half were university-trained. These were the nation's most valuable human resources. If not all of them left with the intention of emigrating permanently, they somehow managed over the years to find ways and reasons not to return. The largest numbers of them apparently were unwilling to put up any longer with the economic disappointments, bureaucratic frustrations, and military dangers of life in the Jewish state. In July 1979, Ronnie Milo, chairman of the Knesset immigration committee, reported that no fewer than 2,000 Israelis were departing monthly.

Earlier, Labor, governments had made serious efforts to address the danger. In 1968 the Eshkol administration established a network of *aliyah* (immigration) centers in American and other Western cities to lure new immigrants and Israeli émigrés alike. For the first time, emigrants who returned were allowed virtually the identical concessions extended to new immigrants. These included travel expenses, cheap housing and business loans, free import of electrical appliances, tax relief, preferential assistance in education for their children. The response to the inducement was not impressive. It was even less so under the Begin government, which had other priorities. In earlier years, seeking Washington's quid pro quo for Israeli diplomatic concessions, Prime Ministers Meir and Rabin frequently had requested American intercession with Moscow on behalf of Soviet Jewish emigration. But in 1981, when Reagan declared his willingness to compensate Israel for the AWACS sale to Saudi Arabia, Begin's price was a "strategic memorandum of understanding" against Soviet penetration in the Middle East. It was a commitment that all but doomed the—already shadowed— possibility of Soviet Jewish departures. Begin's curious indifference to *aliyah* was evinced in the fate of the ministry of absorption. Established by the Eshkol government, it was now subsumed in the ministry of social welfare.

Independently of the Likud government, however, public attention was sharply refocused on the emigration crisis in December 1980. It was in that month that Shmuel Lachis, director-general of the Jewish Agency, reported that the number of Israelis and their children living abroad had recently been calculated at over 500,000. Extensively publicized in the newspapers, the figure evoked astonishment and alarm. When some journalists and government statisticians charged Lachis with exaggeration, the director-general held his ground, insisting that he had acquired his data after extensive consultations with Israeli consuls and other emissaries in the United States and Europe, and with Jewish federation and American immigration officials. More shocking even than the figure itself, Lachis noted, was the quality of the emigrants. A large proportion of them were youthful, well-educated army veterans, many of them heads of families. With some passion, Lachis exhorted the government to confront the problem on an emergency basis, to provide young veterans with a three-year grace period in tax exemptions and subsidized rental apartments following their military discharge. Returnees from abroad might be granted extensive loans, he suggested, and then spared repayment obligations in the event they remained in Israel a minimum of five years. The director-general was aware that these inducements would be costly. Yet he saw no alternative. "Did we pipe water into the Negev because it was economically worthwhile?" he argued. "We must close the emigration faucet before we lose too much of our precious human resources." Lachis warned, too, that a punitive approach would be counterproductive. If their Israeli passports were stripped from them, emigrants would become "permanently alienated from their homeland."

The danger of permanent alienation was not exaggerated. To be sure, the vast Israeli émigré community in New York sponsored Hebrew-language newspapers and journals, conducted its own émigré affairs, organized its own Jewish holiday parties. In the manner of other irredentist ethnic groups, who felt obliged to overcompensate in displays of patriotism, the emigrants tended to be hawkish, calling on Israel to invade Syria, to annex the West Bank, not to bow to American pressure. But the quality of their cultural attachments could better be gauged by their advertisements in the émigré press, offering the services of Hebrew-speaking hairdressers, garage mechanics, sex therapists. Interspersed with their gregarious meetings at Israeli functions in New York and elsewhere, their appearance at Jewish communal gatherings as Israeli heroes, they also found occasion to avoid Israeli taxes and reserve military service, to mock Israel's "inducements" to prospective returnees. With their nexus to the homeland considerably more fragile than their political posturing suggested, they required little additional excuse to sever their ties permanently.

Soon after the Lachis Report, Begin finally was persuaded to chair a special ministerial panel to discuss the document. The group's response continued defensive and bitter. Once again its members attacked Lachis's figures as specious. The actual number of émigrés was placed closer to 350,000 than to 500,000. Lachis himself was characterized as a "messenger of doom," a "negativist." Indeed, resentment against the director-general became so acute that he was obliged to resign his Agency post. On his own, however, Lachis moved vigorously to arouse public support. He organized a committee, Citizens for the Prevention of Emigration, to keep the issue alive in the press. Numerous editorials urged the implementation of at least one of Lachis's suggestions, the provision of cheap rental housing for families of young veterans. Under pressure, then, Begin appointed Dov Shilansky, deputy director of the prime minister's office, as chairman of a committee to deal with the emigration problem. The committee soon proved toothless—largely because it was penniless. Except for his willingness to consult with Lachis, Shilansky's one claim to fame rested on his whirlwind tour of the United States in 1981, where he passed himself off as an Israeli émigré seeking work in the Los Angeles "slave market."

Begin had his own schedule of objectives for government expenditures. Some of these, the evacuation of Israel's military and civilian infrastructure in Sinai, were unavoidable. Others, the renewal of slum neighborhoods and the extension of high school education, were enlightened. But others yet, the subsidization of a formidable Jewish presence on the West Bank and, later, the funding of the Lebanon war, at best were arguable. During this same period, ironically, in August 1981, a ministry of labor study was released that confirmed Lachis's statistics. At least 500,000 Israelis (and probably more) had indeed left the country over the previous three decades. The nation's Jewish population had slipped from 86 percent of the total in 1967 to 84 percent by 1979. With immigration all but frozen, the Jewish fertility rate stagnant, and emigration climbing steadily, even catastrophically, the demographic shift was all but certain to continue. David HaCohen, one of Israel's veteran diplomats and Labor's elder statesman, put the issue succinctly as he addressed his party's convention in 1981. "The crisis of emigration is no longer a question of Zionism," he insisted. "It has become a question of survival."

CHAPTER XI **LIGHTS AND SHADOWS IN ISRAEL'S FUTURE**

CONFRONTING THE STATE OF SIEGE

In the aftermath of the Begin years, Arab hostility remained a brooding presence in Israel's life, so integral a feature of the nation's waking and sleeping hours that it was incorporated into its very collective unconscious. The animus of Syria permitted only the narrowest cease-fire or disengagement with Israel, and even that through third-party—United States—mediation. The April 1983 accord with Lebanon was doomed from the outset by Syrian noncooperation. Even the peace treaty with Egypt lay frozen at dead center, meaningful contacts between the two nations obstructed equally by Menachem Begin's contentious policies in the "territories" and Lebanon and by Hosni Mubarak's determination to end Egypt's pariah status in the Arab world. There remained even an unresolved issue of direct Egyptian-Israeli confrontation. Just south of Eilat, at the Taba inlet, Israeli entrepreneurs recently had completed a 600-room resort hotel. The Taba zone, encompassing less than a square mile, clearly was minuscule in relation to the 22,000 square miles of the Sinai Peninsula that Israel lately had returned to Egypt. Yet the Cairo government was adamant in laying claim to this beach and to the neighboring hill overlooking Eilat to the northeast. The demand was based squarely on the 1979 peace treaty, in which Israel recognized Egypt's international Sinai frontier.

That frontier, as it happened, had first been demarcated in 1906 after lengthy negotiations between Great Britain (acting in Egypt's name but in Britain's interest) and the Ottoman Empire. For its part, Israel now argued that it was specifically the original 1906 demarcation line that was open to question. Recent scholarly investigation had disclosed that Britain at the time had altered Egyptian maps and surveys of the Taba boundary area, doctoring them to move the border still closer to the port of Aqaba at a time when the Turkish garrison there, with its small overspill at Taba, represented a potential threat to the Suez Canal. As matters later developed, Taba proved to be of little defensive value either to Britain or to Egypt. But its offensive usefulness in British hands, and later in Egyptian hands, was another matter. Indeed, Taba's strategic importance became a key issue after the 1979 treaty. Jerusalem alluded to a possible Egyptian offensive springboard on the western shore of the Gulf of Aqaba, in the event of a future violation of the Sinai demilitarized zone. Cairo insisted that its claim was dictated exclusively by considerations of national sovereignty.

Publicly, to be sure, the Israelis sought to pass Taba off as a minor issue that eventually would be resolved through negotiation. Ezer Weizman, the former defense minister, observed jovially early in 1985 that "Taba is not Masada and the Sonesta Hotel does not stand on hallowed ground." Yet the Peres government was concerned that Hosni Mubarak had proclaimed "satisfaction" on Taba to be indispensable for the normalization of Egyptian-Israeli relations, and was invoking the treaty's provision for arbitration. As Peres (and Begin before him) saw the matter, a renegotiation of the Israeli presence in Taba was acceptable only within the context of ongoing normalization—an even more fundamental provision of the treaty. Notwithstanding Israel's accelerated withdrawal from Lebanon, its de facto settlements freeze on the West Bank (p. 241), that context of normalization plainly had not yet materialized.

Altogether, the prognosis for a major Israeli breakthrough to peace remained guarded. Except for the West Bankers and (conceivably) the Jordanians, no Arab nation discerned a compelling need to reach accommodation with the Jewish state. Although the Egyptians favored a respite from war, the friendship they sought was with their Arab neighbors, not with Israel. Syria had emerged from the Lebanon conflict with enhanced prestige in the Arab world. The evidence was palpable, moreover, of ongoing Arab weight and influence in international affairs at large, and thus of continued diplomatic leverage against Israel. The Gulf oil nations were acquiring a network of refineries and shipping facilities that offered a potentially more decisive source of wealth and influence even than their substantial ownership of crude oil. The political impact of this economic might remained everywhere visible. During the 1970s, the Giscard d'Estaing government in France conducted an aggressive pro-Arab campaign within the Common Market, sponsoring declarations of "unequivocal support" for the Palestinian people, lobbying vigorously for a revision of UN Resolution 242. Giscard's courtship of the Arabs soon paid off. Iraq increased production specifically to meet French oil needs, placed $1.9 billion worth of new construction-and-service orders with French corporations, selected France as the primary supplier for its multibillion dollar electronics market. Saudi Arabia signed a contract for nearly $6 billion in French weapons and construction projects, then joined Kuwait, Qatar, and the United Arab Emirates in proclaiming France to be its prime supplier for military equipment through the next decade.

From the 1970s on, too, West Germany quietly emerged as a major economic and cultural center for expatriate Moslem communities. Upon negotiating an extensive scientific and technological cooperation pact with Saudi Arabia, Bonn dropped its former restrictions on investments by foreigners in German strategic industries. Kuwait accordingly purchased 30 percent of Korf Stahl AG, one of Germany's biggest steel firms; 10 percent of Metall-gesellschaftverein, a huge metals and engineering conglomerate; 14 percent of Daimler-Benz, Germany's largest vehicle manufacturer. Kuwait and Qatar purchased a 15 percent stock share of the two giant chemical corporations, Hoechts and BASF; even as the Arab Banking Corporation took over Richard Daus, a major West German bank. By no coincidence, Bonn's position on the PLO became increasingly friendly. Former Chancellor Willi Brandt met with Arafat. Chancellor Helmut Schmidt in 1980 worked with Giscard in a behind-the-scenes effort to tilt the Common Market in favor of the PLO. In turn, during his state visit to the Bundesrepublik some months later, King Khaled of Saudi Arabia announced that West German firms had been awarded a $910 million contract for laying a water pipe system, that the Bundesrepublik also had been granted most-favored-nation status as it prepared to submit bids on desalinization and petrochemical projects.

During this same period, wealthy Arabs were inundating England, buying up choice real estate, giving extensive business to British doctors and solicitors, and transforming London into the most expensive city in Europe. The single largest institutional investor in London was the Kuwait Investment Office, even as Arab money soon became the most important source of British urban funding altogether. The avalanche of Arab capital into the British banking and financial system could hardly have failed to register on the nation's foreign policy. In 1979 Foreign Secretary Lord Carrington officially dissociated his government from Camp David, a summit process, he argued, that "takes no account of the legitimate rights of the Palestinians . . . with the right to a homeland." Well into the 1980s, the British government maintained its original Yom Kippur War embargo on military sales and shipments to Israel, and Prime Minister Margaret Thatcher remained a sharp critic of Begin's settlements policy. Most of the British press followed this line.

Neither could the United States have remained unaffected by the palpable evidence of Arab economic power. As early as the Yom Kippur War, the nation imported a third of its oil, much of it from Muslem countries. The very symbol of the shifting American-Arab economic relationship could have been that preeminent boom city, Houston. Here, until the early 1980s, Arab investors were swallowing major portions of the urban economy, purchasing massive share-blocs of American oil companies, their refineries in Houston and elsewhere, their distribution networks overseas and in the United States itself. The precise figure of Arab investment in the United States was unknown, but estimates ran as high as $200 billion by 1982. It was certain that Arab deposits in American banks represented a sword of Damocles over the nation's financial community. For Israel, this power was evident less in a shift of Washington's policy than in the foreclosure of private American investment in Israel. An antiboycott law had been introduced under the Carter administration. Despite the best supervisory efforts of the commerce department, however, the effect of that measure was not far-reaching. A significant number of America's largest corporations declined to do business in Israel, evading the antiboycott legislation by the simple device of including a few non-American components in the parts and materials they exported. Houston became the center for a new industry of legal professionals (among them, John Connally, an ex-governor of Texas and a major contender for the Republican presidential nomination in 1976) who guided American firms through the complexities of the antiboycott law.

These bleak circumstances notwithstanding, Israel's tunnel was by no means without its promise of light. The collapse of the OPEC pricing structure in the early 1980s foretold a possible reduction of Arab diplomatic leverage—at least for the rest of the decade. The election of the Mitterand government in France in 1981 opened possibilities for a revived Franco-Israeli relationship. Even Begin's precipitous bombing of the Iraqi nuclear reactor would not alter Mitterand's decision to rekindle the Franco-Israeli friendship of the 1950s, to pay a ceremonial state visit to Israel. Thus, fifteen years after de Gaulle had branded the Jews "an elite people, self-assured and domineering," and given Israel the back of his hand, a French president came to Jerusalem and laid a wreath on the tomb of Theodor Herzl, the founder of political Zionism. Elsewhere, too, there was evidence of a revival of Israel's image, even of Israel's presence. In 1986 Spain's Prime Minister Felipe Gonzalez announced his government's decision to establish diplomatic relations with Israel. It was a less than oblique bid to dissipate the memory of the Franco era, to ease Spain's membership into the European Common Market by cultivating Jewish and liberal goodwill in the Western world. Nevertheless, the gesture was a poignant one, awakening memories of the once-historic symbiosis of Iberian and Jewish civilizations.

Israel slowly resumed its earlier foothold in Africa, as well. During the 1973 Yom Kippur War, President Joseph Mobutu of Zaire was the first of nearly thirty Black African leaders to sever diplomatic ties with the Jewish state. In May 1982, Mobutu was the first Black African president to restore them. Liberia followed suit in August 1983; its president, Samuel Doe, became the first Black African head of state to visit Israel in twelve years. But, in fact, Israel's connections with Africa never actually had lapsed, even after the 1973 wave of diplomatic ruptures. Rather, Jerusalem's economic links with Africa continued to expand. Israel's exports there exceeded $100 million in 1981 alone, a quantity three times greater than in 1973. Private Israeli companies operated in twenty-one African nations, performing work under contracts worth $2 billion. Israel also maintained diplomatic "interest sections" in Kenya, Ghana, Gabon, Togo, and the Ivory Coast. In Nigeria, some 500 Israeli families were living and working, operating three Hebrew schools. There was every likelihood that full diplomatic relations soon would be restored in these and other African nations. Israel's troops were out of Sinai by 1982, after all, out of Africa. More importantly, the signature of a peace treaty with Egypt opened the way for other African governments to restore and broaden their former contacts with the Jewish state.

Nor was the hope of Israeli relations in the Arab world necessarily unrealistic. For all its lifelessness, the 1979 treaty with Egypt foreclosed the imminent danger of war between Israel and its largest traditional enemy. In the manner of Germany and France, locked together in NATO, these two Near Eastern military powers between them sterilized their regional front. Under Jordan's moderate King Hussein, too, the Hashemite government evinced no interest whatever in renewed hostilities with its formidable Israeli neighbor, or even in the continuation of the state of war. Well after the unproductive Hussein-Rabin meetings of 1974 and 1975 (Chapter I), contacts between Amman and Jerusalem persisted as both governments sought ways to circumvent the 1974 Rabat Resolution, to restore Hussein's mandate to negotiate on Palestine. The mutual probing continued through the Rabin government and into that of Begin. In August 1978, Moshe Dayan met with Hussein in London. The king was concerned about the role envisaged for him in the current Israeli discussions with Sadat. In fact, Dayan envisaged none. He was certain now that Israel's best chance of a deal on Palestine was through the Egyptian president alone. On his own, therefore, without seeking Begin's approval, the Israeli foreign minister revived the harsher surgical alternative of partitioning the West Bank; and Hussein predictably rejected the notion. Dayan then returned to Jerusalem to inform the cabinet that there was nothing to discuss with the Hashemite ruler.

Nine weeks later, when Hussein telephoned Sadat at Camp David, offering to join the talks, the Egyptian leader proved equally unresponsive. Like Dayan, he feared a last-minute disruption of the fragile understanding then being worked out for a bilateral Egyptian-Israeli agreement. So, apparently, ended a fourteen-year dialogue; and Hussein was left with no choice but to proceed to the Baghdad conference, where he joined the rejectionist front against the Camp David accords. Several years passed then, with Jordan locked out of meaningful Palestine discussions, and Begin intensifying his settlements program on the West Bank. Yet Hussein for his part never abandoned hope of resuming his earlier spokesmanship for the Palestinians. Ironically, Begin himself fortified that hope, by destroying the PLO military-administrative infrastructure during the 1982 Lebanon invasion. Thereafter, with Arafat's position gravely undermined, Hussein again became the senior available Arab negotiator for the occupied territories.

Would the diminutive Hashemite monarch be allowed to exercise that role? The chances were more promising than in earlier years. The likelihood of other Arab nations acting on behalf of the Palestinians was hardly worth discussion, after all. They had been unwilling even to receive evacuees during the Israeli siege of Beirut. For that matter, they scarcely could agree on any position on any substantive issue, from the Reagan initiative in September 1982 to the May 1983 Israel-Lebanon accord, or even to the Gulf War between Iraq and Iran. At long last, Hussein remained without serious competitors. Indeed, no one appreciated the fact better than did Yasser Arafat, who finally bestirred himself to a series of renewed conferences with the Jordanian king in search of a joint Palestine formula. For Hussein, it was indispensable now to secure a forthcoming response from the United States— and from Israel. The possibilities were at least even. Secretary of State Shultz no longer appeared fixated by the Reagan plan for the West Bank, particularly since its initial rejection by Amman, if a more acceptable scenario could be devised between the Jordanians and Israelis themselves. And here it was, in September 1984, that Shimon Peres assumed office as prime minister of Israel's twenty-second government.

AUGURIES OF TERRITORIAL AND
POLITICAL MODERATION

As much as for any reason, Peres had won his narrow electoral mandate to end the debilitating military occupation of Lebanon. The precedent already had been set, the process of withdrawal from the Lebanese quagmire begun, as far back as July 1983, when the Begin government decided to carry out a unilateral retreat to the Awali River line. Now, after settling into office, Peres gave urgent attention to ways of accelerating the evacuation deadline. It was not a simple matter for his hybrid government. The Likud ministers initially opposed further withdrawal in the absence of a reciprocal commitment from Syria. By contrast, several of Peres's Labor colleagues favored an immediate, unqualified departure without any preconditions whatever. On January 14, 1985, however, with the support this time of Rabin, who was serving as defense minister, Peres was able to secure cabinet endorsement of a full, if calibrated, evacuation. Within the next five weeks, Israeli forces began pulling back to a new, interim position south of the Zahrani River, between 12 and 25 miles from the border. Then, remaining long enough only to protect their flanks and evacuate their heavy equipment, the army soon pulled up stakes again. By late June of 1985, after suffering only minimal losses in the carefully staged withdrawal, Israel ended its formal occupation of Lebanon altogether. The bulk of its invading army was across the northern frontier, and home.

The evacuation did not take place without painstakingly devised future safeguards. In the aftermath of Operation Litani in 1978, it is recalled, the Israelis had won tacit Lebanese recognition of a security zone, extending from 3 to 12 miles north of the Galilee border. In 1985 a modified buffer strip once again was to be guarded by the 1,800 militiamen of the predominantly Christian South Lebanon Army, the force now commanded by General Antoine Lahad. Initially, several cabinet members had expressed misgivings about the surrogate SLA, for its earlier incarnation under Major Haddad had been less than effective. But in fact the militia turned out this time to be a reasonably competent buffer against guerrilla penetration. Its success was largely attributable to the military equipment provided by Israel, together with the "advisory" presence of several hundred Israeli officers and men.

Indeed, after several months, the Israelis no longer bothered to disguise their collaborative role. That role extended from training to tactics. The latter resembled a "game" of sorts, worked out obliquely between the Israelis and the Christian militiamen, on the one hand, and Amal, the Shia militia, on the other. Israel, for its part, generally refrained from bombing Shia villages in the south. The Amal in turn was "permitted" to maintain its image as the successor to the PLO in the battle against the "satanic" Israeli presence, to launch attacks on SLA positions inside the security zone. Yet it was not allowed to mount attacks directly against northern Israel or even against the 400 or 500 Israeli troops working in the zone; nor was it to permit attacks against Israel by Palestinians or by "radical" Shia organizations from north of the security buffer, essentially in the old Red Line area. Altogether, the Amal and other Moslem groups knew just how far they could go before Israel's "iron fist" came down. "If we shoot one rocket," explained Daoud Daoud, the senior Amal official in south Lebanon, to a *Newsweek* reporter in December 1985, "Israel will shoot 100 rockets back. We can't win that war."

The very modesty of the Israeli foothold in the buffer zone also served a useful purpose. The Shia militia was less interested in expending its energy against this vestigial Israeli presence in south Lebanon than in mobilizing its strength to defeat other, more radical, Shia groups and in securing its fair share of the political pie in Beirut. Indeed, both the Amal and the Syrians were increasingly preoccupied with the continuing battle to prevent Yasser Arafat from reestablishing an independent base in West Beirut and south Lebanon. By early 1986, then, with the security zone effectively insulating Upper Galilee from the Lebanese conflict, Israeli newspapers and politicians tended to ignore Lebanon altogether. The attitude reflected a collective psychological need. The scars of 1982–84 needed time to heal.

The new government accordingly enjoyed a measure of leeway to turn its attention to other problems. In his earlier years as defense minister, Peres had equivocated on the territorial issue. Once he became Labor's candidate for the prime ministry in 1977, however, and in later elections, he was in a position to adopt a somewhat more conciliatory approach to the West Bank question. "Labor has no interest in becoming masters of 1.3 million Arabs against their will," he declared in 1981, and again in 1984. "Our goal is territorial partition, rather than the 'freedom' to be perpetual rulers over others." Was Peres too late? Had the process of de facto annexation gone too far? The "window" of reversibility by then had been drastically narrowed, after all, and maneuvering room for new initiatives in a national unity government would be constricted even further.

On the other hand, Peres had been elected to the premiership not only to end the Lebanon war but to resolve Israel's metastasizing inflation. To that end, he soon began intervening in economic affairs more vigorously than had any other Israeli prime minister since Levi Eshkol in the late 1960s, and the new government's fiscal and monetary policies soon came to be identified as much with him as with Likud's Yitzchak Modai, the finance minister. Harsh new budgetary restrictions, wage restraints, and tax increases were adopted to curb the overheated economy. In his effort to salvage the economy, moreover, Peres had made clear early on that he would impose a freeze on his predecessor's lavish investment in West Bank settlements and infrastructure. Some 1.5 million dunams in the area already were in Israeli possession, to be sure. But most of this terrain was uncultivated, and had not yet been put to civilian use. As late as 1984, the number of Israelis who had taken up residence beyond the Green Line (except for East Jerusalem) did not exceed 30,000. Many were haunted by recollections of the Sinai evacuation, and were less than sanguine that their presence among an Arab majority was necessarily irreversible. Even the Begin government had been sluggish in creating local employment opportunities for them, in providing them with neighborhood legal and social services, in sparing them the need for repeated trips to various government agencies in Jerusalem— above all, in recognizing their legal status.

Citing Israel's diplomatic obligations under the Camp David accords, too, Begin and his advisers had reminded the settlers that formal steps toward annexation, beyond jeopardizing the peace with Egypt, would provoke Amman into closing the Jordan River bridges; Arab farmers on the West Bank would have had no alternative then but to flood Israeli markets with their agricultural surplus. Coming from a Likud government, the warning was a source of even graver concern. Neither had the settlers forgotten the supreme court's ruling (in the Elon Moreh decision) that the Hague Conventions applied to the West Bank and Gaza. Under a Peres government, that judicial decision conceivably would be invoked to dismantle the settlements altogether. Only a minority of the colonists were devout Gush Emunim adherents. Most had taken up residence beyond the Green Line in response to Ariel Sharon's pragmatic—financial—inducements. What money had made, money presumably could unmake.

Neither could it have escaped even the hard-liners that the post-Lebanon mood altogether was one of fatigue, of a desire to end the bombast and military overreaction that had cost the nation dearly in lives and economic resources. The Peace Now movement had regained its former momentum. Lately, too, it was fortified by N'tivot Shalom, a dovish religious faction, and by such groups as Yesh G'vul, Soldiers Against Silence, and Parents Against Silence. Perhaps the most interesting of the new organizations was HaMizrach l'Shalom—East for Peace. Its members came exclusively from Israel's non-European communities, those that traditionally had been identified with Begin's hawkish policies. HaMizrach l'Shalom's raison d'être, in fact, was specifically to counter the orientals' image as indiscriminate Arab-haters. One of the group's spokesmen, Jules Daniel, a Moroccan-born journalist, reminded the nation that the Gush Emunim by and large was an Ashkenazic movement, that its European-style romanticism originally had been alien to the passive Jewish mores of the Middle East, in some degree, even of North Africa. Anyway, Daniel insisted, the poorer oriental Jews were those who had suffered most from resources diverted to settlements in the West Bank. If Daniel and other members of HaMizrach l'Shalom did not yet speak for a majority of their kinsmen, neither could they be written off as an exotic fringe group. Other Jews of Sephardic-oriental background were making their voices heard, and respected, at the uppermost level of the nation's leadership.

One of these was Yitzchak Navon. Born in 1921 of an old Sephardic family of Jerusalem, Navon studied Arabic literature at the Hebrew University, then served as local director of the Haganah's Arab department during Israel's war of independence. Soon afterward he became political secretary to Foreign Minister Sharett, and subsequently to Prime Minister Ben-Gurion. Operating as Ben-Gurion's right-hand man, Navon in the 1950s developed an effective constituency in his own right. Thus, in 1963, following Ben-Gurion out of office, he was elected as one of the Rafi candidates to the

Knesset, where he rose eventually to the posts of deputy speaker and chairman of the powerful foreign affairs and defense committees. A brilliant linguist, a poet and successful playwright as well as a skilled political negotiator, Navon in 1978 became Labor's candidate for president of Israel, and was elected. Thereafter, his dynamism and charm soon transcended the ceremonial limits of his office. His visit to Cairo in the aftermath of the Egyptian-Israeli peace treaty was a triumph of conciliatory diplomacy. Following the Sabra and Shatila massacres in September 1982, it was Navon's eloquent appeal to the national conscience that played a decisive role in the formation of the Kahan Commission. By the expiration of his presidential term, in 1983, Navon was widely respected as a forthright advocate of Arab-Jewish rapprochement. A year later he took office in the Peres government as minister of education and culture. There, serving as Labor's principal liaison with the non-European population, he was regarded by many as the party's logical eventual chairman, in the event Peres stumbled.

Navon's counterpart in Likud would have appeared an unlikely spokesman for moderation or intercommunal reconciliation. Yet the meteoric rise of David Levy, from an undereducated Cherut party hack to the acknowledged voice of right-wing populism, was one of the authentic political surprises in Israel's recent history. Born in Morocco in 1937, one of twelve children of an impoverished carpenter, Levy was brought to Israel with his family in 1956. In common with tens of thousands of other Jewish immigrants from the Moslem world, the Levys were settled in a development community, in this case Beit Sh'an, a shantytown in the Jordan Valley. Their economic plight, like their housing, was submarginal; and when David Levy married at a young age, he and his wife similarly were obliged to live in a cramped government flat. His own intermittent employment was as a construction worker; his wife's, as a cook and cleaning woman. Often the Levys and their growing brood subsisted on little more than rice and soup. Their frequent dependence on social welfare was soul-destroying. So was the indifference to their ordeal of the neighboring kibbutz settlements. Levy wrote later:

> [W]e lived side by side in the same valley, under the same sun, and they [the kibbutz members] couldn't have cared less. They didn't want to know anything about us and they didn't want to help. . . . They were able to take middle-class Europeans and turn them into productive laborers. But when it came to the immigration from Arab countries the [Ashkenazim] took people, many of them educated and all with strong ethical, moral and religious values, and . . . humiliated them and turned them into welfare cases. I also realized that this was not from evil intent, but from negligence that bordered on the criminal.

A strong, broad-shouldered, articulate young man, Levy was not prepared to accept his lot supinely. On one occasion he broke into the local employment office and wrecked the premises (he was briefly jailed). As a hired laborer in a kibbutz, he organized and won a strike for better drinking water. It was as a young activist, too, that Levy applied his frontal approach to the political process. During an election campaign, he mobilized his fellow Beit Sh'an immigrants to create pandemonium when visiting candidates of both major parties took the rostrum. Cherut's response to this provocation was imaginative. It was to select Levy as its candidate for the Beit Sh'an local council in 1964. He won. Subsequently, from his Beit Sh'an base, he assembled his own Cherut-dominated coalition. Within three years he and his political allies succeeded in excluding Labor from its traditional control of the Jordan Valley development communities. Thereafter, Levy was given increasingly important roles, including a spot on the party's Knesset list. Following the Likud upset victory of 1977, he was rewarded for his work on the hustings with a cabinet appointment, as minister of immigrant absorption. The "minor" portfolio dissatisfied him. He also demanded, and swiftly achieved, the ministry of housing.

Here at last was Levy's opportunity to make his mark on behalf of his non-European constituents. His predecessor, Gideon Patt, a Liberal devotee of free market economics, had left behind a critical shortage of apartments for young working-class families. Approaching the challenge with his characteristic straightforwardness, Levy bulldozed the cabinet into sponsoring remedial legislation. Contractors were offered the inducements of government guarantees. Mortgage interest rates were linked to income. Borrowers in the lower brackets repaid at a scale not to exceed 20 percent of their salaries. Over the next four years, as a result, the nation's chronic low-income housing shortage was noticeably alleviated. The remedy was shockingly expensive for the treasury, of course; yet it was effective both as a social measure for young marrieds and as a political asset for Likud. It was even more of a boon for David Levy's public image. During the 1981 election, campaigning in oriental neighborhoods, he was greeted with rapturous chanting, "David, Melech Yisrael"—David, King of Israel—a sobriquet he shared increasingly with Begin and Sharon. Working-class Israelis adored this forty-four-year-old dynamo, his eloquent speaking style, his frequent references to the modest Beit Sh'an flat where he continued to live with his wife and eleven children.

Levy's Cherut colleagues sensed his drawing power. After the 1981 election, he was awarded the additional title of deputy prime minister. Thereafter, in 1983, when Begin announced his intention to resign, Levy challenged Yitzchak Shamir for the Cherut leadership, and won 40 percent of the vote in the party convention—an impressive showing against a veteran political war horse. Although he did not contest Shamir again before the 1984 Knesset election, he was granted the number two spot on the Likud list, and accordingly retained his deputy premiership in Peres's coalition government. By then the chances were at least even that Levy would be Cherut's candidate for prime minister in a future national election.

If the prospect was unsettling to many educated Ashkenazic Israelis, others noted that Levy was by no means a boor or an uncontrollable demagogue. With the help of private teachers, he was diligently overcoming his limited education, studying English and international affairs. For all his populism, he seemed determined to project a "respectable" image. Thus, in the 1984 election campaign, he sternly warned his followers not to borrow from his own youthful example in embracing violence and demonstrations. The moderate Liberal faction within Likud was grateful for his ongoing support of the Camp David accords and the 1979 peace treaty, for emerging as a minority critic, later, of Begin's and Sharon's conduct of the Lebanon war. Behind the scenes, moreover, Levy allowed it to be known that he favored priority attention for the nation's economic needs over a doctrinaire obsession with an "undivided Land of Israel." If these caveats were expressions of a developing political maturity, David Levy's spectacular rise in Israel's national firmament may have been less a cause for anxiety than for cautious hope.

TOWARD A SCIENTIFIC SOCIETY

In 1984, the gap between Israel's total imports and exports exceeded $4.2 billion. Its accumulated foreign debt hovered at $23 billion. Among the factors responsible for this economic debacle, not the least was the nation's failure to match the competitive output of other developed countries. Its work force was not producing enough. Experts agreed that, for the standard of living merely to stay even, Israel's economy would have to generate a 12 percent annual increase in exports of goods, reaching the level of $11 billion by 1990. There was one method only for achieving this prodigious leap. Indeed, it was a simple matter of eliminating alternatives. Israel's natural resources were minimal, after all. Chemicals from the Dead Sea were a steady earner, but of peripheral importance in the country's trade balance. Polished diamonds, for years a major export, had been depressed since the market for investment gems collapsed in 1980. Although Israel had been reasonably self-sufficient in food production for a decade and a half, the nation's citrus exports were encountering stiff competition in the Common Market from Spain, Italy, Greece. It was evident then that Israel, like Switzerland and Japan, would have to develop an export-oriented industry based essentially on skilled manpower, and in sophisticated items where price was a secondary consideration. "Israel's future heroes," argued Stefan Wertheimer, one of his nation's most successful industrialists, "will not be its soldiers but its pioneers in high technology."

Here there may have been reason for a certain optimism. Well into the 1980s, the country possessed an impressively large reservoir of engineering and scientific talent. Its universities and technical institutes over the years had produced some 80,000 graduates, and the number of its working scientists approached 10,000. In this land, moreover, the tradition of applying science to practical, nation-building ventures was as old as Zionism itself. Many of Palestine Jewry's and Israel's spectacular dry farming achievements were pioneered by local Jewish scientists and engineers. No country proved more imaginative in the development and application of new irrigation techniques. None matched Israel in the exploitation of solar energy. Indeed, solar heating of domestic water was at work in two-thirds of the nation's households, and solar power was being applied increasingly to air conditioning. The world's largest solar pond, driving a turbine-powered 5-megawatt generating plant along the shores of the Dead Sea, was the vanguard of a planned network of solar power stations to be erected throughout southern Israel over the next two decades. The outlay would exceed $30 million, a considerable sum for a small country already foundering under crushing economic burdens. But on matters of water and energy, Israel traditionally enjoyed a political consensus. Applied science, like the military, had always been regarded as the key to national survival.

If the availability of scientific talent offered hope for Israel's economic growth, so did a concomitant pool of entrepreneurial skills. Stefan Wertheimer was a case in point. The son of a German Jewish mercantile family, he arrived in Palestine in 1937, attended school briefly in Tel Aviv, then hired himself out as an apprentice to a series of optics manufacturers. During World War II, enlisting in the British army, he served as a technician, repairing bomb sights and other precision equipment. There he found his métier. In the Haganah after 1945, Wertheimer became one of the Jewish defense force's most valuable tool- and weapon-makers. A brilliant improviser of munitions during Israel's war of independence, endlessly devouring technical manuals, he was soon put in charge of the fledgling government arms research corporation, Rafael, and launched the enterprise on a development program that ultimately would make it one of Israel's industrial giants.

Well before Rafael's eventual "takeoff," however, Wertheimer departed the company to enter business for himself, opening his own machine tool shop in 1951. His meticulous workmanship soon won him a respectable list of customers. Then, in the early 1960s, securing a line of credit from the Israel Discount Bank—a pioneer investor in the nation's industrial development—Wertheimer expanded and modernized his plant. Under the trade name of Iscar, Ltd., his company began to find a market in Europe for its high-quality tungsten carbide cutting tools. Wertheimer's decisive breakthrough came in 1968, however, as a consequence of de Gaulle's embargo on French military replacement parts for Israel. When the Israeli air force found itself in urgent need of turbine blades for jet engines, it asked Wertheimer to manufacture the equipment under license from the French Snecma company. He complied with alacrity. By 1970 his new factory was completed, his force of 200 workers hired and trained. And then, almost at the same moment, Paris banned the shipment of Snecma blueprints to Israel. Unfazed, Wertheimer resorted to his proven technique of improvisation. No one familiar with his record was surprised when the quality of his blades eventually surpassed that of the Snecma originals, when Iscar began receiving orders from well beyond Israel, even from Pratt Whitney in the United States. At the same time, endlessly innovating and refining his techniques, Wertheimer continued to place his major emphasis on the manufacture of sophisticated cutting tools for industry. In the late 1970s, Iscar was selling its products in Western Europe, North America, Japan. Its capital value in 1986 was estimated at $400 million, its annual exports at $50 million.

In these same recent years, Wertheimer viewed his developing success as evidence that Israel was capable of achieving a leadership role in skilled, light industry. The nation's economic future rested with entrepreneurs like himself, he believed, with men who started small, who imaginatively pioneered science-based ventures such as irrigation equipment or electronic software. On his own, then, Wertheimer threw his considerable energy and funds into the selling of his concept. One of his more visionary schemes was an industrial-residential community known as Kfar V'radim—Rose Garden Village. With other private investors, and with some government help, Wertheimer in the late 1970s purchased a 7,400-dunam tract in the Western Galilee, twenty minutes from Naharia and forty-five minutes from Haifa. There, private contractors were invited to build handsome, modern homes for an anticipated nucleus of younger professional families. As an inducement for construction and settlement, Kfar V'radim would boast the finest in educational and recreational facilities. More important, its residents would find their employment nearby in a light industrial park—Gefen—that would function as a kind of Silicon Valley.

It was in the blueprint for the industrial park that Wertheimer and his associates displayed particular imagination. Even highly qualified engineers and scientists, they knew, were often put off by the expense of administrative services. Thus, at Gefen, young entrepreneurs would be provided with a wide array of "start-up" facilities, with low-cost office and workshop space, with secretarial, telex, and computer services. Experienced private businessmen and government export specialists also would be at hand to supply their expertise. After a three-year time limit, the venture scientist would move on to larger accommodations elsewhere, leaving room for newcomers. By 1983 some 500 families, many of them American, already had moved into Kfar V'radim, the husbands embarking upon their start-up projects at the Gefen industrial park. Wertheimer was confident that ten times that many families would arrive within the half-decade. The expectation seemed even more realistic a year later when the supreme technocrat, Shimon Peres, became prime minister, and when extensive public subsidies were less likely to be poured into alternative West Bank bedroom communities.

If not all Israeli businessmen or economists shared Wertheimer's unalloyed faith in private enterprise, none at least doubted the transcendant importance of skilled manpower. Uzia Galil had discovered early the best method of mobilizing that resource. A product of the Technion, an electrical engineer, Galil completed his graduate work at the Massachusetts Institute of Technology, then stayed on briefly in the United States to work for the Motorola Corporation. It was during this American period that he witnessed the vital interaction between university faculties and high-technology industry. The model struck him as ideal for Israel. Upon returning home in 1961 to join the Technion's faculty, Galil set about establishing a laboratory in his small Haifa apartment to produce electronics equipment. With eight other faculty members and graduate students as his co-workers, he found a ready market for his products, and began earning a tidy profit. Two years later Galil outlined his approach to Dan Talkowsky, a former chief of the Israeli air force and currently chairman of the Israel Discount Bank's investment division. Talkowsky was impressed. As it had for Stefan Wertheimer's Iscar, Ltd., his bank advanced Galil a line of credit. In this case, the sum was enough to permit Galil to organize a small corporation, Elron Electronics, which itself would function as a source of seed money—in effect, a holding company—for other promising high-tech ventures. The partnership between Elron Electronics and the Israel Discount Bank was destined to revolutionize Israel's industrial development.

In subsequent years, Galil and Talkowsky examined dozens of proposals for linking the nation's scientists to research-and-production enterprises. Thus, the abundance of doctors and engineers opened out important possibilities in medical engineering. To exploit this talent pool, Elron and the Israel Discount Bank founded a subsidiary company, Elscint, for the development and manufacture of CAT scanners. The equipment proved marvelously effective as a successor to the conventional X-ray machine, and soon was marketed in the United States by General Electric. Within a short period, Elscint assumed a dominant position worldwide in the production of medical imaging devices, and in 1984 became the first Israeli company listed on the New York Stock Exchange. By then, too, it was manufacturing an entire family of integral camera-processors for nuclear medical applications. Elscint's managing director, Dr. Avraham Suhami, spoke enthusiastically of building the company into another Philips Industries, a giant international conglomerate that also maintained its base in a small nation, the Netherlands. Elron itself, meanwhile—Uzia Galil's original parent firm—continued to develop other subsidiaries. Among these were Elron-Iscar Biomedical Instruments, established jointly with Stefan Wertheimer; Bio-Technology, jointly developed with Dr. Isaac Kaplan of Tel Aviv University Medical School, and producing the world's most sophisticated surgical laser equipment; Elmar Medical Systems, Ltd., for pioneering techniques in automated hemodialysis. Not all these subsidiaries were profitable. In 1985, Elscint found itself overextended and obliged to retrench. Yet the mother company's future remained unchallenged as a brilliant alliance of science and industry.

By the late 1970s, high tech had become a talisman for Israel, an apparent panacea for an otherwise modestly endowed nation. Indeed, other competing banks now were seeking to share in the action. Bank Le'umi, the nation's oldest and largest, established its own venture capital division to fund high-tech projects. Bank HaPoalim, owned by the Histadrut Labor Federation, proved notably adept in seeking out joint investors from overseas. Best known among the latter was Ampal, the America-Israel Corporation, which sold debentures in the United States for Israeli high-tech enterprises. Nor was the Israeli government itself laggard in encouraging these projects. As far back as 1968, the ministry of industry and trade engaged a chief scientist to evaluate promising new ventures. Thereafter, at the recommendation of this official, the ministry matched an "approved" company's research and development budget pound for pound. By 1984 the money was going into approximately 600 projects.

Ultimately, a major share even of Israeli government funds depended on overseas help. Here an important source was the Israel-United States Binational Industrial and Research and Development Foundation—"Bird F"—underwritten jointly with $60 million from the United States and Israeli governments to provide low-interest loans for approved ventures. If the investments succeeded, Bird F was entitled to repayment of up to 150 percent. To ensure that payoff, moreover, Bird F in turn required joint ventures between Israeli and American private companies. Among the latter were such giants of American industry as Mennen Medical, Control Data, National Semiconductor, Motorola, Digital Equipment, and ITT. These and other corporations, transforming Herzlia and Rehovot into miniature Silicon Valleys, apparently were impressed by the wealth of Israeli scientific talent, the role played by faculty and graduates from the Technion, from the Weizmann Institute, the Hebrew and Tel Aviv universities and other research centers in lending their skills to high-tech enterprises. They were impressed as well by Israeli wage scales, which still tended to be lower than in the United States. Thus, by 1984, Bird F was funding 121 joint projects, and already receiving repayments from 35 of them. In that year, too, with the help of these local and overseas sources, electronics rose from 7 percent to 33 percent of the value of all Israeli industrial exports. Some economists projected a figure of 60 percent by the end of the decade.

ISRAEL AS ARMORER

A uniquely profitable component of the nation's burgeoning high-tech industry turned out to be military equipment. It was a less than surprising development, in view of Israel's painful circumstances of isolation and beleaguerment. Electronics production in other countries received its start in the consumer market, then later was applied to defense needs. In Israel, the process was reversed. More than any other factor, too, it was the French arms embargo in the aftermath of the Six-Day War that accelerated Israel's military electronics program. From then on, the establishment of a national military-industrial capacity became a matter of hardly less than survival. As local weapons systems of increasing sophistication were developed, moreover, they produced unexpected dividends as currency earners in the export market, for they were battle-proven in Israel's chronic state of warfare.

The field of military electronics was dominated by two conglomerates, Tadiran and Israel Aircraft Industry (IAI). Between them, these firms accounted for three-quarters of all production. Tadiran, jointly owned by Histadrut's Koor subsidiary and the GTE company of the United States, produced almost every kind of military communications equipment. Its research and development budget was the single largest in the nation, except for the aircraft industry, and the reputation of its products was evident in its 17 percent share of the world market. By contrast, IAI was strictly a government enterprise. First organized in the early 1950s, when it was known as Bedek, the firm was operated for a number of years essentially as an aircraft maintenance and repair facility; but later its functions were expanded to the assembly and manufacture of military and civilian aircraft, then of ancillary weapons systems. By 1984, employing 25,000 people, IAI was producing jet fighter planes, executive transports, naval vessels, and sea-to-sea missiles. IAI's subsidiaries, meanwhile, achieved world standards in sophisticated weapons systems. One of these, Eltra-Electronic, manufactured aviation instrumentation. A second, Mabat, produced aeronautical, naval, and ground weapons systems. At the same time, another—separate—company, the government-owned Rafael (like IAI, dating back to the early days of the state), specialized in air-to-air missiles and minicomputers. Still another, private, firm, AEL, produced electronic intelligence equipment, including jamming and electronic radar alarm systems.

It was at the Paris International Air Show, in 1981, that observers first grasped the scope and diversity of Israeli military production. Only then did IAI, Tadiran, Rafael, Beit Shemesh Engines, and other Israeli companies first unveil the full range of their output. Here were displayed the Kfir fighter plane, the Arava and Westwind civil aircraft, the E1/M 2121 intelligence battlefield surveillance radar—this latter an extremely long-range device capable of locating and classifying targets on or near the ground, at daytime or nighttime. Also revealed was an upgraded version of the Gabriel sea-to-sea missile, a brilliant water-skimmer which even in its earlier incarnation had revolutionized the Israeli navy; and the Tactical Command and Control System, a mobile computer providing updated information of the battlefield environment. Still another item was the Barak Advanced Shipborne Point Defense Missile System, a modular, compact, lightweight apparatus, boosted by a solid propellant supersonic engine, capable of all-weather multiple target engagement, and possessing a high electronic countermeasure immunity. IAI also introduced the Spider-11 Air Defense Artillery System, a unit capable of identifying and tracking targets, and automatically aiming and firing six twin-30mm cannon; a range of airborne electronic warfare and countermeasure systems, among them jammers, chaff dispensers, deception, SIGNIT and target acquisition equipment; a series of pilotless drones that only a year later were virtually to paralyze Syria's SAM-missile system in the Lebanon war. The display, with attendant "smart" bombs, rockets, warheads, mines, and other explosives, as well as laser range-finders, artillery computers, and related integrated fire-control systems, left visitors gasping.

By 1984, 112 Israeli companies were exporting approximately $1 billion worth of military equipment each year. IAI and Rafael both marketed their products in some forty nations (the United States sold arms to sixty-seven countries). Israel was not yet among the world's major weapons exporters, to be sure. It ranked fifteenth or sixteenth, behind even North Korea, Brazil, and Belgium—all nations with access to the vast Arab arms market. Yet, by some estimates, weapons comprised an astounding one-fourth of Israel's total industrial exports, and manifestly a disproportionate share of its high-tech exports. Ironically, even Iran (militantly anti-Israel following the Khomeini revolution) became an intermittent purchaser of Israeli arms as it waged its Gulf War against Iraq—the latter an implacable "confrontation" state the Israelis were eager to keep hard-pressed elsewhere. The Israelis excelled, too, in meeting the needs of Third World nations that could not afford costly new weapon systems. Their technicians had accumulated much experience in modernizing older equipment for domestic military purposes. Thus, Israel "updated" fifteen Mirage-55 jet fighters for Colombia, became the single largest infantry-equipment supplier to El Salvador and Guatemala, offered Venezuela and Costa Rica huge stocks of arms captured in Lebanon. With its massive quantities of captured Soviet weapons, Israel actually became the world's second largest exporter of Soviet arms, after the USSR itself. In 1982, the sale of Soviet-built equipment figured in the resumption of diplomatic relations between Israel and Zaire.

No one could have regarded these developments more approvingly than did Shimon Peres, a man whose entire career had been identified with the invigorative possibilities of science-based industrial and military growth. On independence day, 1985, six months after assuming the prime ministry, Peres declared Israel's thirty-seventh year to be the "year of high technology." The proclamation was at once an expression of hope and anxiety. Defining his party's stance only two years earlier, Peres had emphasized that

> Labor conceives Zionism as not only a change of location but also a change in the structure of the Jewish people: a break with *galut* [exile] livelihoods and voluntary transition to a life based on work and creativity. . . . Likud has turned Israel into a stock-market state. Less and less money is concentrated on the development of agriculture and industry. Farms are caving in, and no one comes to their aid. There is virtually no [*sic*] government investment in the industry of the future. And the young joining Israel's work force increasingly turn to finance and to administrative services, where the pay is higher and the work is more comfortable. A renewed *galut* existence is taking shape on the unredeemed soil even before our eyes.

In his preoccupation with a creative, "productive" society, Peres doubtless understated the accomplishments of past years. Yet the prime minister hardly underestimated the obstacles to an educated, science-based work force. As early as 1982, Avraham Suhami, chairman of the board of Elscint, had warned that his company's future might no longer reside in Israel itself. As a consequence of emigration to the United States and other Western countries, Israel was running out of engineers and scientists. Nor did the Technion and the nation's other universities possess the capacity any longer to make up the shortage. Rather, the slack already was driving up wages beyond the level of competitiveness on the world labor market. Suhami noted that the cost of employing an engineer in Israel was often twice as high as in England. Plain and simple inefficiency, the sheer expense of the nation's support services, also threatened the future of science-based industry in Israel. With some bitterness, Suhami observed that he, the chairman of a corporation with sales of $100 million a year, could not get a second telephone installed in his apartment. The government bureaucracy was still clogged by its own weight. Perhaps the Israelis themselves sensed the uncertainty of local conditions. In the early 1980s, the Israel Discount Bank and other financial institutions were more cautious about risking capital in local high tech. By then most of the venture funds for research and development were coming from abroad—in the case of Elscint, 97 percent. During the Begin years, moreover, increasing numbers of Israeli capitalists were investing even their own funds overseas, in the United States and other Western countries. It was not until the Peres coalition cabinet of 1984 that the "free market" in capital transfers out of Israel gradually was halted.

Government directive alone, however, was incapable of resolving the emergent shortage of trained personnel. Less than a month after Peres declared 1985 to be the nation's year of "high technology," the state comptroller issued his annual report. It was a shocker. Far from progressing in science and engineering, Israel, creaking under an inefficient public bureaucracy and falling educational standards, actually was dropping behind. The warning had been issued before (Chapter VII), but never in such massive and overpowering detail. In two particularly frightening chapters, the report debunked the nation's self-satisfied assumption that it was maintaining its qualitative edge over its enemies. One chapter described the inability of Israel's school system to provide the armed forces with technologically trained manpower. The other scrutinized the qualifications of the military's officer corps up to the level of battalion commander. In all, the evidence provided a sobering picture of decline and deterioration. It also confirmed Avraham Suhami's estimate that Israel would experience a shortage of between 18,000 and 23,000 engineers and skilled technicians by 1992. So acute was the danger that the harvest of Israel's renowned scientific and technological achievements, sown ten or twenty years earlier in education and investment, might not be renewable. Here too, then, even in the brightest patches of Israeli progress, sunlight was mottled with ominous shadows.

A REEVALUATED PARTNERSHIP WITH THE DIASPORA

Israel's most dependable ally meanwhile remained the Jewish hinterland. Without that alliance, the little republic very possibly would not have survived the early years of its independence. It was in the 1940s and 1950s, during the initial avalanche of post-statehood immigration, that the Jewish Agency fulfilled its role as the conduit of Diaspora philanthropy by ensuring that not a single newcomer spent even one night without a roof over his head. Dispensing overseas Jewish funds, the Agency established 480 new agricultural villages and furnished them with livestock and equipment—settlements that by 1970 were producing 70 percent of the nation's food. It was the Agency, too, administering Jewish National Fund and United Israel Appeal money, that built Israel's roads and planted its forests, established clinics and schools, and functioned as the government's partner in transforming a colony of barely 600,000 Jews into a viable population six times that size. Altogether, between 1948 and 1968 alone, world Jewry accounted for 60 percent of Israel's $6 billion cumulative import surplus, and 69 percent of all long-term capital transfers to Israel. The sum would nearly double in the ensuing decade.

At no time during that period, however, were American and other Western Jews who contributed these funds "structurally" identified with the vast redemptive effort of ingathering and economy building. Their role by and large was passive, of giving but not of policy determining. The relationship was not healthy, in the view of Louis Pincus, who served as chairman of the Jewish Agency during the 1960s. Thus, in the aftermath of the Six-Day War, Pincus worked closely with Diaspora Jewish leaders to integrate the overseas community more closely with Israel's ongoing economic development. In 1969 the Jewish Agency was reconstituted, its authority henceforth to be shared equally between Israelis, "affiliated" members of the Zionist Organization overseas, and "nonaffiliated" Diaspora leaders. Presumably this equal partnership would link Western Jews more effectively to Israel's economic-humanitarian needs. It did not. Very little changed—for better or worse. As in the earlier years of practical alliance, but of "nonpartnership," the American and other Diaspora members of the Jewish Agency declined to interfere in administrative judgments. Content to be involved in loose, policymaking discussions, they were equally prepared to leave the day-to-day execution of programs, and the disbursement of funds, to the Israeli-staffed departments of the Agency. Neither did their essential response to the nation's humanitarian needs significantly increase or decline, whatever the vicissitudes of Israeli government policy. Even in the early years of the Begin era, their contributions and those of their Diaspora constituencies did not fluctuate significantly.

By and large, then, the Diaspora remained a grateful—but compliant—partner in the Israeli relationship. Influenced by a mixture of ethnic pride and local insecurity, Jews abroad tended to regard Israel as the dynamic champion of Jewish peoplehood worldwide. Over the years, their communal organizations, even such older, nativized institutions as the American Jewish Committee, the Conseil Représentatif des Israélites de France, the Delegación de Asociaciónes Israelitas Argentinas, tacitly recognized Israel's claim to speak for world Jewry in protesting the treatment of Jews in Arab lands or in the Soviet Union. The Israelis were uncommonly effective in this spokesmanship, after all. Their diplomatic emissaries enjoyed important status overseas. Their presence offered a certain mantle of dignity, even of protection, to local Jewish communities. If there were occasional conflicts of interest between Jerusalem and the Diaspora on foreign policy issues, the Diaspora rarely caviled (at least in public). Thus, in the early 1960s, Israel consciously risked the discomfiture of South African Jewry as it joined in United Nations condemnations of apartheid. The protests of the South African Jewish leadership in turn were feeble and largely pro forma. In the case of Soviet Jews, Israel and its loyal Diaspora constituencies stressed the right of their kinsmen to emigrate, rather than the right to enjoy cultural and religious freedom within the USSR itself—a boon that would have been of no particular value to Israel.

This forbearance ultimately may not have been in the interest either of Israel or of the Diaspora. By the second decade of Israel's existence, surely by the third, it was evident that the Zionist "solution" was vulnerable on several levels. A majority of Jews in free, Western nations did not wish to emigrate to the Jewish homeland. The promise of Israel as the sure and certain haven for the Jewish people was itself open to question. Israel was the only Jewish community in the world, after all, that was calling on Jews elsewhere to save it from "another Holocaust," a threat that its very establishment was supposed to avert. Jewish statehood also had been envisaged as a guarantee of Jewish economic self-sufficiency and productivity. The opposite now seemed to be the case. Israel's economy was a shambles, the nation was emerging as one of the "developed" world's chronic mendicants. Neither was the historic assumption borne out that a Jewish state would effect a Jewish vocational transformation. Peres's warning was apt: the business and professional patterns of the Diaspora were increasingly being replicated on Israeli soil. To an extent that would have shocked both Herzl and Achad HaAm, so was the reviving ultra-Orthodoxy of the nineteenth-century *shtetl*.

The militantly Jewish premise, too, upon which Israel was founded, created as many problems as it solved. Thus, not only Israeli Jews but foreign Jews possessed a status in the country that was denied non-Jewish Israelis. Israel's loyal Druze citizens (and surely Israel's Arab citizens) were not permitted to buy homes in the Jewish quarter of Jerusalem's Old City, although Jewish noncitizens were welcome. Meir Kahane could bring his American hoodlums to Israel and aggravate tensions on the West Bank, the Satmar Rebbe's visiting American zealots could join in the stoning of vehicles in Sabbath traffic—all with near impunity—while Arab citizens of Israel risked arrest for demonstrations against government policy. Finally, Israel's authorities experienced no qualms in mobilizing the political and economic support of Jews everywhere, in urging American Jews to petition their government on Israel's behalf, to solicit congressional support for Israel's foreign aid requests. It was the virtual certainty of world Jewish support, or at least of Diaspora acquiescence, that allowed Begin and Sharon to pursue their policies on the West Bank.

In the end, Jewish loyalty worldwide manifestly derived from feelings of pride and solidarity, even of guilt for not living in Israel and exposing themselves to Israel's military and economic pressures. Would these feelings have been as intense if Israel were a peaceful, prosperous, "normal" country? It did not appear likely. The warrior image that traditionally had been disdained by generations of Jews as antithetical to Judaism had become an admired Israeli characteristic in contemporary Jewish eyes. Identifying with it, Jews overseas led a vicarious existence as heroes by proxy, certain that they were casting a more respectable shadow among their Gentile neighbors. The economic price for these emotional compensations was high. But so it was for the Israelis, many of whom unwittingly felt obliged to act out the role for which they were being paid. It was a burden that may have inhibited them from adopting a more imaginative stance on the question of the territories. In some measure, it may have induced the Likud government into the Lebanon expedition.

Would Jews living abroad review more actively the manner in which their support was used? The possibility of change in this compliant partnership first surfaced after 1979, when the Begin cabinet intensified Israel's settlements program in Judea and Samaria. It was then, we recall, that criticism was ventilated in the Jewish hinterland far more openly and forthrightly than was the case in the early and mid-1970s. Concern was voiced that Diaspora funds were being applied to the widening Israeli infrastructure on the West Bank. In response, Jerusalem insisted that only Zionist Organization and government money was being spent beyond the Green Line; that United Jewish Appeal and other Diaspora philanthropic funds were applied exclusively to domestic humanitarian purposes within integral Israel. Yet, for many, the explanation smacked of bookkeeping manipulation. There was growing concern, too, during Begin's intimate partnership with the religious parties, that a disproportionate share of overseas charity was going to yeshivot and other Orthodox institutions. Although the level of philanthropic support did not yet fall dramatically, Diaspora misgivings were mounting. Now, at last, American Jews began seriously exploring alternatives to their traditional United Jewish Appeal contributions, and to the disbursement of those funds through party-staffed Israeli departments in the Jewish Agency.

One of the most intriguing philanthropic alternatives was the Palestine Economic Foundation's Israel Endowment Funds. The PEF was no latecomer on the Jewish scene. It had been established in 1922 by Louis Brandeis specifically to bypass the Zionist-controlled Jewish National Fund. For years afterward, the PEF operated quietly, the recipient of contributions from Brandeis himself and from other affluent American Jews. Then and later, it employed no staff. Its volunteers sought out worthy charitable activities and institutions in Israel. With few exceptions, the latter were small, unpublicized groups that worked in poor neighborhoods, gathering and distributing used clothing or used medical equipment for needy families, organizing visitation societies for the ill or the "socially fallen," operating mother-and-baby clinics, toy-and-game libraries, loan-and-grant funds, workshop shelters for the blind, the crippled, the battered, the mentally or physically disabled. The PEF was as restrained in its promotional efforts as in its administrative expenditures, but during the 1970s and 1980s its activities drew increasing attention and support from American Jews.

Elsewhere in the Diaspora, concern with Israeli policies, if more diffuse, was becoming more vocal. Here too the Lebanon war served as a watershed of sorts. On the one hand, there was widespread resentment at the indiscriminate media criticism of Israel, of the skewed Gentile tendency to equate Lebanon with the Holocaust—palpably a moral exorcism for Western indifference to the European Jewish tragedy. Yet, even as Israel's survival remained central to their lives, many Diaspora leaders now experienced increasing difficulty in coping with Begin's political territorialism and military adventurism. Their confusion and ambivalence was evident in tense gatherings of Jewish communal representatives in London and Paris, in Stockholm and São Paulo, in Montreal and New York. In December 1982, at a meeting of the Zionist Congress in Jerusalem, a majority of delegates for the first time supported a resolution urging Begin to relinquish occupied territory in exchange for peace, to limit Jewish settlements on the West Bank to sparsely populated areas. The proposal evoked such consternation that the chairman halted the proceedings before a vote could be taken. At no time was there doubt that Israel remained the very wellspring of Diaspora loyalty and pride. Yet, at the least, the wounds of bitter contention seemed increasingly likely to affect fund raising for Israeli causes, even for such purely domestic programs as Jewish education, with its markedly Zionist orientation.

THE SEARCH FOR REALISM

From 1982 on, the U.S. government sensed a rare opportunity to exploit these developing Jewish misgivings. For the first time, Washington believed it possible with at least tacit Diaspora support to press for an equitable solution of the Palestine question. If Israel were to be saved, so the argument went, it would have to be saved in spite of itself, under pressure of a sustained American initiative with full presidential endorsement. Had the time already passed for this initiative? Not if Washington managed to disabuse itself of the fashionable intellectual notion that its pressure instantly would mobilize a reflexive Israeli counterreaction, a closing of the nation's ranks around its government. In truth, the largest numbers of the Israeli people were mortally afraid of losing the American connection. More significantly, important elements among them would quietly have welcomed that pressure as a face-saving rationale for compromise.

In 1982, Yehoshafat Harkabi, the former chief of Israeli military intelligence, published a short volume entitled *The Bar Kochba Syndrome*. The allusion was to the second-century Judean uprising against imperial Rome, the climactic revolt in a series that began in 60 A.D., and that eventually destroyed the Temple, eradicated Jewish statehood, and cast the Jewish people to the margins of history. For centuries afterward in the Diaspora, this final rebellion, commanded by the redoubtable warrior Simeon Bar Kochba, was lamented as a gratuitous and costly disaster. It was only the rise of modern Zionism, Harkabi noted, that suddenly transformed the Bar Kochba episode into an act of folkloristic heroism. Thereafter, Israeli schoolbooks described the event as an inspiring talisman of Jewish resistance against all odds. In May 1982, we recall, Prime Minister Begin actually conducted an official "burial" ceremony in the Judean Desert to inter the bones—ostensibly those of Bar Kochba's lieutenants—that had been unearthed by archaeologists some twenty years before.

As Harkabi saw it, the original Bar Kochba uprising was catastrophic enough. Its contemporary metamorphosis into a touchstone of Jewish valor was more tragic yet. "To admire the rebellion," he warned, "is to admire rebelliousness and heroism detached of responsibility for their consequences." It was precisely this "Bar Kochba syndrome" that Harkabi regarded as an endemic characteristic of the Begin government. "Having chosen statehood," he noted, "our destiny is . . . in our hands, more than at any time since Bar Kochba. This new situation demands not myths, but sobriety, much self-criticism, and severe criticism of the historical circumstances in which we find ourselves." Those circumstances, of de facto territorial annexation, of military adventurism, of fiery defiance of international criticism—in short, the Bar Kochba syndrome—represented a plain and simple departure from realism. Their consequences for modern Israel could well be as suicidal as they were for the ancient Judea of Bar Kochba's day.

Harkabi's appeal for realism was echoed by Amnon Rubinstein, dean of Tel Aviv University Law School and founder of the moderate-centrist Shinui party. In his volume, *The Zionist Dream Revisited*, Rubinstein traced the emergence of a kind of fatalistic Israeli apocalypticism to the period of the Six-Day War. It was in that crisis, when the nation stood alone before imminent destruction, and then triumphed on its own, that the illusion rapidly became widespread that Israel, contrary to all Zionist aspirations, was the successor to, and inheritor of, the rejected Jew. By that interpretation, Israel's behavior, its diplomatic or military stance, were irrelevant. The Jewish state, like the Jewish people, was apparently forever doomed to isolation and contumely. In its secular version, this fate often was regarded as the inescapable curse of Jewish peoplehood. To religious fundamentalists, rejection corroborated an ancient truism and validated the continuity of Jewish history, with the Arabs the successors of the Jebusites and Amalekites, and the Christian world the successor of ancient Romans and medieval Crusaders.

For Rubinstein, the analogy was meretricious. Israel's current difficulties in no sense were of the genre that had plagued Jews as minorities in reluctant host societies. Rather, they stemmed from geopolitical circumstances that were wholly unrelated to the religious, historical, and psychological obsessions that had acompanied and dispersed Jews in earlier times. Israel was a state now, with a government, a foreign service, an army of its own. For all its vulnerability of size and resources, it had proven its ability to defend itself, to win economic aid, military equipment, and diplomatic support from nations as powerful as France and the United States. Notwithstanding its financial dependence on the Diaspora, moreover, Israel exerted significant diplomatic leverage, not to mention moral and psychological influence, in protecting Jewish communities in numerous other lands. Plainly, there were not lacking abnormalities to bedevil the tiny Jewish republic and accentuate its built-in neuroses. But Israel also displayed enough of the features of other sovereign states to impose on its policymakers and intellectual leaders the obligations of perspective and responsiblity—the dignity of realism.

From non-Israelis, on the other hand, a certain realism was equally mandatory. This was the realism of compassionate understanding. Expectation of that sentience from Israel's enemies, from hard-line Arab states or from the Soviet Union, no doubt would remain the vainest of hopes. Yet from Israel's millions of Gentile friends, and surely from the Jews of the Diaspora, understanding was a reaction as legitimately to be anticipated as was constructive criticism. Israel had fallen short in their eyes lately, without question, had revealed itself to be a less than ideal compensation for their own insecurities. Even so, disappointment might usefully have been tempered with a reevaluation of this little people's authentic identity. The question then deserved to be asked: who, after all, were the Israelis?

To Westerners, it was the Orthodox bloc that represented the heaviest millstone around Israel's neck. The appraisal very likely was accurate. It was also worth recalling, however, that the first Jews to return to Palestine in modern times were not the Zionists, but precisely these religionists. Their departure from Eastern Europe in the early 1700s, a vanguard of 1,500 devout families under Rabbi Judah the Chasid, the grim ordeal of their odyssey, the desolation that awaited the survivors among them in the Wilderness of Zin—were painful almost beyond description. A half-century later, the largest numbers of the original pilgrim community had perished. Even then, their kinsmen did not stop coming, generation after generation. Unquestionably, they were not a productive group. They had arrived with only one purpose, to live and die within the shadow of the ancient Temple Wall. Yet what was the life, and the death, of this "sleeping settlement?" Subsisting behind the medieval Turkish battlement of the Old City, they froze in winter, roasted in summer, expired by the hundreds of typhoid and typhus. In peacetime, they were taxed, tyrannized, stoned, and lynched by their Arab neighbors to within an inch of their lives. In wartime, enemy blockade reduced them to the edge of starvation. By 1880, nevertheless, it was this Orthodox element that constituted the largest Jewish enclave in Palestine, and, in the city of Jerusalem, the majority of the population altogether. For all their perversity and obscurantism, they were the foothold, the first to resume and maintain the ancestral tradition. But for them, there would have been no Jewish settlement in the Holy Land.

No one claimed, either, that Begin and his Cherut associates were a congenial lot, as they stubbornly insisted upon their right not simply to the State of Israel but to the totality of the "Land of Israel." By the same token, in their original, Etzel, incarnation, it was this group also that risked and gave their lives perhaps more unreservedly than any other in the Zionist cause. Begin himself was a case in point, a man of almost superhuman physical courage, facing the hell of a Soviet prison camp for his Zionist convictions, afterward living in Palestine with a price on his head, orchestrating a campaign that destroyed British installations, slew British personnel, forced the mandatory government to station 80,000 troops in the Holy Land, to "protect" its civilian employees in barbed wire compounds—until by 1948 the sapping guerrilla campaign all but bankrupted the British Exchequer and ultimately played a central role in driving the British from Palestine altogether. "Of course, the Land of Israel is everything to us," Begin and his partisans would insist after 1967. "All of it. Otherwise, why are we here rather than in the United States?" The question was more than rhetorical at a time when tens of thousands of other Israelis had emigrated. They were an abrasive group, without question. As much as any element in Israel, however, they had paid their dues.

Nor was Begin's widest constituency likely at first to evoke much resonance from Western observers. These were the Jews from Moslem lands. By the 1980s, the "Easterners" and their children comprised 60 percent of Israel's Jewish population. The bulk of the right wing's strength, they appeared entirely uncompromising toward Arab claims to self-determination—

and accordingly were a source of despair to those Westerners who had been raised in a more tolerant, humanistic tradition. It may have been worth recalling, nevertheless, how little in the recent experience of these orientals would have impelled them to charity. Nurturing memories of their own fate under Moslem rule, they were quite familiar with the circumstances of authentic oppression: a half-million North African Jews who for hundreds of years had lived among the Berber majority as second-class *dhimmis;* 135,000 Iraqi Jews who had functioned as the commercial and intellectual elite of their country, until they were driven in 1950, stripped and pauperized, from a land in which they had lived for three millennia; 90,000 Egyptian and Syrian Jews, subsisting after 1948 in a rictus of terror, escaping from ghettos and concentration camps with the clothes on their backs and little else. It was legitimate to ask where these people would have learned charity or trust toward Arab good intentions. Unquestionably Israel's oriental-Sephardic Jews were difficult, occasionally semiliterate, at times a volatile lazzarone. But they were also the sweat of the nation's work force, the fodder of its infantry. And, in far greater proportions than Ashkenazic Jews, they remained on in Israel. Those who decried them would have been advised to envisage Israel without them, to contemplate a Jewish population of less than half its current size—in effect, a nonviable state.

The Ashkenazic establishment, surely, could not escape its own share of blame for the nation's shortcomings. More than any other, these were the forefathers responsible for the lockstep of Israeli politics, the dysfunctional party-list system, the stridency of party factionalism, the institutionalization of patronage into a national sociology. Their political behavior may well have been an affront to the Western conception of functional democratic government. Yet, if realism were the key, it was fair to inquire where these political novitiates would have learned the Anglo Saxon art of moderate compromise. The largest numbers of them had come from Eastern Europe, from countries in which they remained almost entirely unexposed to the democratic process, where compromise and restraint hardly figured in their collective experience. They were refugees, or the children of refugees. During the years when Jews in the Western Hemisphere or the British Commonwealth were working through the normal hostilities of adolescence, Israelis of East European extraction were caught up in the sheer glandular struggle for survival—against tsarist pogroms, subsequently against Fascist successor state cabals, later against the Nazis, later yet against Communist police states, and still later, in Palestine, against the British and against the Arabs. It was in Israel alone, during the brief interregna of non-war, that the accumulation of their repressed frustrations at last were vented: in the trivia of daily existence, whether driving on the highway, awaiting attention in a government office, lining up before a box office; or in the responsibilities of leadership, whether sitting in the Histadrut central committee, in the Knesset, or in the cabinet. A seething heterogeneity of tribes and cultures, most of these survivors now tasted authentic independence for the first time in their lives, and were intent on acting out that freedom—every moment of their lives. Here, in sum, was the Israeli people, neither better nor worse.

Was there not some other, special, mission that the Diaspora expected of its beloved Jewish state, some higher spiritual purpose beyond mere existence? Its prophets had always insisted that there was. The first modern Zionist, Moses Hess, set the tone of that expectation as early as 1840. Only a national renaissance, he argued, "can endow the religious genius of the Jews with new strength and raise its soul again to the level of prophetic inspiration." While subliminally craving only the right to live as other peoples lived, each successive wave of Zionist ideologues felt obliged to articulate its own vision of a "special mission" as the public rationale for a private normalcy. Thus, at the First Zionist Congress in 1897, the Mizrachi leader Rabbi Shmuel Mohilever insisted that Jewish settlers bore the obligation of transforming the Holy Land into an arena for "Torah-true Judaism." The Labor Zionist leader, Nachman Syrkin, envisaged Jewish Palestine as a microcosm of "social justice." Not to be outdone, the apolitical statesman Chaim Weizmann anticipated in Israel a land where "God's children would come serving their old country and make it a center for human civilization."

Students of history would recall, however, that all ideologists of modern nationalism, surely not the Jews alone, somehow felt impelled to legitimize their movements by superimposing upon them the afflatus of a "special mission." The Greek patriot Alexandros Ypsilanti defined his people's struggle against the Turks, in the "cradle of democracy," as an effort to cast a new "beacon of freedom to the world." Giuseppe Mazzini, father of the Italian Risorgimento, envisaged his disjointed nation restored to "Roman unity" as a model for other fractionalized peoples. French nationalists sought to romanticize their imperial growth by endowing it with a *mission civilisatrice;* the British, by their selfless willingness to bear the "white man's burden." For Dostoevsky, Pan-Slavism was the spirit of the artless Russian *mouzhik* in contest with the "decadent rationalism" of the Latin West. Fichte's and Treitschke's version of German "moral earnestness" anticipated the Aryan pseudoscience of the kaiser's, and of Hitler's, Pan-Germanism. For that matter, the American founding fathers proclaimed the epochal discovery, in 1776, that their freedom was "self-evident" to human reason.

Eventually all these peoples achieved their sovereignty or their empires. All they lost in the process was their "special mission" to the world. Observers would have been hard-pressed to recognize meaningful democracy in the corruption of twentieth-century Greece; unity in the chaos of modern Italy; the dignity of the simple peasant in the brutality of tsarist and Soviet Russia; France's *mission civilisatrice* in the slaughter and counterslaughter of Algeria; Britain's "white man's burden" in a network of Jim Crow living- and dining-facilities extending from Cairo to Hong Kong. The legacy of *Deutsche Kultur* required no embellishment whatever. Even as the right to freedom was rather less than "self-evident" to 5 million black slaves during the first eighty-seven years of American independence.

An objective appraisal of history would have suggested, rather, that no people or nation was endowed with a special mission. Its statehood's only legitimacy was the security it provided its citizens, the freedom it offered them to live and express their collective identity within protected boundaries, the open immigration it guaranteed their refugees, the passports it issued for travel in safety elsewhere in the world. All the rest was verbalization and bombast, excesses that applied no less to the ideologues of Zionism and Israel. The Christian world, in its theological and psychological effort to cope with Israel as a surrogate for Gentile imperfections, no doubt would continue to expect more of a Jewish state. The Diaspora might have been less inclined to do so. Normally unwilling to live in the realm of illusion, this most experienced and sophisticated of international peoples was under no compulsion to abdicate its traditional acuity of judgment in its relationship to Israel. Jews abroad might yet have found it useful to link ad hoc economic and political support to constructive criticism. But the moral and emotional ties, the commitment to Israel's survival, would never have been questioned—not by a civilization endowed with the Jews' historical memory. In their perspective, the simple fact of a revived and sovereign homeland transcended all further need for ongoing legitimization. Israel existed: a state, an asylum, a touchstone of security and dignity. Without question it was as imperfect in its own way as were older, richer, and more powerful nations in theirs. Its citizens, and its kinsmen abroad, nurtured the hope that it would improve. If asked, however, whether the mere fact of Israel's existence were enough for them now, they would have been tempted to paraphrase Mark Twain's reply to the friends who commiserated with him upon his prolonged illness. "The experience has been a painful one," he admitted, "but somehow it is tolerable—when I consider the alternative."

BIBLIOGRAPHY

The weekend editions of *Ma'ariv* and *HaAretz* have been indispensable to the author. Readers will find these two distinguished newspapers a unique reference source for contemporary Israeli history. As in Volume I, the ensuing compendium is presented topically. Where relevant, some works occasionally are listed under more than one heading.

I. *The Ramifications of October*

Ablin, R. "A Lost Decade of Israeli Growth? Economic Policy since 1973," *Economic Review*, May 1980.

Abu-Ayyash, A. "Israeli Regional Planning Policy in the Occupied Arab Territories," *Journal of Palestine Studies*, Nos. 3–4, 1976.

Alexander, Y. "The Nature of the PLO: Some International Implications," *Middle East Review*, No. 3, 1980.

Alexander, Zvi (Netzer). "Immigration to Israel from the USSR," *Israel Yearbook on Human Rights*, Jerusalem, 1977.

Altshuler, Mordechai. *Ma'amadah shel M'dinat Yisrael ba'Kerev Yehudei B'rit HaMo'atzot* [The Position of Israel among Soviet Jewry]. Jerusalem, 1983.

Amit, M. "Israel: The Bases of Economic Growth," *Washington Quarterly*, Spring 1979.

Avnery, U. "Inside the PLO," *New Outlook*, October–November 1977.

Avruch, K. "Gush Emunim: Politics, Religion, and Ideology in Israel," *Middle East Review*, No. 2, 1978/79.

Barkai, C. "The Israeli Economy in the Past Decade," *Jerusalem Quarterly*, Summer 1984.

Bar-Yaacov, N. "Keeping the Peace Between Egypt and Israel," *Israel Law Review*, April 1980.

Bar-Zohar, Michael, and Eitan Haber. *The Quest for the Red Prince*. New York, 1983.

Ben-Ami (Arieh Eliav). *Between Hammer and Sickle*. Philadelphia, 1967.

Benbassat, J. "Wanted: A New Code of Medical Ethics," *Jerusalem Quarterly*, Winter 1978.

Ben-Ezer, E. "War and Siege in Israeli Literature after 1967," *Jerusalem Quarterly*, Fall 1978.

Ben-Horin. *MaKoreh Sham: Sipuro shel Yehudei MiB'rit HaMo'atzot* [What Is Happening There: The Story of Soviet Jewry]. Tel Aviv, 1970.

Ben-Porat, Yeshayahu, Eitan Haber, and Ze'ev Schiff. *Entebbe Rescue*. New York, 1977.

Bernstein, Burton. *Sinai: The Great and Terrible Wilderness*. New York, 1979.

Borthwick, B. M. "Religion and Politics in Egypt and Israel," *Middle East Journal*, Spring 1979.

Carrère-d'Encausse, Helen. *La politique soviétique au Moyen-Orient, 1959–1975.* Paris, 1976.

Chibwe, Ephraim C. *Afro-Arab Relations in the New World Order.* New York, 1978.

Cobban, Helena. *The Palestinian Liberation Organization.* Cambridge, 1984.

Cohen, Y. "The Implications of a Free Trade Area between the EEC and Israel," *Journal of World Trade Law,* May–June 1976.

Crittendon, A. "Israel's Economic Plight," *Foreign Affairs,* Summer 1979.

Davis, Uri. *Israel Incorporated: A Study of Class, State, and Corporate Kin Control.* London, 1977.

Dubberstein, W. "Sinai II in Retrospect," *Strategic Review,* No. 2, 1977.

Duclos, J. S. "Un essai de rapprochement égypto-israélien," *Maghreb-Machrak,* January 1976.

Eban, A. "A New Look at Partition," *Jewish Frontier,* August–September 1976.

Efrat, E. "Settlement Pattern and Economic Changes in the Gaza Strip, 1947–1977," *Middle East Journal,* Summer 1977.

Eliashivili, Natan. *HaYehudim HaGruzim BaGruziyah u'v'Eretz Yisrael* [Georgian Jews in Georgia and in Israel]. Tel Aviv, 1975.

Elizur, Yuval, and Eliahu Salpeter. *Who Rules Israel?* New York, 1973.

Farago, U. "The Ethnic Identity of Russian Immigrant Students in Israel," *Jewish Journal of Sociology,* December 1978.

Fishelson, Gideon. *The Economic Integration of Israel into the EEC.* Tel Aviv, 1977.

Flapan, S. "Golda Meir in Retrospect," *New Outlook,* January–February 1979.

Forrestral, F. "The Sinai-American Connection," *MERIP Reports,* December 1977.

Friedberg, M. "From Moscow to Jerusalem—and Points West," *Commentary,* May 1978.

Friedlander, Dov, and Calvin Goldscheider. *The Population of Israel.* New York, 1979.

Gerson, Allan. *Israel, the West Bank, and International Law.* London 1978.

Ginsberg, Y. "Rural Urban Migration and Social Networks: The Israeli Case," *International Journal of Comparative Sociology,* September–December 1979.

Gitelman, Zvi. "Baltic and Non-Baltic Immigrants in Israel: Political and Social Attitudes and Behavior," *Studies in Comparative Communism,* Spring 1979.

———. *Becoming Israelis: The Political Resocialization of Soviet and American Immigrants.* New York, 1982.

———. "Moscow and the Soviet Jews: A Parting of the Ways," *Problems of Communism,* January 1980.

Goldmann, N. "Zionist Ideology and the Reality of Israel," *Foreign Affairs,* Fall 1978.

Hallstein, Walter. *Kein Frieden im Israel zur Sozialgeschichte d. Palëstina-Konflikts.* Bonn, 1977.

Isaac, Rael Jean. *Israel Divided.* Baltimore, 1976.

"Israel and the Resources of the West Bank," *Journal of Palestine Studies,* Summer 1979.

Karni, E. "The Israeli Economy, 1973–1976," *Economic Development and Cultural Change,* October 1979.

Khouri, R. G. "Israel's Imperial Economics," *Journal of Palestine Studies,* Winter 1980.

Kleiman, A. "International Guarantees and Secure Borders," *Middle East Review,* No. 2, 1977/78.

Krausz, E., and M. Bar-Lev. "Varieties of Orthodox Religious Behavior," *Jewish Journal of Sociology,* July 1978.

Leitenberg, Milton, and Gabriel Sheffer, eds. *Great Power Intervention in the Middle East.* New York, 1979.

Lesch, A. M. "Israeli Settlements in the Occupied Territories," *Journal of Palestine Studies*, Autumn 1977.

Losman, D. "Inflation in Israel: The Failure of Wage and Price Controls,' *Journal of Social and Political Studies*, No. 1, 1978.

Louvish, M. "Histadrut and Government," *Jewish Frontier*, August–September 1976.

Meerson-Aksenov, M. "The Jewish Exodus and Soviet Society," *Midstream*, April 1979.

Meron, Raphael. *The Economy of the Administered Areas: 1977–1978.* Jerusalem, 1980.

Morraflet, Claude. *Otages à Kampala.* Paris, 1976.

Mushkat, M. "The Socioeconomic Malaise of Developing Countries as a Function of Military Expenditures: The Case of Egypt and Israel," *Co-Existence*, October 1978.

Nachmias, David, and David Rosenbloom. *Bureaucratic Culture: Citizen and Administration in Israel.* New York, 1978.

Nisan, Mordechai. *Israel and the Territories.* Ramat Gan, 1978.

Rabin, Yitzchak. *Rabin M'socheach im Manhigim u'Rashei M'dinot* [Rabin Converses with Leaders and Heads of State]. Ramat Gan, 1984.

———. *The Rabin Memoirs.* New York, 1979.

Rana, A. "The Objectives and Strategy of the Palestine Liberation Organization," *India Quarterly*, April–June 1976.

Raphael, Gideon. *Destination Peace.* New York, 1981.

Rass, Rebecca. *From Moscow to Jerusalem.* New York, 1976.

Rosen, Steven J. "The Occupied Territories," *Jerusalem Quarterly*, Fall 1977.

Sachar, Howard M. *Diaspora.* New York, 1985.

———. *Egypt and Israel.* New York, 1981.

Safran, Nadav. *Israel: The Embattled Ally.* Cambridge, Mass., 1978.

Samarah, Adil. *Iqtisad al-manatiq al-Muh'tallah* [The Economy of the Occupied Territories]. Jerusalem, 1975.

Schnall, D. "Gush Emunim: Messianic Dissent and Israeli Politics," *Judaism*, No. 2, 1977.

Schroeter, Leonard. *The Last Exodus.* Seattle, 1981.

Shuval, J., E. Markus, and J. Dotan. "Age Patterns in the Integration of Soviet Immigrants in Israel," *Jewish Journal of Sociology*, December 1975.

Sprinzak, E. "Gush Emunim: The Tip of the Iceberg," *Jerusalem Quarterly*, Fall, 1981.

"A Survey of Israeli Settlements," *MERIP Reports*, September 1977.

United States. House of Representatives. Subcommittee on Immigration and Naturalization. *Colonization of the West Bank Territories by Israel.* Washington, D.C., October 17, 18, 1977.

Voronel, A. "The Immigration of Russian Intelligentsia," *Economic Review of Problems of Aliyah and Absorption*, October 1975.

Wolffsohn, M. "Israel und der Nachost-Konflikt: Eine einführende Bibliographie," *Aus Politik und Zeitgeschichte*, May 6, 1978.

Yaniv, A., and F. Pascal. "Doves, Hawks, and Other Birds of a Feather: The Distribution of Israeli Parliamentary Opinions on the Future of the Occupied Territories, 1967–1977," *Journal of Middle Eastern Studies*, April 1980.

———. "Jews and Arabs on Campus," *Jerusalem Quarterly*, No. 30, 1984.

II. *Israel Turns to the Right*

Abraham, S. "The Jew and the Israeli in Modern Arabic Literature," *Jerusalem Quarterly*, Winter 1977.

Abu-Mushlib, Ghalib. *Al-Druz fi thil al-Ihtilal al-Isra'ili* [The Druze under the Shadow of Israeli Occupation]. Beirut, 1975.

Adle, A., and R. Hodge. "Ethnicity and the Process of Status Attainment in Israel," *Israel Social Science Research*, No. 1, 1983.

Aloni, S. "Legal Discrimination," *New Outlook*, May–June 1978.

Amir, Y. "Interpersonal Contact between Arabs and Israelis," *Jerusalem Quarterly*, No. 13, 1979.

Amit, M. "Israel: The Bases of Economic Growth," *Washington Quarterly*, Spring 1979.

Bar-Gal, Y., and A. Soffer. *Geographical Changes in the Traditional Arab Village in Israel.* Durham, England, 1981.

Barkai, C. "The Israeli Economy in the Past Decade," *Jerusalem Quarterly*, Summer 1984.

Barmash, I. "How's Business in Israel? How Should It Be?" *Present Tense*, Autumn 1978.

Bar-Nir, Dov. *MiJabotinsky ad Begin* [From Jabotinsky to Begin]. Tel Aviv, 1982.

Barzilay, I. "The Arab in Modern Hebrew Literature: Image and Problems," *Hebrew Studies*, No. 18, 1977.

Begin, Menachem. *BaMachteret* [In the Underground]. 4 vols. Tel Aviv, 1959–61.

———. *The Revolt.* New York, 1951.

———. *White Nights.* New York, 1957.

Ben-Dor, Gabriel. *The Druze in Israel.* Boulder, Colo., 1979.

Ben-Ezer, E. "War and Siege in Israeli Literature after 1967," *Jerusalem Quarterly*, Fall 1978.

Ben-Porat, Y. "Israeli Dilemmas: Economic Relations between Jews and Arabs," *Dissent*, No. 4, 1984.

Ben-Rafael, Eliezer. *The Emergence of Ethnicity: Cultural Groups and Social Conflict in Israel.* Westport, Conn., 1982.

Benyamini, K. "Israeli Youth and the Image of the Arab," *Jerusalem Quarterly*, No. 29, 1981.

Bitan, Arieh. *T'murot Yeshuviyot baGalil HaTachton HaMizrachi, 1800–1978* [Settlement Changes in Eastern Lower Galilee, 1800–1978]. Jerusalem, 1982.

Brinker, M. "A Battle of Interpretations: The Last Campaign in the Israeli Election," *Dissent*, Fall 1977.

Burstein, P. "Political Patronage and Party Choice among Israel's Voters," *Journal of Politics*, No. 4, 1976.

———. "Social Cleavage and Party Choice in Israel," *American Political Science Review*, March 1978.

Caspi, Dan. "Ta'amulat HaB'chirot u'Totza'at HaB'chirot: Hashva'ah bein Ma'arachot b'Yisrael [Electoral Propaganda and the Electoral Decision: A Comparison of Campaigns in Israel]," *M'dina, Memshal, Vichasim Beinle'umiyim*, Nos. 19–20, 1982.

Caspi, Dan, et al. *The Roots of Begin's Success.* New York, 1984.

Cohen, S. "The Sephardic Condition," *New Outlook*, January–February 1979.

Davis, Uri. *Israel Incorporated: A Study of Class, State, and Corporate Kin Control.* London, 1977.

Doron, A. "Public Assistance in Israel," *Journal of Social Policy*, October 1978.

Dror, Yehezkal. "Are the Israel Labor Party and the Alignment Still Alive?" *Middle East Review*, No. 1, 1977.

Dutter, L. E. "Eastern and Western Jews: Ethnic Divisions in Israeli Society," *Middle East Journal*, Autumn 1977.

Elbaz, M. "Oriental Jews in Israeli Society," *MERIP Reports*, November–December 1980.

Eliachar, E. "The Sephardi Non-Presence," *New Outlook*, January–February 1977.

Elizur, Yuval, and Eliahu Salpeter. *Who Rules Israel?* New York, 1973.

Esmar, Fouzi al-. *To Be an Arab in Israel*. Beirut, 1978.

Eyal, Eli. "The Democratic Movement for Change: Origins and Perspectives," *Middle East Review*, No. 1, 1977.

Felsenthal, D. "Aspects of Coalition Payoffs: The Case of Israel," *Comparative Political Studies*, July 1979.

Flink, Salomon J. *Israel—Chaos or Challenge: Politics vs. Economics*. Ramat Gan, 1979.

Freedman, Robert O., ed. *Israel in the Begin Era*. New York, 1982.

Ginor, Fanny. *Socio-Economic Disparities in Israel*. Tel Aviv, 1979.

Globerson, Aryeh. *Higher Education and Employment in Israel*. New York, 1979.

Goering, K. "Israel and the Bedouin of the Negev," *Journal of Palestine Studies*, No. 4, 1979.

Gonem, A. "Geografikat Hitacharut B'chirot bein HaMa'arachah v'haLikud b'arim yehudiyim b'Yisrael, 1965–1981 [The Geography of the Electoral Competition between the Labor Alignment and the Likud in Jewish Cities of Israel, 1965–1981]," *M'dina, Mimshal, Vichasim Beinle'umiyim*, Nos. 19–20, 1982.

Haber, Eitan. *Menahem Begin*. New York, 1978.

Habibi, Emile. *Al-Waqayi al-Gharibah fi Ikhtifa Sa'id Abi al Nahs al Muhtasha'il* [The Strange History of Said Pessoptimist, The Luckless Palestinian]. Beirut, 1974.

Haddad, Hezkel M. *Yehudei Artzot Arav v'Islam* [The Jews of Arab and Islamic Lands]. Tel Aviv, 1983.

Harari, Yechiel. *HaB'chirot HaMunitzipaliot BaMigzar HaArvi* [The Municipal Elections in the Arab Sector]. Givat Havivah, 1978.

Hareven, A., and A. Radian. *The Likud Government and Domestic Policy Change*, *Jerusalem Quarterly*, Winter 1981.

Hirschler, Gertrude, and Lester Eckman. *Menahem Begin*. New York, 1979.

Hoffman, J., and N. Rouhana, "Young Arabs in Israel: Some Aspects of a Conflicted Social Identity," *Journal of Social Psychology*, June 1977.

Horowitz, D. "More Than a Change of Government," *Jerusalem Quarterly*, Fall 1977.

Isaac, E., and R. J. Isaac. "The Impact of Jabotinsky on Likud's Policies," *Middle East Review*, Fall 1977.

Isaac, Rael Jean. *Israel Divided*. Baltimore, 1976.

Israel. Ministry of Education and Culture. *Yehudei HaMizrach b'Sifrutenu* [Oriental Jews in Our Literature]. Jerusalem, 1974.

Israeli, A. "Ma'apechat HaTa'asukah bein HaMiyutim HaLoYehudiyim b'M'dinat Yisrael [The Employment Revolution among the Non-Jewish Minorities of Israel]," *HaMizrach HeChadash*, Nos. 3–4, 1976.

Jörgensen, A. "Der Kampf der Kommunistischen Partei Israels für die nationalen Rechte des palästinesischen arabischen Volkes," *Asien, Afrika, Latinamerika*, No. 2, 1978.

Jurays, Sabri. *The Arabs in Israel*. New York, 1976.

Katz, Elihu, and Michael Gurevitch. *The Secularization of Leisure: Culture and Communication in Israel.* Cambridge, Mass., 1976.

Klein, Claude. *La caractère juif de l'Etat d'Israël: Etude juridique.* Paris, 1977.

"The Koenig Report: Demographic Racism in Israel," *MERIP Reports,* October 1976.

Lewis, A. "Educational Policy and Social Inequality in Israel," *Jerusalem Quarterly,* Summer 1979.

Lustick, Ian. *Arabs in the Jewish State.* Austin, Tex., 1980.

Mar'i, Sami Khalil. *Arab Education in Israel.* Syracuse, 1978.

Malka, Victor. *Menahem Begin.* Paris, 1977.

Mayer, E. "The Druze of Bet Ja'an: Modernization and the Activation of Minorities in Israel," *Middle East Review,* No. 2, 1976/77.

Nachmias, D., and D. Rosenbloom. "The Right-Wing Opposition in Israel," *Political Studies,* September 1976.

Ofer, G. "Israel's Economy: Diagnosis and Cure," *Jerusalem Quarterly,* Summer 1981.

Oz, Amos. *In the Land of Israel.* New York, 1983.

Peled, E. "Equality in Israel's Educational Policy," *Jerusalem Quarterly,* No. 30, 1981.

Penniman, Howard R., ed. *Israel at the Polls: The Knesset Elections of 1977.* Washington, D.C., 1979.

Peretz, D. "The Earthquake—Israel's Ninth Knesset Elections," *Middle East Journal,* Winter 1985.

Perlmutter, A. "Cleavage in Israel," *Foreign Policy,* Summer 1977.

Preuss, Teddy. *Begin baShilton* [Begin in Office]. Jerusalem, 1984.

Radian, A., and I. Sharkansky. "Tax Reform in Israel," *Policy Analysis,* No. 3, 1979.

Rekhess, E. "The Israeli Arab Intelligentsia," *Jerusalem Quarterly,* Spring 1979.

Rosenberg, B. "The Arabs of Israel," *Dissent,* Spring 1980.

Sachar, Howard M. *Diaspora.* New York, 1985.

Sa'id, Abd al-Munim Sa'id. *Wathi gat Koenig wa Arab al-Ard al-Muh tallah* [The Koenig Report and the Arabs of the Occupied Territories]. Cairo, 1977.

Said, Edward. *Al-Ubur i la al-Mustaqbal: Qasa id min Wahi Uktubr* [Modern Arabic Poetry Related to the Arab-Israel War of October 1973]. Cairo, 1975.

Schnall, David. *Radical Dissent in Contemporary Israeli Politics.* New York, 1979.

Schölch, Alexander. *Palestinians over the Green Line: Studies in the Relations between Palestinians on Both Sides of the Armistice Line since 1967.* London, 1983.

Shamir, M., and A. Arian. "HaB'chirah HaEdatit baB'chirot 1977 b'Yisrael [The Ethnic Vote in Israel's 1977 Elections]," *M'dina, Mimshal, Vichasim Beinle'umiyim,* Nos. 19–20, 1982.

Shokeid, M. "The Israel Arab Vote in Transition," *Middle Eastern Studies,* January 1978.

Silver, Eric. *Begin: The Haunted Prophet.* New York, 1984.

Silver, G. A. "Love Is Not Enough: Hadassah and Israel's Medical Care Dilemma," *Midstream,* March 1978.

Smooha, S. "Ethnic Stratification and Allegiance in Israel: Where Do Oriental Jews Belong?" *Il Politico,* No. 4, 1976.

———. "Existing and Alternative Policies toward the Arabs in Israel," *Ethnic and Racial Studies,* January 1982.

———. *Israel: Pluralism and Conflict.* London, 1979.

Smooha, S., and O. Cibulski. *Social Research on Arabs in Israel, 1948–1977.* Ramat Gan, 1978.

Smooha, S., and J. Hofman. "Some Problems on Arab-Jewish Co-Existence in Israel," *Middle East Review*, No. 2, 1976/77.

Sprinzak, E. "Extreme Politics in Israel," *Jerusalem Quarterly*, Autumn 1981.

Stillman, Norman. *The Jews of Arab Lands*. Philadelphia, 1979.

Swirski, S. "The Oriental Jews in Israel: Why Many Tilted toward Begin," *Dissent*, No. 1, 1984.

Syrkin, M. "Labor Opposition to Settlements," *Middle East Review*, No. 2, 1979/80.

Tamarin, G. "Jewish-Arab Relations in Irael Following the 1973 War," *Pluralism and Society*, No. 4, 1976.

Tessler, M. A. "Israel's Arabs and the Palestinian Problem," *Middle East Journal*, Summer 1977.

Treffer, Gerd. *Israels Identitätkrise: Israel zwischen Judaismus, Zionismus, und Israelismus*. Munich, 1975.

Tsirulnikov, Shlomo. *HaDialektika shel Ma'apechat Yisrael* [The Dialectic of the Israeli Revolution]. Tel Aviv, 1982.

Urieh, Nachman. *Aliyatah u'n'filetah shel Dash* [The Rise and Fall of the Democratic Movement for Change]. Tel Aviv, 1982.

Vroman, G. "El Al Fights to Stay Aloft," *Israel Economist*, September 1980.

Wasserstein, David. *The Druze and Circassians of Israel*. London, 1976.

Weiker, W. "Stratification in Israel Society: Is There a Middle Grouping?" *Israel Social Science Research*, No. 2, 1983.

Weingrod, A., and M. Gurevitch. "Who Are the Israeli Elites?" *Jewish Journal of Sociology*, July 1977.

Winter, M. "Arab Education in Israel," *Jerusalem Quarterly*, Summer 1979.

Wistrich, Robert S., ed. *The Left Against Zion: Communism, Israel, and the Middle East*. London, 1979.

Yinon, A. "Nosim b'Sifrut Arvei Yisrael [Themes in the Literature of Israeli Arabs]," *HaMizrach HeChadash*, No. 1, 1975.

Yishai, Y. "Challenge Groups in Israeli Politics," *Middle East Journal*, Autumn 1981.

———. "Factionalism in Israeli Political Parties," *Jerusalem Quarterly*, Summer 1981.

———. "Israel's Right Wing Jewish Proletariat," *Jewish Journal of Sociology*, December 1982.

III, IV. *Egypt's Quest for Peace, The Precarious Embrace*

Abshire, David, and Associates, eds. *Egypt and Israel: Prospects for a New Era*. Washington, D.C., 1979.

"After the Peace Treaty: The Military Situation in the Middle East," *Journal of Palestine Studies*, Winter, 1979.

Ajami, Fuad. "The Struggle for Egypt's Soul," *Foreign Policy*, No. 35, 1979.

Amos, John W., II. *Arab-Israeli Military/Political Relations: Arab Perceptions and the Politics of Escalation*. New York, 1979.

Arad, Ruth, Ze'ev Hirsch, and Alfred Tovias. *The Economics of Peacemaking*. London, 1983.

Aronson, S. "Israeli Views of the Brookings Report," *Middle East Review*, No. 1, 1977.

Aruri, N. H. "Kissinger's Legacy to Carter," *Middle East*, March 1977.

Bar-Yaacov, N. "Keeping the Peace between Egypt and Israel, 1973–80," *Israel Law Review*, April 1980.

Blitzer, Wolf. *Between Washington and Jerusalem: A Reporter's Notebook*. New York, 1985.

Borthwick, B. M. "Religion and Politics in Egypt and Israel," *Middle East Journal*, Spring 1979.

Buheiry, M. "The Saunders Document," *Journal of Palestine Studies*, No. 1, 1978.

Bull, V. "The West Bank—Is It Viable?" *Journal of Developing Areas*, July 1976.

Carter, Jimmy. *Keeping Faith*. New York, 1982.

———. *The Blood of Abraham*. New York, 1985.

Dawisha, A. "Syria and the Sadat Initiative," *World Today*, May 1978.

Dayan, Moshe. *Breakthrough to Peace*. London, 1981.

Derogy, Jacques, and Hesi Carmel. *The Untold History of Israel*. New York, 1979.

Devore, Ronald L., ed. *The Arab-Israeli Conflict: A Historical, Social, and Military Bibliography*. London, 1977.

Din, Amin Az al-. *The Implications of the Camp David Agreement on the Movement of Labour Between Egypt and Israel*. London, 1980.

Eban, A. "A New Look at Partition," *Jewish Frontier*, August–September 1976.

Egypt. Ministry of Foreign Affairs. *Egypt and the Palestinian Question, 1945–1982*. Cairo, 1982.

"Egyptian Oil in 1977," *Middle East Economic Survey*, July 19, 1978.

Flapan, Simha, ed. *When Enemies Dare to Talk: An Israeli-Palestinian Debate*. London, 1979.

Friedlander, Melvin A. *Sadat and Begin*. Boulder, Colo., 1983.

Gilboa, E. "Educating Israeli Officers in the Process of Peacemaking in the Middle East," *Journal of Peace Research*, No. 2, 1979.

Golan, Matti. *Shimon Peres*. New York, 1982.

Goldmann, Nahum. "Zionist Ideology and the Reality of Israel," *Foreign Affairs*, Fall 1978.

Haber, Eitan. *Menahem Begin*. New York, 1978.

Haber, Eitan, Ze'ev Schiff, and Ehud Ya'ari. *The Year of the Dove*. New York, 1979.

Hareven, A. "Can We Learn to Live Together?" *Jerusalem Quarterly*, Autumn 1979.

Hattis, S. "Jabotinsky's Parity Plan for Palestine," *Middle Eastern Studies*, January 1977.

Hirschler, Gertrude, and Lester Eckman. *Menahem Begin*. New York, 1979.

Hirst, David, and Irene Beeson. *Sadat*. London, 1981.

Hottinger, A. "Friedenspolitik im Nahen Osten," *Europa-Arachiv*, May 10, 1979.

Kanovsky, E. "Economic Aspects of Peace between Israel and the Arab Countries," *New Outlook*, October 1978.

Katz, Shmuel. *The Hollow Peace*. Jerusalem, 1981.

Kreinin, M. "On the Potential Trade Pattern between Egypt and Israel," *Middle East Review*, No. 3, 1980.

Kubursi, Atif. *The Economic Consequences of the Camp David Agreements*. Kuwait, 1979.

Lapidoth, R. "The Camp David Agreements: Some Legal Aspects," *Jerusalem Quarterly*, No. 19, 1979.

Lipschitz, Ora. *Sinai*. Tel Aviv, 1978.

Luttwak, E. "Strategic Implications of the Camp David Accords," *Washington Quarterly*, No. 19, 1979.

Mansour, A. "Getting to Know You," *New Outlook*, May–June 1979.

Markus, Yoel. *Camp David: Petach l'Shalom* [Camp David: Gateway to Peace]. Tel Aviv, 1979.

Meyer, Gail E. *Egypt and the United States*. Cranbury, N.J., 1980.

Nes, D. "Egypt Breaks the Deadlock," *Journal of Palestine Studies*, No. 2, 1978.

Neumann, R. G. "The Middle East after Camp David: Perils and Opportunities," *Washington Quarterly*, No. 19, 1979.

Peres, S. "Dampening the Fire in the Middle East: A Strategy for a Transition Period," *International Security*, No. 3, 1978.

Peric, M. "Camp David-Damascus-Baghdad," *Review of International Affairs*, November 20, 1978.

Preuss, Teddy. *Begin baShilton* [Begin in Office]. Jerusalem, 1984.

Quandt, William B. *Camp David: Peacemaking and Politics*. Washington, D.C., 1986.

Rabin, Yehoshua. *The Arab-Israeli Military Balance*. Tel Aviv, 1980.

Rabinovich, I. "Israel: The Impact of the Treaty," *Current History*, January 1980.

Rida, Adil. *Al-Riban al-Isra'ili ala Junoob al-Sudan* [The Israeli Role in Southern Sudan]. Alexandria, 1975.

Rondot, P. "Après la signature du traité de paix égypto-israélien. Les limits du refus arabe," *Politique Etrangère*, No. 2, 1979.

Ron-Feder, Galilah. *Moshe Dayan*. Vol II. Jerusalem, 1984.

Roy, D. "Egyptian Liberalization: The First Decade, 1973–1983," *Middle East Executive Reports*, December 1983.

Sachar, Howard M. *Egypt and Israel*. New York, 1981.

Sadat, Anwar al-. *In Search of Identity*. New York, 1977.

"Sadat's Visit to Israel Is an American Plot," *Arab Palestinian Resistance*, February 1978.

Salem-Babikian, N. "The Sacred and the Profane: Sadat's Speech to the Knesset," *Middle East Journal*, Winter 1980.

Savin, P. "Israël vu d'Egypte," *Politique Etrangère*, June 1981.

Sayegh, Fayez. *Palestine and the Camp David Accords*. London, 1979.

Segev, Shmuel. *Sadat, HaDerech l'Shalom* [Sadat, the Road to Peace]. Tel Aviv, 1978.

Shamay, M. "La communauté juive en Egypte," *Peuples Méditerranéens*, No. 16, 1981.

Shamir, Shimon. *Mitzrayim b'Hanhagat Sadat* [Egypt under the Leadership of Sadat]. Tel Aviv, 1978.

Shlain, A., and A. Yaniv, "Domestic Politics and Foreign Policy in Israel," *International Affairs*, Spring 1980.

Shoukri, Ghali. *Egypt: Portrait of a President. Sadat's Road to Jerusalem*. London, 1981.

Shufani, E. "Sadat's Initiative: The Reaction in Israel," *Journal of Palestine Studies*, No. 1, 1979.

Sid-Ahmed, Mohamed. *After the Guns Fall Silent: Peace or Armageddon in the Middle East*, London, 1978.

Silver, Eric. *Begin: The Haunted Prophet*. New York, 1984.

Smith P. A. "European Leaders Unhappy with Sadat-Begin Treaty," *MERIP Reports*, September 1979.

Tariq, Muhammad al-. *Masi rat al-Sadat min Salzburg hatta al-Kinaysit* [Sadat's Journey from Salzburg to the Knesset]. Beirut, 1977.

Taub, Y. "Israel's Economy: The Challenge of Peace," *Jerusalem Quarterly*, Fall 1979.

Torgovnik, E. "Accepting Camp David: The Role of Party Factions in Israeli Policy-Making," *Middle East Review*, No. 2, 1978/79.

Waterbury, John. *Egypt: Burdens of the Past, Options for the Future*. Bloomington, Ind., 1978.

Weizman, Ezer. *The Battle for Peace.* New York, 1981.

Yariv, A. "Strategic Depth," *Jerusalem Quarterly,* Spring 1980.

Yaniv, A., and F. Pascal. "Doves, Hawks, and Other Birds of a Feather: The Distribution of Israeli Parliamentary Opinions on the Future of the Occupied Territories, 1967–1977," *Journal of Middle Eastern Studies,* April 1980.

Yinon, Moshe. *Misva'at HaShalom* [The Peace Equation]. Haifa, 1981.

Zelnikar, Shimshon, and Zaki Shalom. *Cooperation Between Israel and Egypt: Positions and Trends.* Tel Aviv, 1981.

Zion, S., and U. Dan. "The Untold Story of the Mideast Talks," *New York Times Magazine,* January 21, 28, 1979.

v. Fading Hopes on Palestine

Alexander, Y. "The Nature of the PLO: Some International Implications," *Middle East Review,* No. 3, 1980.

Ali, M. "Jordan's Stance toward the Camp David Agreements," *Pakistan Horizon,* No. 31, 1978.

Antonius, S. "Fighting on Two Fronts: Conversations with Palestinian Women," *Journal of Palestine Studies,* No. 3, 1979.

Ashrawi, H. M. "The Contemporary Palestinian Poetry of Occupation," *Journal of Palestine Studies,* No. 1, 1978.

Ayal, Eli. "Economic Factors Behind United Actions," *Middle East Review,* No. 2, 1980/81.

Bailey, C. "Changing Attitudes toward Jordan in the West Bank," *Middle East Journal,* Spring 1978.

Ball, G. "The Coming Crisis in Israeli-American Relations," *Foreign Affairs,* Winter 1979–80.

Ballas, J. "Le personnage israélien dans la literature arabe," *Nouveaux Cahiers,* No. 44, 1976.

Bar-Yaacov, N. "Keeping the Peace between Egypt and Israel, 1973–1980," *Israel Law Review,* April 1980.

Bar-Zohar, M., and E. Haber. *The Quest for the Red Prince.* New York, 1983.

Bassiouni, M. C. "An Analysis of Egyptian Policy toward Israel," *New Outlook,* January 1981.

Benvenisti, Meron. *The West Bank Data Bank Project: A Survey of Israel's Policies.* Washington, D.C., 1984.

Carter, Jimmy. *Keeping Faith.* New York, 1982.

———. *The Blood of Abraham.* New York, 1985.

Cobban, Helena. *The Palestinian Liberation Organization.* Cambridge, 1984.

Cohen, S. "Jerusalem's Unity and West Bank Autonomy—Paired Principles," *Middle East Review,* Nos. 3–4, 1981.

Cohen, Steven M. *The Attitude of American Jews to Israel and Israelis.* New York, 1983.

Cohn, A. "The Changing Patterns of West Bank Politics," *Jerusalem Quarterly,* Fall 1977.

Dajani, al-, A. "The PLO and the Euro-Arab Dialogue," *Journal of Palestine Studies,* No. 3, 1980.

Darin-Drabkin, H. "From Settlement to Colonization," *New Outlook,* February–March 1978.

Davies, P. "The Educated West Bank Palestinians," *Journal of Palestine Studies,* No. 3, 1979.

Davis, U. "Settlements and Politics under Begin," *MERIP Reports*, June 1979.

Don-Yehiya, E. "The Origins and Development of the Agudah and Mafdal Parties," *Jerusalem Quarterly*, Summer 1981.

Egypt. Ministry of Foreign Affairs. *Egypt and the Palestinian Question, 1945–1982*. Cairo, 1982.

Elazar, Daniel J., ed. *Judea, Samaria, and Gaza*. Washington, D.C., 1982.

Eliav, Arieh. *Tabot Edut* [Required Testimony]. Tel Aviv, 1983.

Elon, A. "Mad Settlers on the West Bank," *Nation*, April 21, 1979.

Elrazik, A., A. R. Amin, and U. Davis. "Problems of Palestinians in Israel: Land, Work, Education," *Journal of Palestine Studies*, No. 3, 1978.

Flapan, Simha. *Zionism and the Palestinians*. New York, 1979.

Freedman, Robert, ed. *Israel in the Begin Era*. New York, 1982.

Friedman, M. "Mifleget HaMafdal b'Mashber [The National Religious Party in Crisis]," *M'dina, Mimshal, Vichasim Beinle'umiyim*, Nos. 19–20, 1982.

Golan, Gila. *The Soviet Union and the Palestine Liberation Organization*. New York, 1980.

Goldmann, N. "Grim Prospects for Israel," *New Outlook*, October 1980.

Halabi, Rafik. *The West Bank Story*. New York, 1981.

Halevi, Ilan. *Sous Israël, la Palestine*. Paris, 1978.

Harkabi, Yehoshafat. *Palestinians and Israel*. New York, 1977.

Harris, William W. *Taking Root: Israeli Settlement in the West Bank, the Golan, and Gaza Sinai, 1967–1980*. New York, 1980.

Heller, M. "Begin's False Autonomy," *Foreign Policy*, No. 37, 1979–80.

Hertzberg, A. "Begin and the Jews," *New York Review of Books*, February 18, 1982.

Hillal, Jamil. *Al-Diffah al Gharbiyah: al-Tarqiba al-Ijtima iya wa-al-Iqtisadiya* [West Bank Economic and Social Structure. 1948–74]. Beirut, 1977.

———. "Class Transformation in the West Bank and Gaza," *MERIP Reports*, December 1976.

Isaac, E., and R. J. Isaac. "The Impact of Jabotinsky on Likud's Policies," *Middle East Review*, Fall 1977.

Isaacs, Stephen. *Jews and American Politics*. Garden City, N.Y., 1974.

Issa, Mahmoud. *Je suis un fedayin*. Paris, 1976.

Jabra, J. "The Palestinian Exile as Writer," *Journal of Palestine Studies*, No. 2, 1979.

Johnson, Nels. *Islam and the Politics of Meaning in Palestinian Nationalism*. London, 1982.

Kalter, D. "The Truth about Elon Moreh," *New Outlook*, July–August 1979.

Khalili, Ghazi al-. *Al Mar'ah al-Filastiniyah wa-al-Thawrah* [A Sociological Study of Palestinian Women and the Revolution]. Beirut, 1977.

Komisar, Lucy, "The West Bank as Bantustan," *Nation*, May 29, 1982.

Korey, William. "The PLO's Conquest of the U.N.," *Midstream*, November 1979.

Kornberg, J. "Zionism and Ideology: The Breira Controversy," *Judaism*, No. 1, 1978.

Kuroda, Alice K., and Yasumasa Kuroda. *Palestinians Without Palestine: A Study of Political Socialization among Palestinian Youths*. Washington, D.C., 1978.

Lapidoth, R. "The Autonomy Talks," *Jerusalem Quarterly*, No. 24, 1982.

Lerman, Eran. *The Palestinian Revolution and the Arab-Israel Conflict: A New Phrase*. London, 1982.

Lesch, A. M. "The Politicization of the Occupied Territories, 1967–1977," *New Outlook*, December 1977–January 1978.

Litani, Y. "Leadership in the West Bank and Gaza," *Jerusalem Quarterly*, Winter 1980.

Lustick, Ian. *Israel and Jordan*. Berkeley, 1974.

————. "Israel and the West Bank after Elon Moreh: The Mechanics of De Facto Annexation," *Middle East Journal*, Autumn 1981.

Ma'oz, Moshe. *Palestine Arab Politics*. Jerusalem, 1975.

Ma'oz, Moshe, and Mordechai Nisan. *Palestinian Leadership in the West Bank*. London, 1984.

Marx, E. "Changes in Arab Refugee Camps," *Jerusalem Quarterly*, Summer 1978.

Meron, Raphael. *The Economy of the Administered Areas: 1977–78*. Jerusalem, 1980.

Migdal, Joel S., et al. *Palestine Society and Politics*. Princeton, 1980.

Milson, M. "The Palestinians and the Peace Process," *Forum*, 42/43, 1981.

Nachmias, David. *The West Bank and Gaza: Toward the Making of a Palestinian State*. Washington, D.C., 1979.

Nachmias, David, and E. Zureik, eds. *The Sociology of the Palestinians*. New York, 1980.

Negbi, M. "The Israeli Supreme Court and the Occupied Territories," *Jerusalem Quarterly*, No. 27, 1983.

Nevo, J. "Is There a Jordanian Entity?" *Jerusalem Quarterly*, Summer 1983.

Peleg, I. "Jabotinsky's Legacy and Begin's Foreign Policy," *Israel and the World*, October 1983.

Peres, S. "A Strategy for Peace in the Middle East," *Foreign Affairs*, Spring 1980.

Peretz, D. "Palestinian Social Stratification," *Journal of Palestine Studies*, Autumn 1977.

"Polling Palestinians," *Society*, September–October 1982.

Qurah, Nazih. *Ta'alim al-Filastiniyah: al-Waqf wa-al-Mushkilat* [The Education of the Palestinians]. Beirut, 1975.

"The Rise and Fall of the Camp David Agreements," *Arab Perspectives*, June 1980.

"The Rise of Palestinian Revolutionary Cinema," *Palestine*, April 1983.

Ron-Feder, Galilah. *Moshe Dayan*. Vol. II. Jerusalem, 1984.

Ross, R. "West Bank 'Facts' Are Hitting Home," *Nation*, March 10, 1984.

Sayegh, Fayez. *Palestine and the Camp David Accords*. London, 1979.

Sayigh, Rosemary. *The Palestinians: From Peasants to Revolutionaries*. London, 1979.

Shadid, Mohammed. *The United States and the Palestinians*. New York, 1981.

Shalev, Areh. *Kav Haganah b'Yehudah u'v'Shomrom* [The Defense Line in Judea and Samaria]. Tel Aviv, 1982.

————. "Security Dangers from the East," *Jerusalem Quarterly*, No. 27, 1983.

Shemesh, M. "Egypt's Commitment to the Palestinian Cause," *Jerusalem Quarterly*, No. 34, 1985.

————. "The West Bank: Rise and Decline of Traditional Leadership," *Middle Eastern Studies*, July 1984.

Shiloach Institute, ed. *Ta'alich Nirmul HaYachasim bein Israel l'vein Mitzrayim* [The Process of Normalization of Relations between Israel and Egypt]. Tel Aviv, 1981.

Shimoni, Yaacov. "Israel and Europe," *Jerusalem Quarterly*, Spring 1981.

Shuaibi, Issa al-. "The Development of Palestinian Entity-Consciousness," *Journal of Palestine Studies*, Autumn 1979.

Sicherman, H. "The Politics of Dependence: Western Europe and the Arab-Israeli Conflict," *Orbis*, No. 4, 1980.

Smith, P. A. "European Leaders Unhappy with Sadat-Begin Treaty," *MERIP Reports*, September 1979.

"Special Report on Jordan," *Arab Economist*, August 1979.

Stone, I. F. *Underground to Palestine and Reflections Thirty Years Later*. New York, 1978.

Talmon, Y. L. "The Homeland Is In Danger: An Open Letter to Menahem Begin," *Dissent*, Fall 1980.

Tarbush, S. "Palestinian Groups: The Divided Front," *Middle East Economic Digest*. September 23, 1978.

Tawil, Raymonda. *My Home, My Prison*. New York, 1979.

Tillman, S. "Israel and Palestinian Nationalism," *Journal of Palestine Studies*, Autumn 1979.

Tuma, Elias H., and Haim Darin-Drabkin. *The Economic Case for Palestine*. London, 1978.

United States. House of Representatives. Subcommittee on Immigration and Naturalization. *Colonization of the West Bank Territories by Israel*. Washington, D.C., October 17, 18, 1977.

Weitz, R. "Peace and the Palestinian Problem," *Jewish Frontier*, March 1982.

Weizman, Ezer. *The Battle for Peace*. New York, 1981.

Yaniv, A., and Y. Yishai. "Israeli Settlements on the West Bank: The Politics of Intransigence," *Journal of Politics*, November 1981.

Yariv, A. "Strategic Depth," *Jerusalem Quarterly*, Winter 1981.

VI. *The Politics of Militancy*

Adler, A., and R. Hodge, "Ethnicity and the Process of Status Attainment in Israel," *Israel Social Science Research*, No. 1, 1983.

Allen, D., and A. Pijpers. *European Foreign Policy-Making and the Arab-Israel Conflict*. The Hague, 1984.

Arian, A. "Elections 1981: Competitiveness and Polarization," *Jerusalem Quarterly*, Fall 1981.

———, ed. *The Elections in Israel, 1981*. Tel Aviv, 1983.

Aronoff, Myron J., ed. *Cross-Currents in Israeli Culture and Politics*. New Brunswick, 1984.

Avineri, S. "Letter from Israel," *Dissent*, Winter 1980.

Anmon, Y. "The 1981 Elections and the Changing Fortunes of the Israeli Labour Party," *Government and Opposition*, No. 4, 1982.

Bauberot, J., et al. *Palestine et Liban: Promesses et mensonges de l'Occident*. Paris, 1977.

Benvenisti, M. "Dialogue of Action in Jerusalem," *Jerusalem Quarterly*, Spring 1981.

———. "First Steps to a Solution," *New Outlook*, May 1981.

———. *Jerusalem: The Torn City*. Minneapolis, 1976.

———. "Some Guidelines for Positive Thinking on Jerusalem," *Middle East Review*, Spring–Summer 1981.

Ben-Zadok, E., and G. Goldberg. "Voting Patterns of Oriental Jews in Development Towns," *Jerusalem Quarterly*, Summer 1984.

Blaisse, Mark. *Anwar Sadat: The Last Hundred Days*. London, 1981.

Blitzer, Wolf. *Between Washington and Jerusalem: A Reporter's Notebook*. New York, 1985.

Campbell, J. "The Middle East," *Foreign Affairs*, Fall 1981.

Caplan, Gerald, and Ruth Caplan. *Arab and Jew in Jerusalem: Explorations in Community Mental Health*. Cambridge, Mass., 1980.

Caradon, Lord (Hugh Foot). *The Future of Jerusalem: A Review of Proposals for the Future of the City*. Washington, D.C., 1980.

Caspi, D. "Ta'amkoulat HaB'chirot v'Totza'at HaB'chirot: Hasva'a bein Ma'rachot b'Yisrael [Electoral Propaganda and the Electoral Decision: A Comparison of Electoral Campaigns in Israel]," *M'dina, Memshal, Vichasim Beinle'umiyim*, Nos. 19–20, 1982.

Cattan, Henry. *Palestine and International Law: The Legal Aspects of the Arab-Israeli Conflict*. London, 1976.

———. *The Question of Jerusalem*. London, 1980.

Cohen, P. "Ethnicity, Class, and Political Alignment in Israel," *Jewish Journal of Sociology*, December 1983.

Cohen, S. "Geopolitical Bases for the Integration of Jerusalem," *Orbis*, No. 2, 1976.

———. "Jerusalem's Unity and West Bank Autonomy—Paired Principles," *Middle East Review*, Nos. 3–4, 1981.

Dawisha, A. "Syria under Assad, 1970–78," *Government and Opposition*, No. 3, 1978.

Doron, A. "Social Policy for the Eighties," *Jerusalem Quarterly*, Spring 1981.

Efrat, Elisha. *Israel Towards the Year 2000—Social Aspects of Planning and Development*. Tel Aviv, 1978.

Elazar, D. "Israel's New Majority," *Commentary*, March, 1983.

Etzioni-Halevy, Eva. *Political Culture in Israel*. New York, 1977.

Ferrari, S. "The Vatican, Israel, and the Jerusalem Question (1943–1984)," *Middle East Journal*, Spring 1983.

Foer, P. M. "The War Against Breira," *Jewish Spectator*, Summer 1983.

Freedman, L. "Israel's Nuclear Policy," *Survival*, May–June 1975.

Golan, Matti. *Shimon Peres*. New York, 1982.

Gonem, A. "Geograficat Hitcharut B'chirot bein HaMa'arachah v'haLikud b'arim yehudiyim b'Yisrael, 1965–1981 [The Geography of the Electoral Competition between the Labor Alignment and the Likud in Jewish Cities of Israel, 1965–1981]," *M'dina, Mimshal, Vichasim Beinle'umiyim*, Nos. 19–20, 1982.

Halabi, Rafik. *The West Bank Story*. New York, 1981.

Harkavy, Robert C. *Spectre of a Middle East Holocaust: The Strategic and Diplomatic Implications of the Israeli Nuclear Weapons Program*. Denver, 1977.

Heller, P., ed. *The Middle East Military Balance*. New York, 1984.

Heikal, Mohamed. *Autumn of Fury: The Assassination of Sadat*. New York, 1983.

Hoffman, S. A. "Candidate Selection in Israel's Parliament: The Realities of Change," *Middle East Journal*, Summer 1980.

Hoffmitz, L. "North African Jewry in Israel," *Jewish Frontier*, February 1976.

Horton, Frank B., Anthony C. Rogerson, and Edward L. Warner, eds. *Comparative Defense Policy*. Baltimore, 1984.

Inbar, E. "American Arms Transfers to Israel," *Middle East Review*, Nos. 1–2, 1982–83.

Israel. Ministry of Defense. *Tsahal b'Helo* [The Armed Forces in Readiness]. Tel Aviv, 1981.

———. Ministry of Foreign Affairs. *T'udot l'Mediniyut HaChutz shel Medinat Yisrael* [Documents on Israeli Foreign Policy]. Vols. 1–3. Jerusalem, 1981–84.

"Israeli Society and its Defense Establishment," *Journal of Strategic Studies*. September 1983.

Jacobs, Paul. "Let My People Go—But Where?" *Present Tense*, Winter 1979.

Jaffe, Eliezer B. *Pleaders and Protesters*. New York, 1980.

Katz, Elihu, and Michael Gurevitch. *The Secularization of Leisure: Culture and Communication in Israel*. Cambridge, Mass., 1976.

Khalidi, Rashid, and Camille Mansour, eds. *Palestine and the Gulf*. Beirut, 1982.

Kievel, Gershon R. *Party Politics in Israel and the Occupied Territories.* Westport, Conn., 1983.

Kollek, Teddy. *For Jerusalem.* New York, 1978.

———. "Jerusalem," *Foreign Affairs,* July 1977.

Kraemer, Joel L. *Jerusalem: Problems and Prospects.* New York, 1980.

Landau, J. M. "Bittersweet Nostalgia: Memoirs of Jewish Emigrants from the Arab Countries," *Middle East Journal,* Spring 1981.

Leshem, El'azar. *Noshrim v'Olim baKerev Yotsei B'rit HaMo'atsot* [Dropouts and Immigrants Among Soviet Jews]. Jerusalem, 1980.

Louvish, M. "Co-Existence in Jerusalem," *Jewish Frontier,* January 1975.

Lustick, I. "Israeli Politics and American Foreign Policy," *Foreign Affairs,* Winter 1981–82.

Mendes, Meir. *HaVatikan v'Yisrael* [The Vatican and Israel]. Jerusalem, 1983.

Montoisy, J. "Israël-Vatican: Le nouveau dialogue," *Studia Diplomatica,* No. 6, 1981.

Nesvisky, M. "Ariel Sharon—Super-General, Super-Hawk," *Present Tense,* Winter 1984.

Norton, A. "Nuclear Terrorism and the Middle East," *Military Review,* April 1976.

Orland, N. "Die deutsch-israelischen Begrehungen aus der Beurteiling von Begin," *Orient,* No. 3, 1983.

Oz, Amos. *In the Land of Israel.* New York, 1983.

Peretz, Don. "Israel's Tenth Knesset Elections," *Middle East Journal,* Autumn 1981.

Peri, Y. "Fall from Favor: Israel and the Socialist International," *Jerusalem Quarterly,* No. 24, 1981.

———. "Mushroom over the Middle East," *New Outlook,* May 1982.

Perlmutter, A. "Begin's Rhetoric and Sharon's Tactics," *Foreign Affairs,* Fall 1982.

———, Michael Handel, and Uri Bar-Joseph. *Two Minutes Over Baghdad.* London, 1982.

Pincus, J. "Syria: A Captive Economy," *Middle East Review,* Fall 1979.

Plaut, S. E. "Perverse Suburbanization in Jerusalem," *Israel Social Science Research,* No. 1, 1983.

Pollock, D. "The Politics of Pressure: American Arms and Israeli Policy since the Six-Day War," *Middle East Studies Asociation,* July 1983.

Prittie, Terence. *Whose Jerusalem?* London, 1981.

Quandt, W. B. "The Middle East Crisis," *Foreign Affairs,* Fall 1980.

Quester, G. H. "Nuclear Weapons and Israel," *Middle East Journal,* Autumn 1983.

Rabin, Yehoshua. *The Arab-Israel Military Balance.* Tel Aviv, 1980.

Ramati, Y. "Analyzing Israel's Election," *Midstream,* October 1981.

Ramberg, W. "Attacks on Nuclear Reactors: The Implications of Israel's Strike on Osiraq," *Political Science Quarterly,* Winter 1982–83.

Ramazani, R. K. "Iran and the Arab-Israel Conflict," *Middle East Journal,* Summer 1978.

Rapoport, Louis. *Redemption Song: The Story of Operation Moses.* San Diego 1985.

Rosen, S. "Nuclearization and Stability in the Middle East," *Jerusalem Journal of International Relations,* No. 3, 1976.

Ross, J. A. "The Relationship between the Perception of Historical Symbols and the Alienation of Jewish Emigrants from the Soviet Union," *World Political Quarterly,* June 1979.

Rouleau, E. "Egypt: A Cumbersome Inheritance," *New Outlook,* April 1982.

Roumani, M. "Jews and Arabs in Jerusalem," *Jerusalem Quarterly,* No. 19, 1981.

Rubinstein, A. "To Lose Both the Elections and the Peace: No," *New Outlook,* September–October 1980.

Said, Edward. *The Question of Palestine*. New York, 1980.

Saint-Prot, Charles. *La France et le renouveau arabe. De Charles de Gaulle à Giscard d'Estaing*. Paris, 1980.

Sevela, Ephraim. *Farewell Israel*. South Bend, Ind., 1977.

Shadid, Mohammed. *The United States and the Palestinians*. New York, 1981.

Shamir, Y. "Israel's Role in a Changing Middle East," *Foreign Affairs*, Spring 1982.

Shimoni, Y. "Israel and Europe," *Jerusalem Quarterly*, Spring 1981.

Shokeni, Moshe, and Shlomo Desken. *Distant Cousins: Ethnicity and Politics among Arabs and North African Jews in Israel*. South Hadley, Mass., 1982.

Shoubachi, F. "The Suicide of Anwar Sadat," *Monthly Review*, March 1982.

Slonim, S. "Israel, the U.S., and China," *Midstream*, June–July 1973.

Snyder, J. C. "The Road to Osiraq: Baghdad's Quest for the Bomb," *Middle East Journal*, Autumn 1983.

Swirski, S. "The Oriental Jews in Israel: Why Many Tilted toward Begin," *Dissent*, No. 1, 1984.

Tabbarah, R. B. "Background to the Lebanese Conflict," *International Journal of Comparative Sociology*, May–June 1979.

Tal, I. "The Concept of National Security," *New Outlook*, March–April 1983.

Tessler, Mark A. *Post-Sinai Pressures in Israel and Egypt*. Hanover, N.H., 1982.

Ungar, S. "Washington: Jewish and Arab Lobbyists," *Atlantic*, March 1978.

United States. House of Representatives. Committee on Foreign Affairs. *The Multinational Force and Observers (MFO) for the Sinai*. Washington, D.C., May 1982.

Weissman, Steve, and Herbert Krosney. *The Islamic Bomb*. New York, 1981.

Yaniv, A. "The French Connection: A Review of French Policy toward Israel," *Jerusalem Journal of International Relations*, No. 3, 1976.

Yegar, M. K. "Israel in Asia," *Jerusalem Quarterly*, No. 18, 1981.

Yegner, T. "Saudi Arabia and the Peace Process," *Jerusalem Quarterly*, No. 18, 1981.

Yiftah, S. "The Middle East Nuclear Race: An Israeli View," *New Outlook*, May 1982.

Yishai, Y. "Israeli Annexation of East Jerusalem and the Golan Heights: Factors and Processes," *Middle Eastern Studies*, January 1985.

Zadok, C. "A Law That Did Nothing But Damage," *New Outlook*, January–February 1982.

VII. *"Begin II": The Djinn Out of the Bottle*

Alexander, E. "Terror at Home: Letter from Jerusalem," *Midstream*, August–September 1984.

Amad, A., ed. "Israeli League for Human and Civil Rights," *Journal of Palestine Studies*, Nos. 1–2, 1975–76.

Aprosch, S. "Israeli Defense of the Golan: An Interview with Brigadier General Avigdor Kahalani," *Military Review*, October 1979.

Aruri, N. H. "Repression in Academia: Palestinian Universities versus the Israeli Military," *Arab Prespectives*, April 1981.

———. "Resistance and Repression," *Journal of Palestine Studies*, Summer 1978.

Aviad, Jane. *Return to Jerusalem: Religious Renewal in Israel*. Chicago, 1983.

Avineri, S. "Beyond Camp David." *Foreign Affairs*, Spring 1982.

Benvenisti, M. "The Price of Extremism," *New Outlook*, August 1980.

———. *The West Bank Data Project: A Survey of Israel's Policies*. Washington, D.C., 1984.

Ben-Zvi, Abraham. *Alliance Politics and the Limits of Influence: The Case of the U.S. and Israel.* Tel Aviv, 1984.

"Bir Zeit: Palestinian University Confrontations," *Palestine Bulletin,* January 1979.

Blitzer, Wolf. *Between Washington and Jerusalem: A Reporter's Notebook.* New York, 1985.

Butros-Ghali, B. "The Foreign Policy of Egypt in the Post-Sadat Period," *Foreign Affairs,* Spring 1982.

Bryan, S. "Advancing U.S.-Israeli Strategic Cooperation," *Middle East Review,* February 1984.

Capitanchik, D. "Religion and Politics in Israel," *Jewish Journal of Sociology,* July 1983.

Davis, Uri. *The Golan Heights under Israeli Occupation, 1967–81.* Durham, England, 1983.

Dawisha, A. "Syria under Assad, 1970–1978," *Government and Oppoosition,* No. 3, 1978.

"The Death of a Plan," *New Republic,* May 2, 1983.

Deshen, S. "Israeli Judaism: Introduction to a Major Pattern," *International Journal of Middle Eastern Studies,* May 1978.

Don-Yehilia, E. "Jewish Orthodoxy, Zionism, and the State of Israel," *Jerusalem Quarterly,* No. 31, 1984.

———. "Religious Leaders in the Political Arena: The Case of Israel," *Middle Eastern Studies,* April 1984.

Dowty, A. "Religion and Politics in Israel," *Commonweal,* July 15, 1983.

Elazar, Daniel J., ed. *Judea, Samaria, and Gaza.* Washington, D.C., 1982.

Eliav, Arieh. *Tabot Edut* [Required Testimony]. Tel Aviv, 1983.

Falk, G. H. "Israeli Public Opinion: Looking for a Palestinian Solution," *Middle East Journal,* Summer 1985.

Friedman, M. "Mifleget HaMafdal b'Mashber [The National Religious Party in Crisis]," *M'dina, Mimshkal, Vichasim Beinle'umiyim,* Nos. 19–20, 1982.

Goldstein, M. "Israeli Security Measures in the Occupied Territories: Administrative Detention," *Middle East Journal,* Winter 1978.

Goren, Dina. *Secrecy and the Right to Know.* Ramat Gan, 1979.

Halabi, Rafik. *Yesh G'vul* [There's a Limit]. Jerusalem, 1983.

Harkabi, Y. "Foreign Policy Reassessed: Reflections on National Defense Policy," *Jerusalem Quarterly,* No. 19, 1981.

Hirschfeld, R. "The Village Leagues," *Christian Century,* April 6, 1983.

Hitchens, C. "The Golem on the West Bank: Israel's Pieds Noirs," *Nation,* September 12, 1981.

Inbar, E. "American Arms Transfers to Israel," *Middle East Review,* Nos. 1–2, 1982/83.

International Center for Peace in the Middle East. *Human Rights in the Occupied Territories.* Tel Aviv, 1984.

International Committee of the Red Cross. *Report on Toulkarm Prison.* Geneva, February 25, 1968.

Isaac, Rael Jean. *Party and Politics in Israel.* New York, 1981.

Kievel, Gershon R. *Party Politics in Israel and the Occupied Territories.* Westport, Conn., 1983.

Krausz, E., and M. Bar-Lev. "Varieties of Orthodox Religious Behavior," *Jewish Journal of Sociology.* July 1978.

Langer, Felicia. *With My Own Eyes: Israel and the Occupied Territories. 1967–1973.* London, 1975.

Ledeen, B. and M. "The Temple Mount Plot," *New Republic,* June 18, 1984.

Liebman, Charles L., and Eliezer Ben-Yehuda. *Civil Religion in Israel.* Berkeley, 1983.

Lustick, I. "Is Annexation a Fact?" *New Outlook,* April 1982.

———. "Israeli Politics and American Foreign Policy," *Foreign Affairs,* Winter 1981/82.

Ma'oz, Moshe. "Hafez al-Asad: A Political Profile," *Jerusalem Quarterly,* Summer 1980.

Ma'oz, Moshe, and M. Nisan. *Palestinian Leadership in the West Bank.* London, 1984.

Merrian, J. G. "Egypt under Mubarak," *Current History,* January 1983.

Milson, M. "Arabs and Israeli Attitudes," *Dissent,* No. 4, 1976.

———. "How to Make Peace with the Palestinians," *Commentary,* May 1981.

Minkowich, Avram, Dan Davis, and Joseph Baskin. *Success and Failure in Israeli Elementary Education.* New Brunswick, 1982.

Monteil, Vincent. *Dossier secret sur Israël: Le terrorism.* Paris, 1979.

Naamani, I. "The Secret of Assad's Success," *Jewish Frontier,* June–July 1977.

"The Najah National University in Nablus-Occupied West Bank," *Journal of Social Science,* July 1978.

National Lawyers Guild. *The Treatment of Arabs in the Israeli-Occupied West Bank and Gaza.* New York, 1978.

Oren, M. "A Horseshoe in the Glove: Milson's Year on the West Bank," *Middle East Review,* Fall 1983.

Pattir, D. "After the Sinai Withdrawal," *Public Opinion.* April–May 1982.

Plaut, S. E. "Perverse Suburbanization in Jerusalem," *Israel Social Science Research,* No. 1, 1983.

Reich, W. "A Stranger in My House: Jews and Arabs in the West Bank," *Atlantic,* June 1984.

Ross, R. "West Bank 'Facts' Are Hitting Home," *Nation,* March 10, 1984.

Rubinstein, D. "The Occupied Territories 15 Years Later," *New Outlook,* June–July 1982.

Sabatello, E., "Patterns of Illegitimacy in Israel," *Jewish Journal of Sociology,* June 1979.

Schiff, Gary S. *Tradition and Politics: The Religious Parties of Israel.* Detroit, 1977.

Schiff, Z. "The Spectre of Civil War in Israel," *Middle East Journal,* Spring 1985.

Segre, D. V. "Religion in Israel," *Midstream,* August–September 1980.

Sharkansky, I. "Religion and State in Begin's Israel," *Jerusalem Quarterly,* Spring 1984.

Shemesh, M. "Egypt's Commitment to the Palestinian Cause," *Jerusalem Quarterly,* No. 34, 1985.

Silver, Eric. *Begin: The Haunted Prophet.* New York, 1984.

Spiegel, S. "Israel as a Strategic Asset," *Commentary,* June 1983.

Stock, Ernest. *HaZ'ramim HaDatiyim baGolah u'shluchoteihem b'Yisrael* [The Religious Trends in the Diaspora and their Constituent Groups in Israel]. Jerusalem, 1983.

Stone, R. A. *Social Change in Israel: Attitudes and Events.* New York, 1982.

United States. Department of State. *Human Rights Report.* Washington, D.C., May 1977.

———. Department of State. *Sinai Multinational Force and Observers Established.* September 1981.

Vatikiotis, P. J. "After Sadat," *Policy Review,* Winter 1982.

Waxman, C. "American Israelis in Judea and Samaria," *Middle East Review,* No. 2, 1982/83.

Weiss, S. "Israeli Non-Violent Khomeinism," *New Outlook*, July–August 1979.

Wigoder, G. "Interfaith in Israel," *Midstream*, February 1980.

Witten, E. "The Status of West Bank Universities," *New Outlook*, February 1984.

Yaniv, A., and Y. Yishai. "Israeli Settlements on the West Bank: The Politics of Intrasigence," *Journal of Politics*, November 1981.

Yehoshua, A. B. *Between Right and Right*. New York, 1981.

Yishai, Y. "Abortion in Israel: Social Demand and Political Responses," *Policy Studies Journal*, No. 2, 1978.

Zelnikar, S., and M. Kahan, "Religion and Nascent Cleavages: The Case of Israel's National Religious Party," *Comparative Politics*, October 1976.

VIII, IX. *The Lebanon War*

Allen, D., and Alfred Pijpers. *European Foreign Policy-Making and the Arab-Israel Conflict*. The Hague, 1984.

Banks, Judith M. *Anti-Israel Influences in American Churches*. New York, 1979.

Bard, M. "Israel's Standing in American Public Opinion," *Commentary*, April 1985.

The Beirut Massacre. New York, 1983.

Bulloch, Keith. *Final Conflict: The War in the Lebanon*. London, 1983.

Caplan, N., and I. Black. "Israel and Lebanon: Origins of a Relationship," *Jerusalem Quarterly*, No. 27, 1983.

Chafetz, Z. "Beirut and the Great Media Coverup," *Commentary*, September 1984.

Cobban, Helena. *The Palestinian Liberation Organization*. Cambridge, Mass., 1984.

Cooley, J. J. "The War Over Water," *Foreign Policy*, Summer 1979.

Davidson, L. "Lebanon and the Jewish Conscience," *Journal of Palestine Studies*, No. 2, 1983.

————. "The Tragic Dilemma: American Liberal Jews and Israel," *New Outlook*, September–October 1980.

Dawisha, A. "The Motives of Syria's Involvement in Lebanon," *Middle East Journal*, Spring 1984.

Dishon, D. "The Lebanese War—An All-Arab Crisis," *Midstream*, January 1977.

Eban, A. "The Trauma in Beirut," *Jewish Frontier*, August–September 1982.

Gabriel, Philip L. *In the Ashes: A Story of Lebanon*. Ardmore, Penn., 1978.

Gabriel, Richard A. *Operation Peace for the Galilee*. New York, 1984.

Gamson, Josh. "The Limits of Conscience," *Nation*, June 9, 1984.

Gavron, Daniel. *Israel After Begin*. Boston, 1984.

Gordon, David C. *Lebanon: The Fragmented Nation*. Stanford, Calif., 1980.

Haig, Alexander. *Caveat*. New York, 1984.

HaKibutz HaMe'uhad. *Milchemet Levanon* [The Lebanon War]. Tel Aviv, 1983.

Heller, P., ed. *The Middle East Military Balance*. New York, 1984.

————. "The Syrian Factor in the Lebanese Civil War," *Journal of South Asian and Middle Eastern Studies*, No. 1, 1980.

Hertzberg, A. "The Tragedy and the Hope," *New York Review of Books*, October 21, 1982.

"How Special Is Israel?" *America*, August 21, 1982.

Hudson, Michael C. "The Palestinian Factor in the Lebanese Civil War," *Middle East Journal*, Summer 1978.

Israel. Ministry of Foreign Affairs. *Te'udot l'M'diniyut HaChutz shel M'dinat Yisrael* [Documents on Israeli Foreign Policy]. Vols. 2–3. Jerusalem, 1981–84.

————. Ministry of Information. *Mivtzah "Shalom l'Galil"* [Operation Peace for the Galilee]. Jerusalem, 1983.

Israeli, Raphael. *The PLO in Lebanon.* New York, 1983.

Jansen, Michael. *The Battle of Beirut: Why Israel Invaded Lebanon.* Boston, 1983.

Jones, M. "Verdict in London," *Present Tense,* Summer 1983.

Kapeliuk, Amnon. *Enquête sur un massacre.* Paris, 1982.

Karpel, C. "Onward Christian Diplomats," *New Republic,* September 3, 1984.

Kenan, A. "Stupid War, Stupid Peace," *Nation,* October 15, 1983.

Khalidi, R. "The Palestinians in Lebanon: Social Repercussions of Israel's Invasion," *Middle East Journal,* Spring 1984.

Lanir, Zvi, ed. *Israeli Security Planning in the 1980s.* New York, 1986.

Lissak, Moshe, ed. *Israeli Society and its Defense Establishment: The Social and Political Impact of a Protracted Violent Conflict.* London, 1984.

Lubrani, U. "Israel, Syria, and Lebanon—An Israeli Perspective," *Middle East Insight,* March 1985.

MacBride, Sean, Richard Falk et al. *Israel in Lebanon: The Report of the International Commission.* London, 1983.

Ma'oz, M. "Israel and the Arabs after the Lebanese War," *Jerusalem Quarterly,* No. 26, 1983.

Megged, M. "On Wishing Israel a Happy Birthday," *Nation,* June 11, 1983.

Morris, R. "Beirut—and the Press—under Siege," *Columbia Journalism Review,* November–December 1982.

Muir, J. "Lebanon: Arena of Conflict, Crucible of Peace," *Middle East Journal,* Spring 1984.

Nesvisky, M. "Ariel Sharon—Super-General, Super-Hawk," *Present Tense,* Winter 1984.

Norton, A. "Occupational Risks and Planned Retirement: The Israeli Withdrawal from South Lebanon," *Middle East Insight,* March 1985.

Pail, M. "Lebanon: A Military Analysis," *New Outlook,* August–September 1982.

Pallis, E. "The Widening Rift between Israel and the Diaspora," *Middle East,* August 1983.

Peri, Y. "Fall from Favor: Israel and the Socialist International," *Jerusalem Quarterly,* No. 24, 1981.

Pipes, D. "The Media and the Middle East," *Commentary,* January 1984.

———. "The Real Problem," *Foreign Policy,* Summer 1983.

Preuss, Teddy. *Begin baShilton* [Begin in Office]. Jerusalem, 1984.

Quandt, W. "Reagan's Lebanon Policy: Trial and Error," *Middle East Journal,* Spring 1984.

Raab, E. "Is the Jewish Community Split?" *Commentary,* November 1982.

Rabinovich, Itamar. *The War for Lebanon, 1970–1983.* Ithaca, N.Y., 1984.

Randal, Jonathan C. *Going All the Way.* New York, 1983.

Rasler, K. "International Civil War: A Dynamic Analysis of the Syrian Intervention in Lebanon," *Journal of Conflict Resolution.* September 1983.

Ringler, S. "Public and Private: American Jewish Views on Israel," *Jewish Frontier,* June 1984.

Ron, Moshe. *Milchemet Levanon* [The Lebanon War]. Tel Aviv, 1983.

Rubenberg, C. "The Civilian Infrastructure of the Palestine Liberation Organization," *Journal of Palestine Studies,* No. 3, 1983.

Sadowski, Y. "The Sands Run Out in Lebanon," *Nation,* June 18, 1983.

Salem, E. A. "Lebanon's Political Maze," *Middle East Journal,* Autumn 1979.

Salpeter, E. "The Geva Case," *New Leader,* September 6, November 29, 1982.

———. "The Taste of Ashes," *World Press Review,* September 1982.

Schiff, Z. "Dealing with Syria," *Foreign Policy,* Summer 1984.

————. "Green Light, Lebanon," *Foreign Policy*, Autumn 1983.
————, ed. *Israelis Speak*. New York, 1977.
Schiff, Z., and Ehud Ya'ari. *Israel's Lebanon War*. New York, 1984.
Schueftan, D. "The PLO after Lebanon," *Jerusalem Quarterly*, No. 28, 1983.
Segev, Shmuel. *Mivtsat "Yachsin"* [Operation "Yachsin"]. Tel Aviv, 1984.
Sharett, Moshe. *MiCheter l'Hitarvut Tsva'it b'Levanon?* [From Indirect to Direct Intervention in Lebanon?]. Tel Aviv, 1983.
Sheffer, G. "The Uncertain Future of American Jewry-Israel Relations," *Jerusalem Quarterly*, Summer 1984.
Shiffer, Shalom. *Kadur Sheleg* [Snowball—The Lebanon War]. Jerusalem, 1984.
Shlain, A. "Conflicting Approaches to Israel's Relations with the Arabs: Ben-Gurion and Sharett," *Middle East Journal*, Spring 1983.
Simon, R. "The Print Media's Coverage of the War in Lebanon," *Middle East Review*, No. 1, 1983.
Smith, P. A. "European Reactions to Israel's Invasion," *Journal of Palestine Studies*, No. 1, 1982.
Tabbarah, R. B. "Background to the Lebanese Conflict," *International Journal of Comparative Sociology*, May–June 1979.
Tal, I. "The Concept of National Security," *New Outlook*, March–April 1983.
Timerman, Jacobo. *The Longest War*. New York, 1982.
Travis, T. "The Dilemma of the Jewish American," *New Outlook*, July–August 1978.
Tucker, R. W. "Lebanon: The Case for the War," *Commentary*, October 1982.
Weinberger, N. "Peacekeeping Operations in Lebanon," *Middle East Journal*, Summer 1983.
Ya'ari, E. "Israel's Dilemma in Lebanon," *Middle East Insight*, April–May 1984.
Ya'ari, E., and R. Lieber. "Personal Whim or Strategic Imperative? The Israeli Invasion of Lebanon," *International Security*, No. 2, 1983.
Yermiya, Dov. *My War Diary: Lebanon, June 5–July 1, 1982*. Boston, 1984.
Zagorin, A. "A House Divided," *Foreign Policy*, Fall 1982.
Zamir, M. "Smaller and Greater Lebanon: Squaring the Circle?" *Jerusalem Quarterly*, No. 23, 1982.

x, xi. *The Transition from the Begin Era, Lights and Shadows in Israel's Future*

Adeyufe, A. "Nigeria and Israel," *International Studies*, No. 4, 1981.
Amir, Menachem. *P'shiah, Avarkanut, Meniah, v'Tikun b'Yisrael* [Crime, Delinquency, Prevention, and Solution in Israel]. Jerusalem, 1979.
"Arab Banks in the United States," *U.S.-Arab Commerce*, January–February 1983.
Aronoff, Myron J., ed. *Cross-Currents in Israeli Culture and Politics*. New Brunswick, 1984.
Avishai, B. "Looking over Jordan," *New York Review of Books*, April 28, 1983.
Avneri, Arieh. *David Levy*. Tel Aviv, 1983.
Ayal, E. "Economic Factors Behind United Nations Actions," *Middle East Review*, No. 2, 1978–79.
Bahbah, B. "The United States and Israel's Energy Security," *Journal of Palestine Studies*, No. 2, 1982.
Barkai, C. "The Israeli Economy in the Past Decade," *Jerusalem Quarterly*, Summer, 1984.
Belguedh, M. "The Arab Oil Industry," *OEPEC*, June 1980.
Benbassat, J. "Wanted: A New Code of Medical Ethics," *Jerusalem Quarterly*, Winter 1978.

Benjamin, G. "Israel: Too Much Democracy?" *Midstream*, April 1977.

Bin Talal, Al-Nasser [Crown Prince of Jordan]. "Jordan's Quest for Peace," *Foreign Affairs*, Spring 1982.

Bowden, Tom. *Army in the Service of the State*. Tel Aviv, 1976.

Caspi, D. "How Representative Is the Knesset?" *Jerusalem Quarterly*, Winter 1980.

Chazan, N. "Israel in Africa," *Jerusalem Quarterly*, No. 18, 1981.

Clark, B. "Israel's Science-Based Industries Come of Age," *Kidma*, No. 3, 1983.

Cohen, Y. "The Implications of a Free Trade Area between the EEC and Israel," *Journal of World Trade Law*, May–June 1976.

Crittendon, A. "Israel's Economic Plight," *Foreign Affairs*, Summer 1979.

Curtis, Michael, and Susan A. Gitelson, eds. *Israel in the Third World*. New Brunswick, 1976.

Davis, U, and W. Lehn. "And the Fund Still Lives," *Journal of Palestine Studies*, Summer 1978.

Decter, M. "The Arms Traffic with South Africa," *Midstream*, February 1977.

Doron, A. "Public Assistance in Israel," *Journal of Social Policy*, October 1978.

———. "Social Policy for the Eighties," *Jerusalem Quarterly*, Spring 1981.

Dorot, Aliza M., and Daniel J. Elazar. *Understanding the Jewish Agency*. Jerusalem, 1984.

"The Economic War: Israel Is Losing," *U.S. News and World Report*, August 30, 1982.

Elimelech, A. "Israel Bonds," *Israel Economist*, February 1981.

Eytan, E. "Le lobby arabe," *L'Arche*, August 1975.

Felsenthal, D. "Aspects of Coalition Payoffs: The Case of Israel," *Comparative Political Studies*, July 1979.

Feuth, O. "The Oil Weapon De-Mystified," *Policy Review*, No. 15, 1981.

Finger, S. "The U.N. Security Council and the Arab-Israeli Conflict, 1968–1977," *Middle East Review*, No. 4, 1977.

Fishelson, Gideon. *The Economic Integration of Israel into the EEC*. Tel Aviv, 1977.

Fishman, G. "Youthful Offenders in Israel," *Journal of Social Policy*, January 1977.

Friedlander, Dov, and Calvin Goldscheider. *The Population of Israel*. New York, 1979.

Galnoor, I. "Water Policymaking in Israel," *Policy Analysis*, No. 3, 1978.

Galnoor, I., et al., eds. *Can Planning Replace Parties? The Israeli Experience*. The Hague, 1980.

Gavron, Daniel. *Israel After Begin*. Boston, 1984.

Giniewski, P. "Israel: Un nouveau 'grand' de l'industrie des armaments," *Stratégie*, Spring 1975.

Globerson, Arye, *Higher Education and Employment in Israel*. New York, 1979.

Golan, Matti. *Shimon Peres*. New York, 1982.

Goldin, M. "Why They Give: American Jews and their Philanthropies," *Jewish Frontier*, August–September 1977.

Grinwald, Zvi. *Mayim b'Yisrael* [Water in Israel]. Tel Aviv, 1980.

Hareven, A. "Are We Still the People of the Book?" *Jerusalem Quarterly*, Winter 1984.

Harkabi, Yehoshafat. *The Bar Kochba Syndrome*. Chappaqua, N.Y., 1982.

Harmon, L. "costi: Israel's Centre of Scientific and Technological Information," *Kidma*, No. 4, 1980.

Hartshorn, J. E. *The Objectives of the Petroleum Exporting Countries*. Nicosia, 1978.

"Investment in Israel," *Israel Economist*, March 1981.

Israel. Central Bureau of Statistics. *Energya b'Yisrael* [Energy in Israel]. Jerusalem, 1981.

"Israel-EEC Trade Relations," *Israel Economist*, February 1981.

"Israel's Chemical Industry," *Israel Economist*, August 1977.

"Israel's Electronics Industry," *Israel Economist*, November–December 1980.

"Israel's Security Products Industry," *Israel Economist*, November–December 1980.

Jacobson, J. "Israel's Oil Challenge," *Israel Economist*, May 1981.

Jaffe, Eliezer D. *Givers and Spenders: The Politics of Charity in Israel*. Jerusalem, 1985.

Kahane, M. "The Arab Problem in Israel: Emigration Is the Only Solution," *Judaism*, No. 7, 1977.

Kaikati, J. "The Arab Oil Boycott," *California Management Review*, No. 3, 1978.

Katz, Elihu, and Michael Gurevitch. *The Secularization of Leisure: Culture and Communication in Israel*. Cambridge, Mass., 1976.

Kaufman, Edy, Yorman Shapira, and Josef Barromi. *Israel-Latin American Relations*. New Brunswick, 1979.

Kimche, J. "Taba or Masada?" *Midstream*, April 1985.

King, E. "Solar Energy and Israel," *Jewish Frontier*, December 1980. *Kiyunim b'Michkar Hallidrologit b'Yisrael* [Approaches to Hydrologic Research in Israel]. Tel Aviv, 1981.

Kislev, Y., and M. Hoffman, "Research and Productivity in Wheat in Israel," *Journal of Development Studies*, January 1978.

Kleiman, Aaron. *Israel's Global Reach: Arms Sales as Diplomacy*. New York, 1985.

Lake, J. "The Israel Defense Industry: An Overview," *Defense Electronics*, September 1983.

Land, T. "Black Africa Poised to Restore Relations with Israel," *New Outlook*, March 1980.

Law, John D. *Arab Investors: Who They Are, What They Buy, and Where*. New York, 1981.

Lokat, Y. "The Problem of Yerida," *Jewish Frontier*, August–September 1981.

Levenfeld, D. "Israel and Black Africa," *Midstream*, February 1984.

Levins, Hoag. *Arab Reach*. Garden City, N.Y., 1983.

Levy, W. "Oil and the Decline of the West," *Foreign Affairs*, No. 5, 1980.

Liebman, Charles L. *Pressure Without Sanctions: The Influence of World Jewry on Israeli Policy*. Rutherford, N.J., 1977.

Louvish, M. "Reorganizing the Absorption Machinery," *Jewish Frontier*, November 1976.

Lumer, H. "Israel and South Africa," *Jewish Affairs*, May–June 1976.

Mackie, A. "Oil Companies Wait for Israel to Move Out," *Middle East Economic Digest*, December 8, 1979.

Mahler, Gregory S., ed. *Readings on the Israeli Political System*. Washington, D.C., 1982.

Maslow, W. "Jewish Political Power: An Assessment," *American Jewish Historical Quarterly*, December 1976.

Melamed, Aharon. *No'ar b'Metsukah* [Youth in Distress]. Haifa, 1983.

Meyer, Lawrence. *Israel Now*. New York, 1982.

Minerbi, S. "The Accession of Spain to the EEC and its Implications for Mediterranean Third Countries: The Case of Israel," *Jerusalem Journal of International Relations*, No. 3, 1982/83.

Mor, S. "A Troubled Economy," *World Press Review*, July 1983.

Nacklemann, E. "Israel and Black Africa: A Rapprochement," *Journal of Modern African Studies*, June 1981.

Ne'eman, Y. "The Settling of Eretz Israel," *Midstream*, January 1984.

Oded, A. "Africa, Israel, and the Arabs: On the Restoration of Israeli-African Diplomatic Relations," *Jerusalem Journal of International Relations*, No. 3, 1982/83.

Orfalea, G. "The Arab American Lobby," *Arab Perspectives*. July 1980.

Patiel, K. "The Israeli Coalition System," *Government and Opposition*, No. 5, 1975.

Peres, Shimon. *From These Men*. London, 1979.

———. "The Vision of Labor," *Jewish Frontier*, April 1982.

Peri, Yoram. *Between Battles and Ballots: The Israeli Military in Politics*. London, 1983.

Phillips, Charlotte A. *The Arab Boycott of Israel*. Washington, D.C., 1979.

Pilch, J. "Aliyah In Reverse," *Jewish Frontier*. January 1976.

Plaut, S. "Is Israel Economically Viable?" *Midstream*, January 1982.

Pryce-Jones, A. "The Timid King [Hussein]," *New Republic*, January 3, 1983.

Richmond, A. "A Look at Israel's Institute of Desert Research," *Kidma*, August–October 1977.

Room, J. "Science and Development in Israel," *Kidma*, January–April 1979.

Rubinstein, Amnon. *The Zionist Dream Revisited*. New York, 1974.

Sanders, S.W. "Begin Leaves Behind an Economic Mess," *Business Week*, September 26, 1983.

Schnall, D. "Yored Is Also a Noun," *Midstream*, February 1978.

Sheffer, G. "The Uncertain Future of American Jewry-Israel Relations," *Jerusalem Quarterly*, Summer 1984.

Sherman, Arnold, and Paul Hirschhorn. *Israel High Technology*. Jerusalem, 1984.

Shoham, Shlomo. *Israel Studies in Criminology*. Vols II, III. New Brunswick, 1973, 1976.

Steiner, H. "International Boycotts and Domestic Order: American Involvement in the Arab-Israel Conflict," *Texas Law Review*, November 1976.

Stock, R. "Israel's Tourist Industry Development System," *Kidma*, August–October 1979.

"Surplus Arab Funds at Home and Abroad," *Arab Economist*, September 1980.

Tabor, M. "Solar Energy Development in Israel," *Kidma*, May–August 1979.

Tamarin, G. "Israeli Migratory Processes Today: Does Israel Really Want All Its Immigrants?" *Plural Society*, Nos. 3–4, 1977.

"The Technion Looks to the Future," *Israel Economist*, June 1983.

Toren, N. "The Return of Emigrants to Israel," *Jerusalem Quarterly*, Winter 1981.

Urofsky, M. I. "Fifty Years of the Jewish Agency," *Midstream*, November 1979.

———. *We Are One: American Jewry and Israel*. New York, 1978.

Van Leer Institute. *Echad miKol Shishah Yisraelim* [One of Every Six Israelis]. Jerusalem, 1981.

Vardi, J. "Energy in Israel," *Kidma*, September–December 1978.

Weiner, J. "Ampal [America Israel Corporation]: Mobilizing Capital," *Israel Economist*, May–June 1980.

Winn, I., and A. Peranio. "Israel's Energy Dilemma," *Bulletin of the Atomic Scientists*, April 1980.

Ya'alon, D. "Soil Sciences in Israel, Past and Present," *Kidma*, September–December 1978.

Zamir, D. "Generals in Politics," *Jerusalem Quarterly*, No. 20, 1981.

Zussman, Pinchas. *Ma'arachat HaBitachon v't'rumatah la'Kidmah HaTechnologit* [The Defense System and Its Contribution to Technological Advancement]. Tel Aviv, 1983.

INDEX

Abuhatzeira, Aharon, 46, 111, 138, 230
Abuhatzeira, Baruch, 229, 230
Abu Ghazala, Abd al-Halim, 135
"Abu Nidal," *see* Banna, Sabri al-
"Abu Iyad," *see* Khalef, Saslah
Abu Rudeis, 6, 82
Achad HaAm, 254
Achdut HaAvodah party, 5, 13, 21, 221
Adam, Yekutiel, 183
Africa, 104, 173, 208; *see also individual African countries*
Agranat, Shimon, 3
Agranat Commission, 3, 198, 219
Agron, Gershon, 112
Agudat Israel party: votes nonconfidence in Rabin government, 19; in 1977 election, 21, 22; joins Likud government, 29; theologically wedded to West Bank occupation, 93; in 1981 election, 132; in Begin coalition, 29, 137–8; hard core following of, 138; wins concessions from Begin government, 138 ff.; organizes protests on Ramot road issue, 144–5; in 1984 election, 216; *see also* religion, religious establishment (Jewish, in Israel)
al-Aqsa Mosque, 51, 114
Al-Arish, 52, 57, 70, 71, 80–1, 82
al-Azar University, 67
Aldoraty, Zvi, 37
Algeria, 46, 92, 101, 148, 207, 260
Al HaMishmar (Tel Aviv), 37, 197
Allon, Yigal: represents Labor left wing as Rabin's foreign minister, 5; ideas of, incorporated into Labor administration of Palestine, 12; a minister in Meir and Rabin cabinets, 12, 15; Rabin close to partitionist approach of, 14; proposes linking Jericho with Jordan, 15; converses with Hussein, 15; insists on eviction of Gush Emunim squatters, 17; helps paralyze Rabin cabinet, 19; favors substantial withdrawal from territories, 21; rejects pluralistic state for Israeli Arabs, 35; opposes withdrawl from territories, 21; rejects pluralistic state for Israeli Arabs, 35; opposes withdrawal from Golan, 149; meets with, favors supporting, Lebanese Maronites, 168; deputy premier and foreign minister under Rabin, 221

Allon Plan: implemented by Labor governments in Palestine, 12, 13; evokes no dithyrambs from Arab world, 12; Likud abandons, for extensive settlement, 92; Begin transplants Jews beyond configuration of, 96; Drobles Plan transcends, 98; in Labor's 1981 election campaign, 131; in Labor's 1984 campaign, 216
Alma, 82
Alon, Amos, 22
Aloni, Shulamit, 140, 145
al-Quneitra, 6, 150
Altalena (vessel), 27
Amal, 240, 241
America (journal), 188
American Jewish Congress, 188
American Jewish Committee, 188, 253
Amin, Idi, 10, 11
Ampal (America-Israel Corporation), 248
Anti-Defamation League, 199
Aqaba, 235
Aqaba, Gulf of, 48, 54, 64, 127
Arab League, 46, 78, 135, 167
Arabs, Arab states: attack unprepared Israel, 3; broken after Six-Day War, 4; Rabin seeks military disengagement with, 5; Sadat fears noncompromise by, in Geneva Conference, 6; Egyptian victory would threaten U.S. influence among, 7; Idi Amin a partisan of, 10; Entebbe rescue surmounts hostility of, 11; at Khartoum Conference, reject peace with Israel, 12; uninterested in Allon Plan, 12; rejuvenated after Yom Kippur War, 20; oriental Jewish suspicion of, 24; Begin hectors Labor for appeasing, 27; have many states, Allon explains, 25; Syria demands single delegation for, at Geneva Conference, 43; to pledge peace with Israel under Brookings plan, 43; Sadat to oppose single delegation for, at Geneva, 43; Israeli intelligence warns conservatives among, 46; Sadat contemplates, rejects, inviting leaders of, to Jerusalem, 49; Sadat's planned visit to Jerusalem angers, 50; Sadat seeks wider peace between Israel and, 51; after Jerusalem visit, Sadat vainly seeks international conference with, 52; convene rejectionist con-